BLOCKCHAIN REVOLUTION

ALSO BY

Don Tapscott

Paradigm Shift:
The New Promise of Information Technology (1993)
Coauthor, Art Caston

The Digital Economy:
Promise and Peril in the Age of Networked Intelligence (1995)

Growing Up Digital:
The Rise of the Net Generation (1997)

Who Knows:
Safeguarding Your Privacy in a Networked World (1997)
Coauthor, Ann Cavoukian

Digital Capital:
Harnessing the Power of Business Webs (2000)
Coauthors, David Ticoll and Alex Lowy

The Naked Corporation:
How the Age of Transparency Will Revolutionize Business (2003)
Coauthor, David Ticoll

Wikinomics:
How Mass Collaboration Changes Everything (2006)
Coauthor, Anthony D. Williams

Grown Up Digital:
How the Net Generation Is Changing the World (2008)

Macrowikinomics:
New Solutions for a Connected Planet (2010)
Coauthor, Anthony D. Williams

BLOCKCHAIN REVOLUTION

HOW THE TECHNOLOGY BEHIND BITCOIN AND OTHER CRYPTOCURRENCIES IS CHANGING THE WORLD

Don Tapscott
and **Alex Tapscott**

Portfolio / Penguin

Portfolio/Penguin
An imprint of Penguin Random House LLC
375 Hudson Street
New York, New York 10014

ISBN 9781101980149 (trade paperback)

Library of Congress Cataloging-in-Publication Data

Names: Tapscott, Don, 1947– author | Tapscott, Alex, author.
Title: Blockchain revolution : how the technology behind bitcoin is changing money, business, and the world / Don Tapscott and Alex Tapscott.
Description: New York : Portfolio, [2016] | Includes bibliographical references and index.
Identifiers: LCCN 2016008427 (print) | LCCN 2016014933 (ebook) | ISBN 9781101980156 (ebook) | ISBN 9781101980132 (hardcover) | ISBN 9780399564062 (international edition)
Subjects: LCSH: Electronic funds transfers. | Bitcoin. | Electronic commerce. | Mobile commerce. | Banks and banking—Technological innovations. | Financial institutions—Technological innovations.
Classification: LCC HG1710 (ebook) | LCC HG1710.T385 2016 (print) | DDC 332.1/78—dc23
LC record available at https://lccn.loc.gov/2016008427

First Portfolio/Penguin hardcover edition: May 2016
First Portfolio/Penguin trade paperback edition: June 2018

Printed in the United States of America
1 3 5 7 9 10 8 6 4 2

Book design by Amy Hill

To Ana Lopes and Amy Welsman for enabling this book,
and for understanding that "it's all about the blockchain."

"A masterpiece. Gracefully dissects the potential of blockchain technology to take on today's most pressing global challenges."

—Hernando De Soto, Economist and President, Institute for Liberty and Democracy, Peru

"The blockchain is to trust as the Internet is to information. Like the original Internet, blockchain has potential to transform everything. Read this book and you will understand."

—Joichi Ito, Director, MIT Media Lab

"In this extraordinary journey to the frontiers of finance, the Tapscotts shed new light on the blockchain phenomenon and make a compelling case for why we all need to better understand its power and potential."

—Dave McKay, President and CEO, Royal Bank of Canada

"Deconstructs the promise and peril of the blockchain in a way that is at once accessible and erudite. *Blockchain Revolution* gives readers a privileged sneak peak at the future."

—Alec Ross, author, *The Industries of the Future*

"If ever there was a topic for demystification, blockchain is it. Together, the Tapscotts have achieved this comprehensively and in doing so have captured the excitement, the potential, and the importance of this topic to everyone."

—Blythe Masters, CEO, Digital Asset Holdings

"This is a book with the predictive quality of Orwell's *1984* and the vision of Elon Musk. Read it or become extinct."

—Tim Draper, Founder, Draper Associates, DFJ, and Draper University

"Blockchain is a radical technological wave and, as he has done so often, Tapscott is out there, now with son Alex, surfing at dawn. It's quite a ride."

—Yochai Benkler, Berkman Professor of Entrepreneurial Legal Studies, Harvard Law School

"If you work in business or government, you need to understand the blockchain revolution. No one has written a more thoroughly researched or engaging book on this topic than Tapscott and Tapscott."

—Erik Brynjolfsson, Professor at MIT; coauthor of *The Second Machine Age*

"An indispensable and up-to-the-minute account of how the technology underlying bitcoin could—and should—unleash the true potential of a digital economy for distributed prosperity."

—Douglas Rushkoff, author of *Present Shock* and *Throwing Rocks at the Google Bus*

"Technological change that used to develop over a generation now hits us in a relative blink of the eye, and no one tells this story better than the Tapscotts."

—Eric Spiegel, President and CEO, Siemens USA

"Few leaders push us to look around corners the way Don Tapscott does. With *Blockchain Revolution* he and his son Alex teach us, challenge us, and show us an entirely new way to think about the future."

—Bill McDermott, CEO, SAP SE

"*Blockchain Revolution* is a brilliant mix of history, technology, and sociology that covers all aspects of the blockchain protocol—an invention that in time may prove as momentous as the invention of printing."

—James Rickards, author of *Currency Wars* and *The Death of Money*

"*Blockchain Revolution* serves as an atlas to the world of digital money, masterfully explaining the current landscape while simultaneously illuminating a path forward toward a more equitable, efficient, and connected global financial system."

—Jim Breyer, CEO, Breyer Capital

"*Blockchain Revolution* is the indispensable and definitive guide to this world-changing technology."

—Jerry Brito, Executive Director, Coin Center

"Incredible. Really incredible. The Tapscotts' examination of the blockchain as a model for inclusion in an increasingly centralized world is both nuanced and extraordinary."

—Steve Luczo, Chairman and CEO, Seagate Technology

"Makes a powerful case for blockchain's ability to increase transparency but also ensure privacy. In the authors' words, 'The Internet of Things needs a Ledger of Things.'"

—Chandra Chandrasekaran, CEO and Managing Director, Tata Consultancy Services

"The epicenter of trust is about to diffuse! The definitive narrative on the revolutionary possibilities of a decentralized trust system."

—Frank D'Souza, CEO, Cognizant

"Identifies a profound new technology movement and connects it to the deepest of human needs: trust. Thoroughly researched and provocatively written. Every serious businessperson and policy maker needs to read *Blockchain Revolution*."

—Brian Fetherstonhaugh, Chairman and CEO, OgilvyOne Worldwide

"*Blockchain Revolution* sets the table for a wave of technological advancement that is only just beginning."

—Frank Brown, Managing Director and Chief Operating Office, General Atlantic

"A must read. You'll gain a deep understanding of why the blockchain is quickly becoming one of the most important emerging technologies since the Internet."

—Brian Forde, Director of Digital Currency Initiative, MIT Media Lab

"Blockchain technology has the potential to revolutionize industry, finance, and government—a must read for anyone interested in the future of money and humanity."

—Perianne Boring, Founder and President, Chamber of Digital Commerce

"When generational technology changes the world in which we live, we are truly fortunate to have cartographers like Don Tapscott, and now his son Alex, to explain where we're going."

—Ray Lane, Managing Partner, GreatPoint Ventures; Partner Emeritus, Kleiner Perkins

"Don and Alex have written the definitive guidebook for those trying to navigate this new and promising frontier."

—Benjamin Lawsky, Former Superintendent of Financial Services, State of New York; CEO of The Lawsky Group

"*Blockchain Revolution* is an illuminating, critically important manifesto for the next digital age."

—Dan Pontefract, author of *The Purpose Effect*; Chief Envisioner, TELUS

"The most well-researched, thorough, and insightful book on the most exciting new technology since the Internet. A work of exceptional clarity and astonishingly broad and deep insight."

—Andreas Antonopoulos, author of *Mastering Bitcoin*

"*Blockchain Revolution* beautifully captures and illuminates the brave new world of decentralized, trustless money."

—Tyler Winklevoss, Cofounder, Gemini and Winklevoss Capital

"A fascinating—and reassuring—insight into a technology with the power to remake the global economy. What a prize. What a book!"

—Paul Polman, CEO, Unilever

CONTENTS

PART II: Transformations

ACKNOWLEDGMENTS

This book came from the meeting of two minds and two life trajectories. Don had been leading a $4 million syndicated research program called Global Solution Networks (GSN) at the Rotman School of Management, University of Toronto. The initiative was investigating new, networked models of global problem solving and governance. He researched how the Internet was governed by a multistakeholder ecosystem and became interested in digital currencies and their governance. Meanwhile, Alex was an executive with the investment bank Canaccord Genuity. He noticed the growing enthusiasm for early-stage bitcoin and blockchain companies in 2013 and began leading his firm's efforts in the space. During a father-son ski trip to Mont-Tremblant in early 2014, we brainstormed over dinner about collaborating on this topic, and Alex agreed to lead a research project on the governance of digital currencies, culminating in his white paper, titled *A Bitcoin Governance Network*. The more we dug into the issues, the more we concluded that this could be the next big thing.

Meanwhile our agent, Wes Neff at the Leigh Bureau, along with Don's publisher Adrian Zackheim at Portfolio/Penguin (*Wikinomics*, *Macrowikinomics*), was encouraging Don to formulate a new book concept. When Alex's paper became widely recognized as leading thinking in this area, Don approached Alex to be his coauthor. Adrian, to his credit, made us an offer we couldn't refuse and the book never went to auction, as is normally the case.

We then made what in hindsight was a smart decision. We approached the best book editor we knew, Kirsten Sandberg, formerly of Harvard Business School Press, and asked her to edit our book proposal. She did a spectacular job and our collaboration was so effortless that we asked her to be a full-time member of the book research team. Kirsten participated with us in more than one hundred interviews and collaborated in real time as we tried to understand the myriad issues on the table and develop helpful formulations to explain this extraordinary set of developments to a nontechnical audience. She helped us bring the story to life. In that sense, she was our coauthor and this book would not have appeared, at least in its current comprehensible form, without her. For that, and for all the stimulation and laugh lines, we are very grateful.

Our heartfelt thanks to the people below who generously shared their time and insights with us and without whom this book would not be possible. In alphabetical order:

Jeremy Allaire, Founder, Chairman, and CEO, Circle
Marc Andreessen, Cofounder, Andreessen Horowitz
Gavin Andresen, Chief Scientist, Bitcoin Foundation
Dino Angaritis, CEO, Smartwallet
Andreas Antonopoulos, Author, *Mastering Bitcoin*
Federico Ast, CrowdJury
Susan Athey, Economics of Technology Professor, Stanford Graduate School of Business
Adam Back, Cofounder and President, Blockstream
Bill Barhydt, CEO, Abra
Christopher Bavitz, Managing Director, Cyberlaw Clinic, Harvard Law School
Geoff Beattie, Chairman, Relay Ventures
Steve Beauregard, CEO and Founder, GoCoin
Mariano Belinky, Managing Partner, Santander InnoVentures
Yochai Benkler, Berkman Professor of Entrepreneurial Studies, Harvard Law School
Jake Benson, CEO and Founder, LibraTax
Tim Berners-Lee, Inventor, World Wide Web
Doug Black, Senator, Canadian Senate, Government of Canada
Perriane Boring, Founder and President, Chamber of Digital Commerce
David Bray, 2015 Eisenhower Fellow and Harvard Visiting Executive in Residence
Jerry Brito, Executive Director, Coin Center
Paul Brody, Americas Strategy Leader, Technology Group, EY (formerly IoT at IBM)
Richard G. Brown, CTO, R3 CEV (former Executive Architect for Industry Innovation and Business Development, IBM)
Vitalik Buterin, Founder, Ethereum
Patrick Byrne, CEO, Overstock
Bruce Cahan, Visiting Scholar, Stanford Engineering; Stanford Sustainable Banking Initiative
James Carlyle, Chief Engineer, MD, R3 CEV
Nicolas Cary, Cofounder, Blockchain Ltd.
Toni Lane Casserly, CEO, CoinTelegraph
Christian Catalini, Assistant Professor, MIT Sloan School of Management
Ann Cavoukian, Executive Director, Privacy and Big Data Institute, Ryerson University

Vint Cerf, Co-creator of the Internet and Chief Internet Evangelist, Google
Ben Chan, Senior Software Engineer, BitGo
Robin Chase, Cofounder and Former CEO, Zipcar
Fadi Chehadi, CEO, ICANN
Constance Choi, Principal, Seven Advisory
John H. Clippinger, CEO, ID3, Research Scientist, MIT Media Lab
Bram Cohen, Creator, BitTorrent
Amy Cortese, Journalist, Founder, Locavest
J-F Courville, Chief Operating Officer, RBC Wealth Management
Patrick Deegan, CTO, Personal BlackBox
Primavera De Filippi, Permanent Researcher, CNRS and Faculty Associate at the Berkman Center for Internet and Society at Harvard Law School
Hernando de Soto, President, Institute for Liberty and Democracy
Peronet Despeignes, Special Ops, Augur
Jacob Dienelt, Blockchain Architect and CFO, itBit and Factom
Joel Dietz, Swarm Corp
Helen Disney, (formerly) Bitcoin Foundation
Adam Draper, CEO and Founder, Boost VC
Timothy Cook Draper, Venture Capitalist; Founder, Draper Fisher Jurvetson
Andrew Dudley, Founder and CEO, Earth Observation
Joshua Fairfield, Professor of Law, Washington and Lee University
Grant Fondo, Partner, Securities Litigation and White Collar Defense Group, Privacy and Data Security Practice, Goodwin Procter LLP
Brian Forde, Former Senior Adviser, The White House; Director, Digital Currency, MIT Media Lab
Mike Gault, CEO, Guardtime
George Gilder, Founder and Partner, Gilder Technology Fund
Geoff Gordon, CEO, Vogogo
Vinay Gupta, Release Coordinator, Ethereum
James Hazard, Founder, Common Accord
Imogen Heap, Grammy-Winning Musician and Songwriter
Mike Hearn, Former Google Engineer, Vinumeris/Lighthouse
Austin Hill, Cofounder and Chief Instigator, Blockstream
Toomas Hendrik Ilves, President of Estonia
Joichi Ito, Director, MIT Media Lab
Eric Jennings, Cofounder and CEO, Filament
Izabella Kaminska, Financial Reporter, *Financial Times*
Paul Kemp-Robertson, Cofounder and Editorial Director, Contagious Communications
Andrew Keys, Consensus Systems

Joyce Kim, Executive Director, Stellar Development Foundation
Peter Kirby, CEO and Cofounder, Factom
Joey Krug, Core Developer, Augur
Haluk Kulin, CEO, Personal BlackBox
Chris Larsen, CEO, Ripple Labs
Benjamin Lawsky, Former Superintendent of Financial Services for the State of New York; CEO, The Lawsky Group
Charlie Lee, Creator, CTO; Former Engineering Manager, Litecoin
Matthew Leibowitz, Partner, Plaza Ventures
Vinny Lingham, CEO, Gyft
Juan Llanos, EVP of Strategic Partnerships and Chief Transparency Officer, Bitreserve.org
Joseph Lubin, CEO, Consensus Systems
Adam Ludwin, Founder, Chain.com
Christian Lundkvist, Balanc3
David McKay, President and Chief Executive Officer, RBC
Janna McManus, Global PR Director, BitFury
Mickey McManus, Maya Institute
Jesse McWaters, Financial Innovation Specialist, World Economic Forum
Blythe Masters, CEO, Digital Asset Holdings
Alistair Mitchell, Managing Partner, Generation Ventures
Carlos Moreira, Founder, Chairman, and CEO, WISeKey
Tom Mornini, Founder and Customer Advocate, Subledger
Ethan Nadelmann, Executive Director, Drug Policy Alliance
Adam Nanjee, Head of Fintech Cluster, MaRS
Daniel Neis, CEO and Cofounder, KOINA
Kelly Olson, New Business Initiative, Intel
Steve Omohundro, President, Self-Aware Systems
Jim Orlando, Managing Director, OMERS Ventures
Lawrence Orsini, Cofounder and Principal, LO3 Energy
Paul Pacifico, CEO, Featured Artists Coalition
Jose Pagliery, Staff Reporter, CNNMoney
Stephen Pair, Cofounder and CEO, BitPay Inc.
Vikram Pandit, Former CEO, Citigroup; Coinbase Investor, Portland Square Capital
Jack Peterson, Core Developer, Augur
Eric Piscini, Principal, Banking/Technology, Deloitte Consulting
Kausik Rajgopal, Silicon Valley Office Leader, McKinsey and Company
Suresh Ramamurthi, Chairman and CTO, CBW Bank
Sunny Ray, CEO, Unocoin.com
Caterina Rindi, Community Manager, Swarm Corp
Eduardo Robles Elvira, CTO, Agora Voting

Keonne Rodriguez, Product Lead, Blockchain Ltd.
Matthew Roszak, Founder and CEO, Tally Capital
Colin Rule, Chairman and CEO, Modria.com
Marco Santori, Counsel, Pillsbury Winthrop Shaw Pittman LLP
Frank Schuil, CEO, Safello
Barry Silbert, Founder and CEO, Digital Currency Group
Thomas Spaas, Director, Belgium Bitcoin Association
Balaji Srinivasan, CEO, 21; Partner, Andreessen Horowitz
Lynn St. Amour, Former President, The Internet Society
Brett Stapper, Founder and CEO, Falcon Global Capital LLC
Elizabeth Stark, Visiting Fellow, Yale Law School
Jutta Steiner, Ethereum/Provenance
Melanie Swan, Founder, Institute for Blockchain Studies
Nick Szabo, GWU Law
Ashley Taylor, Conensys Systems
Simon Taylor, VP Entrepreneurial Partnerships, Barclays
David Thomson, Founder, Artlery
Michelle Tinsley, Director, Mobility and Payment Security, Intel
Peter Todd, Chief Naysayer, CoinKite
Jason Tyra, CoinDesk
Valery Vavilov, CEO, BitFury
Ann Louise Vehovec, Senior Vice President, Strategic Projects, RBC Financial Group
Roger Ver, "The Bitcoin Jesus," Memorydealers KK
Akseli Virtanen, Hedge Fund Manager, Robin Hood Asset Management
Erik Voorhees, CEO and Founder, ShapeShift
Joe Weinberg, Cofounder and CEO, Paycase
Derek White, Chief Design and Digital Officer, Barclays Bank
Ted Whitehead, Senior Managing Director, Manulife Asset Management
Zooko Wilcox-O'Hearn, CEO, Least Authority Enterprises
Carolyn Wilkins, Senior Deputy Governor, Bank of Canada
Robert Wilkins, CEO, myVBO
Cameron Winklevoss, Founder, Winklevoss Capital
Tyler Winklevoss, Founder, Winklevoss Capital
Pindar Wong, Internet Pioneer, Chairman of VeriFi
Gabriel Woo, Vice President of Innovation, RBC Financial Group
Gavin Wood, CTO, Ethereum Foundation
Aaron Wright, Professor, Cardozo Law School, Yeshiva University
Jonathan Zittrain, Harvard Law School

Also special thanks to a few people who really rolled up their sleeves to help. Anthony Williams and Joan Bigham of the GSN project worked closely with

Alex on the original digital currencies governance paper. Former Cisco executive Joan McCalla did deep research for the chapters on the Internet of Things and also Government and Democracy. We received a lot of familial support. IT executive Bob Tapscott spent many days downloading and getting under the hood of the entire bitcoin blockchain to give us firsthand insights on some of the technical issues. Technology entrepreneur Bill Tapscott came up with the revolutionary idea of a blockchain-based personal carbon credit trading system, and technology executive Niki Tapscott and her husband, financial analyst James Leo, have been great sounding boards throughout. Katherine MacLellan of the Tapscott Group (conveniently a lawyer) tackled some of the tougher issues around smart contracts as well as managing the interview process. Phil Courneyeur was on the lookout daily for juicy material, and David Ticoll provided helpful insights about the state of the digital age so far. Wes Neff and Bill Leigh of the Leigh Bureau helped us craft the book concept (how many books is this, guys?). As always (now more than twenty years), Jody Stevens flawlessly managed the administration for the entire project including databases, finances, and document management, as well as the proofreading and production process—a full-time job, in addition to her other full-time jobs at the Tapscott Group.

Special thanks to Dino Mark Angaritis, the CEO of blockchain company Smartwallet; Joseph Lubin, CEO of the Ethereum development studio Consensus Systems; and Carlos Moreira of fast-growing security company WISeKey—who each spent considerable time with us brainstorming ideas. They are each brilliant and so kind to help us out. Now we get to enjoy witnessing the success of each of their businesses in this space. Also big thanks to the great team at Penguin Random House led by our editor Jesse Maeshiro and overseen by Adrian Zackheim.

Most important, we'd like to give our heartfelt thanks to our wives, Ana Lopes (Don) and Amy Welsman (Alex), who more than tolerated our obsession with cracking this big nut over the better part of a year. We are both very fortunate to have such wonderful life partners.

Writing this book has been a joyous experience for both of us and it's fair to say that we loved every minute of it. As someone famous once said, "If two people agree on everything, one of them is unnecessary." We challenged each other daily to test our beliefs and assumptions, and this book is living proof of that healthy and vigorous collaboration. Mind you, collaborating does seem effortless when you share so much DNA and have a shared thirty-year history of exploring the world together. We do hope you find the product of this collaboration important and helpful.

Don Tapscott and Alex Tapscott, January 2016

ACKNOWLEDGMENTS FOR THE PAPERBACK EDITION

As with the original edition, our most important collaborator was our editor-in-chief Kirsten Sandberg, who, in addition to leading our BRI editorial and production team, contributed substantively to the content and helped to make the words come alive.

We acknowledge the following people whose wisdom and insights shaped this new material: Brian Behlendorf, executive director, Hyperledger; Ethan Buchman, cofounder and CTO, Tendermint; Anthony Diiorio, cofounder, Ethereum, founder, Decentral; Olga Feldmeier, CEO, Smart Valor; David Jaffray, executive vice president, the University Health Network; Min Kim, cofounder, ICON; Trevor Koverko, CEO, Polymath; Jae Kwon, Founder and CEO, Tendermint; Joe Lubin, founder, Consensus Systems; Pete Martin, founder and CEO, Votem; Charlie Morris; Andrey Petrov; and Matthew Spoke, founder, Aion.

We thank our founding members—Accenture, Barrick, Bell, Capgemini, Centrica, CIBC, City of Toronto, FedEx, Fujitsu, Government of Canada, IBM, Interac, KPMG, Liberty Global, MKS, Nasdaq, PepsiCo, Province of Ontario, Raiffeisen Bank, SAP, Tencent, Thomson Reuters, TMX, and the University Health Network—and our affiliate organizations, namely the Chamber of Digital Commerce, COALA, Enterprise Ethereum Alliance, and Hyperledger, and the Illinois Blockchain Initiative.

We also thank our BRI contributors: Chami Akmeemana, Nolan Bauerle, Tom Baumann, Oliver Bussmann, Davide Cargnello, Michael Casey, Stephen Caswell, Soumak Chatterjee, COALA, Alan Cohn, Chloe Desmonet, Jeremy Epstein, Tommy Gardner, Vlad Gheorghiu, Bill Gillies, Sergey Gorbunov, Dom Guinard, Jennifer Han, Stefan Hopf, Tom Isaacson, Matt Jackson, Reshma Kamath, Christian Keil, Henry Kim, Marek Laskowski, Alan Majer, Massimo Morini, Bob Morison, Michele Mosca, Bill Munsun, Vineet Narula, Marcus O'Dair, Iliana Oris Valiente, Lawrence Orsini, Andreas Park, Mary-Jane Pilgrim, Abhishek Punia, Rachel Robinson, Jill Rundle, Axel Schumacher, Michael Scott, Tony Scott, Tom Serres, Prema Shrikrishna, Usman Sheikh, Mayank Singhal, Chase Smith, Andy Spence, Nick Szabo, Bob Tapscott, Joel Telpner, Tomicah Tillemann, Mark van

Rijmenam, Bettina Warburg, Anthony Williams, Irving Wladawsky-Berger, and Alan Wunsche.

Last but not least, we thank the BRI staff—Carl Amorim, Daria Apelsinova, Joan Bigham, Luke Bradley, Hilary Carter, Roya Hussaini, Genia Mikhalchenko, Jenna Pilgrim, Jane Ricciardelli, Yuliya Samoylova, and Jody Stevens—and the chair of the BRI advisory board, Sanjay Tugnait.

—Don Tapscott and Alex Tapscott, April 2018

PREFACE TO THE PAPERBACK EDITION

Don Tapscott and Alex Tapscott

CONTENTS

THE BIG IDEAS

When we wrote *Blockchain Revolution*, we got off to a good start by characterizing blockchain—the underlying technology of cryptocurrencies—as the Internet of value. We explained that, for nearly four decades, we've had the Internet of information. It vastly improved the flow of data within and among firms and people, but it hasn't transformed how we do business. That's because the Internet was designed to move information—not value—from person to person. When we e-mail someone a document, photograph, or audio file, we're really sending a copy of our original. This information is abundant, unreliable, and perishable. Anyone else can copy, change, and send it to somebody else. In many cases, it's legal and advantageous to share these copies.

In contrast, to expedite a business transaction, we cannot e-mail money directly to someone—not just because copying money is illegal but because we can't be 100 percent sure our recipient is who he says he is. Information about identity needs to be scarce, permanent, and unchangeable. So we go through powerful intermediaries to establish trust and maintain integrity. Banks, governments, and even big technology companies confirm our identities and enable us to transfer assets; they clear and settle transactions and keep records of these transfers. But the limitations of these intermediaries—their operational opacity and their vulnerability to hackers, rogue employees, and equally vulnerable suppliers—are becoming more apparent. We need a new way forward.

Blockchain solves the double-spend problem, as cryptographers call it. Now for the first time ever we have a native digital medium for value, through which we can manage, store, and transfer any asset—from money and music to votes and Stradivarius violins—peer to peer in a secure and private way. Trust is achieved not necessarily by intermediaries but by cryptography, collaboration, and clever code. We almost titled the book *The Trust Protocol.*

It seems that *Blockchain Revolution* was a clearer title—it's still a best seller, as of this writing. The response to it has both encouraged and delighted us. It received widespread coverage from such respected media as the *Financial Times*, *Forbes*, *Fortune*, *The Guardian*, *Harvard Business Review*, *Newsweek*, NPR's *All Things Considered*, *Reuters*, *Time*, and *The Wall Street Journal* and was a feature article in *The New York Review of Books* and the subject of a PBS television special.

It's gone global, too—translated into fifteen languages so far and, as of now, a best seller in five Asian languages alone. Don's second TED talk

(TED's first on blockchain) has received well over three million views. At 2017 TEDxSanFrancisco, Alex spoke on blockchain and financial services; his has become one of the most watched talks on the topic, too.

When we first published in May 2016, ours was one of a handful of serious books about the topic. Now several important new works have entered the market such as Michael Casey and Paul Vigna's *The Truth Machine*, Chris Burniske and Jack Tatar's *Cryptoassets*, and Primavera De Filippi and Aaron Wright's *Blockchain and the Law*, to name a few.

Our book continues to hold its own as the bestselling book on blockchain. We receive positive comments on a number of its big ideas:

1. The book underscores the importance of identity and the end of digital feudalism. What some called "surfing the Internet," we viewed as "serfing the Internet," throwing off our data for the Internet landowners to expropriate and monetize. The notion of a self-sovereign identity for each of us, with our personal data stored in a virtual black box, is one of the most foundational concepts of our time. Realizing this "Virtual You" through blockchain technologies could restore our control over our own identities, the data we create, and the rest of our rights. No serf surfing, we say.

2. As a thought experiment, we tried to get inside Satoshi's mind and tease out his design principles for blockchain. It turns out there were seven. That chapter (chapter 2) was technical, appealing more to technologists and business engineers. We applied these seven principles to seven domains—financial services (chapter 3), the architecture of the firm (chapter 4), business model innovation (chapter 5), the Internet of Things (chapter 6), economic inclusion (chapter 7), government and democracy (chapter 8), and the creative industries (chapter 9)—and argued that blockchain would create seven new substructures for a distributed economy.

3. We dubbed the financial services industry a Rube Goldberg contraption, a ridiculously complex system that actually performs eight basic functions. That taxonomy has proven helpful for industry executives and regulators alike. Do take a look at chapter 3 and the Golden Eight. Smart contracts (aka distributed applications) on a blockchain could, in theory, do each of these eight to disintermediate incumbents. Conversely, incumbents could transform their businesses for the better, if they embrace blockchain.

4. Nobel Prize–winning economist Ronald Coase's theory on the firm proved quite applicable to an analysis of blockchain's impact on corporate architecture. We explained how blockchain would radically reduce the transaction costs of search, coordination, contracting, and building trust in an open market. Inexorably, this efficiency will lead to more decentralized models for orchestrating the capabilities needed to create new products, services, and wealth. The new "blockchain business models" that we described hold up well, and many new ones have emerged since the book's publication. Decentralized business models are subject to network effects so that, when the number of nodes increases, so does the network. This in part explains the rapid growth of cryptoassets.

5. Blockchain can help us solve the prosperity paradox, where developed economies grow but the middle class and prosperity for most stagnates. Rather than the usual solution—the redistribution of wealth through taxation—we explained how blockchain could help us *pre*distribute wealth by including billions of people in the global economy. For example, we could protect property rights through immutable land titles, create a true sharing economy through shared, open, and distributed platforms, empower diasporas to remit funds through low-fee mobile payment systems, and endow entrepreneurs with the same capabilities as large companies.

6. Soon most transactions will occur between things, not people. We can instill intelligence into our infrastructure by adding smart devices—sensors, cameras, microphones, global positioning chips, gyroscopes—that reconfigure themselves according to availability of bandwidth, storage, or other capacity, and therefore resist interruption. Blockchain is critical. This Internet of Things depends on a Ledger of Things to track every node, ensure its security and reliability, record its production and consumption, and schedule and pay for its maintenance or replacement. There are potential applications across every sector.

7. Our work on blockchain applications in government, democracy, and culture has received much attention. Since Donald Trump's inauguration as U.S. president, our insights seem even more prescient. Engaged citizens and dedicated public servants everywhere are exploring how blockchain can help them reinvent government, protect the free press, restore legitimacy to democratic institutions, and find common ground in public discourse on the Internet. The technology also helps not only journalists to quash claims of "fake news" but also creators of such cultural assets as songs and art to receive fair compensation for their work.

8. We were reluctant to include a chapter on leadership and governance, but we're glad we did. The space is full of formal and informal leaders, that is, those with executive roles in start-ups, blockchain consortia, and regulatory bodies, and those whose vision and talent are both compelling and influential. That said, concerted effort to transform obstacles into opportunities has been the most important factor in the blockchain's success thus far. So crucial is blockchain stewardship that the World Economic Forum asked us to write a special report on governance and launched important programs based on that work.

We also cofounded the Blockchain Research Institute (BRI), a think tank on distributed ledger technology, to investigate blockchain use cases, transformative thought leadership, and implementation challenges. The multimillion-dollar program includes some seventy-five projects across ten industry verticals and seven C-suite roles in both public and private sectors. Many of the quotes in this new preface come from the leaders of these projects.

BRI membership consists of large corporations, governments, nonprofits, and members of the start-up community. Some of our founding members include IBM, Accenture, Capgemini, SAP, NASDAQ, CIBC, PepsiCo, Liberty Global, Tencent, Fujitsu, FedEx, Thomson Reuters, and Centrica, along with the governments of several countries. To our delight, our institute's editor-in-chief is Kirsten Sandberg, who was the original editor of *Blockchain Revolution*.

Notwithstanding all this goodness, a lot of water has gone under the bridge. While the book holds up well, we wanted to report on our latest discoveries in this new edition. Rather than revise the whole manuscript, we are consolidating our findings in this new preface and an afterword. This new material derives from our ongoing research, investments in the space, and speaking engagements around the world. We welcome your feedback (www.blockchainresearchinstitute.org/contact-us).

CRYPTOASSETS AND THE NEW REVOLUTION IN FINANCIAL SERVICES

When *Blockchain Revolution* went to print in May 2016, the entire cryptoasset market had a value of $9 billion. Ethereum had just crossed $1 billion in network value, becoming the second blockchain unicorn (after bitcoin). These were early days. Had the cryptoasset market been a public company, it would barely have cracked the S&P 500 index.[1] Fewer than two years later, the cryptoasset market is $420 billion in size.[2]

This explosion of value in cryptoassets has captured the imagination of developers, entrepreneurs, nongovernment organizations, and the media, not to mention governments, central banks, the investing public, and regulators. It has also thrust these digital assets (and the underlying blockchain technology), once the domain of a few passionate technologists, into mainstream interest. It has made enthusiasts euphoric, Nobel laureates skeptical, and old-school billionaires dyspeptic.[3] Charlie Munger of Berkshire Hathaway went so far as to call bitcoin "noxious poison."[4] (Is there any other kind of poison?)

Vitalik Buterin, Ethereum's inventor, captured the dissonance in late 2017 when the cryptoasset market cap hit half a trillion dollars. He tweeted, "Have We *Earned* It?"[5] "How many unbanked people have we banked?"[6] "How much value is stored in smart contracts that actually do anything interesting?"[7] Buterin pointed out that the level of activity is positive, but perhaps not significant enough to warrant the size of the market. "The answer to all of these questions is definitely not zero, and in some cases, it's quite significant," he added. "But not enough to say it's $0.5T levels of significant. Not enough."[8]

To be sure, there is a lot of hype in this market. For every cryptoasset that succeeds, many fail. Scammers have an outsized negative effect on the space as a whole. According to Reuters, "Twitter Inc. will start banning cryptocurrency advertising . . . joining Facebook and Google in a clampdown that seeks to avoid giving publicity to potential fraud or large investor losses."[9] Moreover, the industry must confront serious challenges. How will these technologies scale? How will incumbents react? What will governments and regulators do? We have good reason to believe this industry urgently needs sound regulation to protect investors and thwart fraudsters, or at least hold them accountable for their crimes. Moreover, to continue investing and building in this technology, market participants need to understand the rules of the road. On the other hand, bad regulations (even with the best intentions) can have unintended consequences and stifle innovation. In some countries, multiple regulators with overlapping mandates are sending conflicting messages. Regulators are not in an easy position. Some jurisdictions, such as Switzerland and Singapore, have emerged as favorable locations for companies to locate and operate with positive outcomes for the local economy. By one (informal) estimate, three thousand jobs have been created in the so-called "Crypto Valley" around Zug and Zurich in the past few years. The Crypto Valley Association has over six hundred members. Smaller and nimbler, these jurisdictions have been able to capitalize on a new industry, though they remain the exception, not the norm. For now, the lack of regulatory clarity in general has created uncertainty.

These are such important issues that we dedicated all of chapter 10 to them, "Overcoming Showstoppers: Ten Implementation Challenges." We continue to view them as implementation challenges to overcome. If we look beyond the hype and mania (not to mention fear, uncertainty, and doubt), we see something profound happening. Bitcoin was the first move in a long campaign to create an entirely new technology stack for the Internet, enabling the first native digital medium for value. That's what blockchain is, and it's limited only by our imagination. Some inventors have imagined a whole new asset class with what we think are at least seven types:

- Cryptocurrencies (bitcoin, Zcash, Monero, and Dash)
- Protocol tokens (ether, ICON, Aion, COSMOS, NEO)
- Utility tokens (Golem, BAT, Spank)
- Securities tokens (cryptoequities, cryptobonds)
- Natural asset tokens

- Crypto collectibles (CryptoKitties, Rare Pepe)
- Crypto fiat currencies and stablecoins (Fedcoin proposal, Singapore's Project Ubin, MakerDAO)

We are witnessing one of the largest transformations of wealth in human history, from paper-based analog assets to digital ones. To be sure, $265 billion is a lot of money. But in terms of all the assets in the world—from stocks, bonds, and mortgages to carbon, land, and water—we have barely scratched the surface of what we can create with crypto.

Is this all a bubble? Possibly. Joseph Lubin, CEO of Consensys and cofounder of Ethereum, says, "We will see bubble after bubble in our space, each one with higher highs and higher lows. I think that's perfectly reasonable. People claim that the dot-com era of boom and bust was destructive, but I would call it creatively destructive."[10] It may have harmed those looking only to make a quick buck, but it otherwise sorted out the sustainable business models from the unsustainable ones, and it weeded out inefficient operations. Perhaps more important, talent shifted to this new area of the economy, and the excitement of the Internet era precipitated billions of dollars of investment in new technology infrastructure.

However, blockchain differs from the Internet in two important ways. First, where the Internet was a free utility built by a diverse group of stakeholders, many of them volunteers with little financial incentive, blockchain provides huge financial rewards for those who can build successful, scalable, and widely used technology through the appreciation of underlying cryptoassets. The early Internet pioneers probably would have appreciated some upside from building a utility worth trillions of dollars, but that was impossible.

Blockchain is different—creators and early adopters can participate *directly* and *financially* in the growth of the second era of the Internet. As a result, there is no "one blockchain" but an explosion of competing, overlapping, complementary platforms, all driven by incentives.

Second, blockchain is tackling value industries such as financial services and supply chains, far larger than information industries like media and publishing. So not only will the impact be greater but the aggregate value will be, too. The excitement is indeed palpable. But, as the saying goes, sometimes we need a little irrational exuberance to build the future.

1. Cryptocurrencies

When *Blockchain Revolution* came out, bitcoin was worth around $7 billion. Today it's more than twenty-two times that. Bitcoin is the workhorse of the cryptocurrency world and the cryptocurrency that launched a thousand ships. Bitcoin has become: a store of hundreds of billions of dollars of value on the most robust computer network ever formed (and entirely bootstrapped), a secure payment system that enables billions of dollars in daily on-chain transactions, a reserve currency for the burgeoning cryptoasset world, a final settlement layer when it's time to cash out, and a favorite punching bag for every armchair analyst in the world. Paradoxically, bitcoin's meteoric price rise makes it easier, not harder, for new investors to justify stepping in because it has become an asset class too big to ignore. Moreover, the bigger it gets, the more utility it has. With the launch of the Lightning Network and other scaling solutions in 2018, bitcoin may also fulfill the promise of its most ardent supporters and obliterate the need for traditional financial intermediaries (chapter 3).

To wit, consider the recent shift in tone of some of the biggest banks. When *Blockchain Revolution* went to print, most banks were tactfully supporting the potential for blockchain but dismissing bitcoin (and its crypto brethren) out of hand. "Bitcoin bad, blockchain good" became cliché. As late as 2017, Jamie Dimon, CEO of JPMorgan Chase, was calling bitcoin a fraud. (He has subsequently changed his mind.) Times have changed. In February 2018, Goldman Sachs–backed Circle acquired Poloniex, one of the world's largest cryptocurrency exchanges, suggesting that it sees risks and opportunities in cryptoassets. In its 2017 annual report, JPMorgan echoed Bank of America in acknowledging that cryptocurrencies could pose a risk to its business: "Both financial institutions and their non-banking competitors face the risk that payment processing and other services could be disrupted by technologies, such as cryptocurrencies, that require no intermediation."[11]

Taken alone, bitcoin's impact on culture and the economy has been extraordinary. Its endowment to the world will continue to be profound. More recently, an emphasis on privacy has shaped newer entrants in the currency use case for cryptoassets. New cryptocurrencies such as Zcash and other "privacy coins" have emerged that build upon bitcoin's principles but add this new functionality.[12] This is not just the domain of cypherpunks and other Internet communities: JPMorgan integrated Zcash's core anonymity technology (zero-knowledge proofs) into its own Quorum blockchain for use cases in a range of asset classes and business functions.[13] That JPMorgan was spending time, energy, and capital pushing the boundaries of this technology's wildest frontier, while its CEO was simultaneously denouncing it, suggests that (at least until recently) the bank's technologists understood the potential of blockchain more than its management did. Another intriguing new entrant is Metronome, which can be "imported and exported across chains," with the initial issuance happening on the bitcoin, Ethereum, Ethereum Classic, and Qtum networks.[14] As we will see in the next section on platforms, interoperability is a big challenge and opportunity in this space. Zcash and Metronome join Dash, Monero, and others vying for market share in the cryptocurrency sphere of this market. But currencies as a use case are the beginning of this story. Consider Ethereum.

2. Platforms

To the outside world looking in, Ethereum and bitcoin could be mistaken for two sides of the same coin—cryptocurrencies designed to function as cash for the Internet. This view couldn't be further from the truth. Whereas bitcoin serves such a purpose, Ethereum is a platform technology, designed from the outset to enable *distributed applications* (DApps), what Nick Szabo calls "an application that runs in a distributed and trust-minimized manner on a blockchain."[15] At the core of distributed applications are smart contracts, software that mimics the logic of a business agreement. Because they are decentralized and running on blockchains, they minimize the need for intermediaries (banks, brokers, lawyers, courts, escrow agents, corporations) to guarantee execution.

The promise of Ethereum was basically theoretical when we were writing the book: it launched only weeks before our first draft had gone to the editor. Yet today, Ethereum's native token (ether) has a market value of $70 billion. More important, Ethereum emerged as the leading platform for ICOs, where a project can raise millions of dollars peer to peer from a global community of investors and supporters. To date, dozens of new distributed applications have been launched on the Ethereum network. In aggregate, some $3 billion have been raised on Ethereum using its ERC-20 protocol, making Ethereum the proto–investment bank for the digital economy. By some estimates, 70 percent of all distributed applications now run on the Ethereum blockchain, giving it powerful network effects that will be hard to dislodge. Ethereum has also galvanized such large enterprises as Microsoft, JPMorgan, and BP, which collectively established the Enterprise Ethereum Alliance in 2017.

As expected, some of these distributed apps have made a great deal of progress, while many others have floundered. For every great start-up that changes the world, countless others fail and most are forgotten. Platforms like Ethereum, however, are largely agnostic to the success of any one distributed application, so long as the next big thing is built on them. The more DApps built on the network, the more demand for the associated platform token, ether. If Ethereum is the city grid, and the DApp is the car, then ether is the fuel, or "gas" in crypto parlance. We pay in ether to use the network for running the smart contract that powers the DApp. But will Ethereum be the platform for the next generation of distributed applications? Will it be one of the core protocols of the new Internet of Value, or will something else take its place? It's currently the best candidate for a "flippening," the point at which an alternative blockchain displaces bitcoin as the network with the most participants and most capital.[16] Massive work is under way to expand Ethereum's capability, including Casper, sharding, and a shift to proof of stake.

A number of other emerging platforms could challenge or complement it. DApp-focused platforms such as NEO (China), ICON (South Korea), and other regional leaders have emerged. Protocols such as Aion designed with large-scale enterprise applications in mind—a huge but generally untapped market—have also emerged, while some of the biggest hype is reserved for still-unreleased protocols like Polkadot and Cosmos, which promise to eliminate scalability and interoperability bottlenecks and unite all blockchains into a giant seamless web of blockchains. All protocols will not succeed, but some will, and those that remedy the showstoppers (chapter 10) will form the backbone of the next era of the Internet.

3. Utility Tokens (App Coins)

In chapter 3, we wrote about Augur, a prediction market designed to harness the wisdom of crowds in order to make markets in virtually anything.

To us, Augur illustrated the potential power of blockchain technology. It was also among the first projects to issue funds in a crowdsale on the blockchain. (We dubbed it the "blockchain IPO," but the term never took off. Instead, people latched on to "initial coin offering," a misnomer if ever there was one.) Augur proved a harbinger of what was to come. In 2016, roughly $165 million was raised in ICOs, which was interesting but not really enough to raise eyebrows outside the blockchain community. By 2017, the figure had reached at least $3 billion, perhaps as much as $7 billion. Joe Lubin believes this new fund-raising mechanism is "democratizing the ability for projects to fund themselves either via tokenized securities issued in a global context or by selling utility tokens that provide consumer membership, consumer access to services, or access to scarce resources . . . basically preselling something and using those proceeds to build what you need or to take it from a rudimentary stage to a more sophisticated stage."[17]

Augur's native token is not equity but a utility token required by users to interact with the network—in effect, a programmable blockchain asset that has functionality in the distributed application. Most ICOs in 2018 were "utility tokens," though many were probably also securities. Consider Golem, a decentralized alternative to today's centralized clouds run by such digital conglomerates as Amazon and Apple. Golem aims to harness the power of the billions of devices used daily to distribute computation. For its model to work, it needs an incentive for participation. So in 2017, Golem issued a utility token that allows users to pay and get paid on its platform. If Golem works, it could disrupt cloud computing as we know it.

Another example is Sweetbridge, which originated the concept of a "discount token," where users receive a monthly discount on goods and services as long as they hold the token in a Sweetbridge wallet. "The amount of the discount is controlled by the revenue in the network and the number of discount tokens held in their wallet. This means that discount tokens have an intrinsic value that increases as more customers use the network, says Scott Nelson of Sweetbridge. "Discount tokens change the business from driven by shareholder value to [driven by] customer value, making the customer the center of the focus of a business."[18] Others are pioneering myriad so-called cryptoeconomic models for utility tokens in virtually every industry.

Utility tokens are usually not stand-alone blockchains. Rather, they run on top of platforms like Ethereum, ICON, and EOS. To be clear, the borders between utility tokens and the underlying platform tokens can be porous. After all, protocol tokens also have utility, as with the ether used to pay transaction fees on the Ethereum network. Some protocols today have only one application. Tomorrow, they may have a lot more. Filecoin, a distributed file sharing system, completed its own ICO in the summer of 2017. However, because it is an open network, developers will ultimately be able to build any number of applications on it. Exceptions notwithstanding, we believe most utility tokens will be application-based and run on networks such as Ethereum.

4. Security Tokens

Though not insignificant, the $265 billion cryptoasset market is a small fraction of the value of virtually any other major asset class. The global equity market, for example, is more than $100 trillion. However, the underlying technology of cryptocurrencies, blockchain, is broadly applicable to basically any asset in the world.

The next ten years will see today's cryptoassets lose their monopoly as securities, particularly nonphysical securities like stocks and bonds, migrate to this technology and increasingly dominate the market. After all, why should a stock trade settle T+3 and involve a handful of intermediaries when buyer and seller can conduct the same transaction peer to peer and settle T+0 on a decentralized exchange? Why shouldn't all stocks, bonds, dividends, futures, forwards, swaps, options, and other financial assets exist in purely digital form on blockchains? An "equity token," for example, is not merely a thumbprint on a blockchain representing some off-chain asset but a native digital asset that we can trade peer to peer without custodians, clearinghouses, brokers, exchanges, and banks.

ICOs have already upended venture capital. Wall Street could be next. To wit, Fidelity, Wellington, and other giants of asset management have taken steps to prepare themselves for this brave new world.

While projects and companies like Polymath, Overstock's tZero, the Jibrel Network (a platform for security token offerings using ERC-20), and the Canadian Securities Exchange build out the technology infrastructure for such a historic transformation, the industry awaits the regulatory infrastructure to give it clarity. This gulf between technology and rule setting creates what legal scholar and blockchain expert Primavera De Filippi calls a "regulatory lag" or "governance gap," which "has resulted in the destabilization of traditional mechanisms of adjudication and rule-making, and the erosion of public confidence in the 'state of play'—that is, what is permissible and what is not."[19] Security tokens could help bridge this divide by defining themselves by what they are *not*. They are *not* cryptocurrencies, protocols, or utility tokens but "digital bearer assets" (securities) native to blockchains. The offspring of ICOs, *security token offerings* (STOs), will become ubiquitous in venture capital and financial services more generally.

This great migration of value from analog to digital will transform the roles of markets and intermediaries as we know them.

5. Natural Asset Tokens and Commodity Tokens

Natural assets such as water, carbon, and air are foundational to the economy and essential for life on earth. However, with the exception of some nascent carbon trading schemes, these assets have largely remained immune to market-based forces. This has led to overuse and exploitation of these resources, with costs borne by society in the form of what economists call "negative externalities." Sociologist Garrett Hardin describes this as the tragedy of the commons—a situation where a shared common resource is depleted because there is no system to govern its use or consumption.

Michael Casey, coauthor of *The Truth Machine*, uses the work of Hardin as a jumping-off point to examine the role of blockchain in helping to solve this problem of governance. He writes, "With the advent of blockchain technology and the cryptocurrencies, cryptotokens, and other digital assets that it has engendered, we may be moving toward a model of programmable money that can deliver a more automated system of internal governance over common resources." Indeed, in much the same way that we can tokenize technology protocols, applications, and securities, so too can we tokenize physical assets in the real world. "The great promise of the token economy is that it might solve the Tragedy of the Commons," writes Casey.[20]

Mostly, entrepreneurs and enterprises have applied this concept to traditional commodities with established markets, such as gold, oil, natural gas, etc. Indeed, it's true that we can apply the same principle of security tokens to physical commodities like these. Replicating the business logic of an oil-futures contract on the blockchain is feasible as blockchain start-up Nuco demonstrated with the TMX Group, owner of the Toronto Stock Exchange. Even though someone still needs to take physical possession when the contract expires, we can still simplify the mechanism of clearing and settling a trade in an underlying physical asset. In many ways, we could use a token backed by gold as a less volatile and more liquid medium of exchange (see section 7). For example, the Royal Mint partnered with the Chicago Mercantile Exchange to create Royal Mint Gold, a digital gold token backed by physical gold held in the Royal Mint's vaults.[21]

To be sure, we have opportunities to streamline and simplify existing markets. However, as with all technologies, the bigger opportunities are in new and previously impossible use cases. To wit, today's carbon trading schemes create a marketplace for carbon and reward companies for good behavior, allowing them to earn credits for reducing their carbon footprint. If companies can be rewarded for good behavior, why can't people? As it exists today, the market is weighed down by a lack of standards and highly fragmented and regional marketplaces.

Blockchain could change that by aligning incentives with a common and collective goal, such as reducing carbon emissions.[22] Companies like CarbonX (Canada) and Veridium (United States) are tackling this market by tokenizing carbon into fungible liquid tokens. By reducing their footprint, individuals can earn carbon credits redeemable for real value. Compared with cryptocurrencies, utility tokens, and even security tokens, natural asset tokens are a tiny market. Most of what has been proposed is theoretical, and there are real challenges such as government policy and regulations that blockchain alone cannot hope to solve. However, with a massive and untapped underlying market and pressing social, economic, and environmental reasons to move forward, it is only a matter of time before this becomes one of the largest cryptoasset types.

6. Crypto Collectibles: Virtual and IRL

In December 2017, the crypto world caught CryptoKitty fever. CryptoKitties are unique, tradable virtual pets that people can purchase, raise, and even breed with other CryptoKitties. As of January 2018, CryptoKitties had more than 235,000 users and had processed $52 million in transactions. CryptoKitties became so popular that the Ethereum network, on which this particular DApp was running, initially struggled to keep up, surely a sign of both the powerful network effects of popular apps and the current limitations of the underlying platform technology. At its peak, the dearest CryptoKitties were selling for more than $100,000. The phenomenon became personal when a close friend told us that he and a new girlfriend were considering taking their relationship to the next step by breeding their CryptoKitties to create a cryptobaby—surely a novel and modern spin on "Let's get a dog." Such "silly and fun things are important" in engaging people with breakthrough technology, said Elon Musk as he blasted his sports car into space.[23] So it is with CryptoKitties, an example of crypto collectibles.

There are two kinds of crypto collectibles. The first are native digital assets that have no equivalent in real life. CryptoKitties and virtual trading cards (such as Rare Pepe) spring to mind. So, too, do in-game purchases of unique assets of virtually any kind. Artists are applying cryptoeconomics to their virtual art. Art derives much of its value because it is scarce. But the Internet of information allowed us to copy free forms of expression, such as images and songs, ad infinitum, reducing the value to zero and losing track of the original. The blockchain connects the creative work to a unique and scarce token. In *The New York Times*, Scott Reyburn recently wrote, "Will cryptocurrencies be the art market's next big thing?" He explored a number of artists working solely in the virtual world.[24] The opportunities are tantalizing. As art and other forms of expression increasingly begin as a digital medium, whole new categories of virtual art, collectibles, and other unique assets could explode in value. The second kind of crypto collectible represents a claim on something tangible. Whereas we will eventually have 21 million bitcoin in circulation, each CryptoKitty is *unique*, as is every Rothko, Picasso, Monet, and Pollock. In chapter 9, we wrote about a company called Artlery, which employs an art-backed cryptocurrency called the CLIO to register physical artwork in the real world. More have joined the fray, including Dada.nyc.[25] While virtual art is a growing market, the existing art market is enormous. Total sales in 2016 of fine art and antiques was $45 billion.[26] Notoriously opaque, this market is beginning to benefit from the disinfecting sunlight of blockchain. Artwork can get a digital fingerprint through a cryptoasset that allows us to trace, track, and authenticate it.

7. Crypto Fiat Currencies and Stablecoins

In 2017, Venezuela announced that it was launching a new cryptocurrency, dubbed "The Petro," backed by its vast oil reserves. The reaction from the cryptocurrency community was a mix of dumbfoundedness and anger. Why would a corrupt and antidemocratic government, which had plunged its own currency into a hyperinflationary death spiral, co-opt this technology if not to exploit its association with trust, security, and immutability? According to analysts, it has three strikes against its credibility: there is no evidence that the Petro is actually backed by oil, there is little technical information online about how it works or which blockchain it runs on, and it is controlled by the same people who collapsed the bolívar.[27]

Unbowed by the criticism, the government moved ahead and raised $735 million, according to officials but not corroborated by any other evidence. News of the Petro was quickly followed by reports that lawmakers in Iran and Russia were also considering their own fiat cryptocurrency. All of these countries have three traits in common: they are authoritarian (or deeply undemocratic), they have a lot of oil, and they are under sanctions. So necessity is the mother of invention after all.

Why does this matter? Most obviously, it shows how rogue governments could use their cryptocurrencies to undermine international law, treaties, and sanctions and further destabilize their already weak economies. The Brookings Institution wrote that the Petro would harm other legitimate cryptocurrencies and undermine international sanctions.[28]

More important, it demonstrates that governments can actually do this. In chapter 11, we ruminated on the idea of a government-backed cryptocurrency, though none existed at the time. Indeed, the most promising candidates—august institutions like the Bank of England, Bank of Canada, and Federal Reserve—have made little headway in this regard, with some even backtracking. They should reconsider.

Crypto fiat currencies will probably not be fully decentralized and censorship resistant, like bitcoin. However, implemented properly, they can still make markets more efficient through real-time settlement, improve inclusion by reducing barriers to entry, improve transparency into our institutions, and make central bank policy more effective by improving responsiveness. Consider the example that Bitt is setting in the Caribbean. The company is working with the region's financial heads to create a digital dollar standard that has a number of benefits to the economy. CEO Gabriel Abed explains, "This is what the Caribbean needs. It's the entire world in one little melting pot, yet there is no cross-border system for payments. . . . The goal is to enable movement of money between two central banks using smart contracts and digital dollars built by Bitt or others that follow a digital dollar standard."[29] There are economic and social reasons to make this happen. Abed says, "Remittances are expensive because interregional settlement is not existent. Forty percent of Caribbeans don't have access to banking. Three percent fees are being taxed by foreign bankers on merchant charges using credit cards." A digital dollar standard for the region could help alleviate these problems.

Another benefit is price stability. Media of exchange are generally not as volatile as bitcoin has been historically. A crypto fiat currency could help solve this. Some crypto diehards will balk. So be it. We still believe that bitcoin (or something like it) will continue to be a legitimate alternative to fiat currencies.

Stablecoins—cryptocurrencies that try to maintain the same value over time by pegging themselves to some underlying asset, such as a fiat currency or gold, or by managing price through an ever-changing supply—could emerge as a hybrid. Mostly these have been the brainchild of entrepreneurs running private companies. The largest of these today is Tether (USDT). Its creators say that Tether is backed dollar for dollar with USD reserves, though analysts have openly questioned this assertion.[30] Others such as MakerDao, BitCNY, and basecoin (backed by Andreesen Horowitz and other prominent VCs) have also emerged. Stablecoins could gain traction if we assume that existing cryptocurrencies such as bitcoin will remain highly volatile and governments will not create their own fiat currencies. At least for now, both conditions exist, and so stablecoins will continue to be an interesting area of innovation. Still, doubts linger. Stablecoins like Tether "decentralize the dollar but centralized the issuance. You have to trust a single entity who now becomes the monetary authority." Abed asks, "Are you better than the Federal Reserve?"[31]

Ultimately, however, we think governments will move into this market and that the future reserve currencies of the world will likely be a mix of crypto fiat currencies (digital dollars and such) and decentralized cryptocurrencies like bitcoin. Regional hybrids like the digital dollar standard in the Caribbean are likely to succeed, too. Don't count on the Petro joining their ranks.

PERMISSIONED NETWORKS

As we were submitting the final manuscript, forces were coalescing not only around the concept of the Fourth Industrial Revolution but also around special-purpose blockchains for industries such as the Industrial Internet of Things.

Ripple, typically one of the three largest cryptoassets by market cap, is an enterprise-friendly alternative to bitcoin, geared toward displacing SWIFT and other global payment networks. Ripple's architecture—relying on a handful of trusted nodes rather than on miners to secure the blockchain—gives it the ability to process more transactions but also makes it more centralized, which, in the eyes of some critics, makes it more vulnerable to attack and capricious and arbitrary behavior. Still, Ripple has been very successful in courting large banks and other potential enterprise users to employ their products and services and, to a lesser extent, use the native token of the network, called XRP.

The Linux Foundation, famous for building ecosystems around open source projects, had been looking at distributed ledger technology for a while. After hearing from several leaders in the space, executive director Jim Zemlin decided that the time had come for Linux to start a blockchain project.

In December 2015, the foundation announced Hyperledger, positioned as a collaborative project to "develop an enterprise grade, open source distributed ledger framework" so that developers could "focus on building robust, industry-specific applications, platforms, and hardware systems [that] support business transactions."[32] The project had "technical and organizational governance structure and 30 founding corporate members," notably IBM, Fujitsu, DTCC, and Accenture.[33] The Linux Foundation is a good home for open, transparent governance of the software development process and the management of intellectual property provenance and safeguards.

Our book covers public blockchains such as bitcoin and Ethereum, which remain two of the most important platforms today. They are open, meaning that anyone reading this book can conduct transactions, verify transaction data, race to create blocks, and develop distributed applications without anyone else's permission. Upgrades to the codebase are reached by consensus. Users who don't agree with particular upgrades (such as the increase of a bitcoin block size) can choose not to adopt it, and the blockchain forks in two. Both bitcoin and Ethereum have forked since we first wrote about them. Hyperledger pioneered the notion of the "consortium" model, which formalizes the governance of such upgrades and consolidates industry expertise around the formulation of standards.

Unlike bitcoin and Ethereum, Hyperledger's focus is on permissioned blockchains, networks in which verified, nonanonymous nodes can post transactions to the ledger and confirm other transactions. Such networks are typically also read-limited to that same network of verified nodes, but the network could allow a larger audience to read the data. A subset of that network could allow further control over read and write access so that it could use a much simpler form of consensus, one based loosely on a "supermajority vote" of the nodes, rather than on the more CPU-intensive proof of work (PoW) that bitcoin, ether, and most other coins have used. Such a network could also accommodate a much higher transaction volume than typically provided by PoW blockchains. Many of those building distributed ledger applications for the financial industry and the industrial Internet of Things, for example, prefer this model. It may also prove more valuable for those building certain public-facing applications, such as educational credentialing, carbon emissions monitoring, or fiat currency administration.

Hyperledger is not alone in building blockchain platforms that allow permissioned use cases and separate the need for a native token or cryptoasset (at least for now). Hashgraph, developed by computer scientist Leemon Baird, does not rely on miners to validate transactions. Instead of bundling transactions into blocks, Hashgraph uses *directed acyclic graphs* to time-sequence transactions on an ongoing basis. Theoretically, this means far faster transaction times, something many enterprises are emphasizing as they embrace this technology. Bloomberg reported that Hashgraph is working with twenty-plus enterprises as well as a number of credit unions in the United States.[34] Whether these new systems will succeed remains to be seen, but the progress of Hashgraph is very encouraging.

Some critics argue that permissioned blockchains are the equivalent of the intranets of the mid-1990s, many of which faded over time as the public Internet grew more robust, secure, and ubiquitous. But this time, blockchain technology enables transactions and management of value—assets owned by persons—and will necessarily have many public and more private forms.

Hyperledger is the fastest-growing project ever hosted by the Linux Foundation. More than two hundred member companies that span numerous industries make up the project and support five blockchain frameworks and four tools/modules.[35] Hyperledger membership is also quite global with 39 percent in Asia Pacific (25 percent in China), 20 percent in Europe, the Middle East, and Africa, and 41 percent spread across North America.

Hundreds of active pilots and proofs of concept (PoCs) are under way with Hyperledger technologies. Many will see production deployment in 2018. Industries as diverse as agriculture, finance, health care, real estate, energy, and diamonds will see blockchain applications disrupt their value chains. Walmart is currently testing blockchain technology for supply chain management. Specifically, it is using Hyperledger Fabric to track and trace pork in China and produce in the United States.[36] In May 2017, the Danish shipping enterprise Maersk announced completion of its first live blockchain trial, also using Hyperledger Fabric. The PoC aimed to simplify how it sends trillions of dollars' worth of products around the world. Deutsche Börse Group selected three use cases to be based on Hyperledger Fabric, relevant for its core business: cross-border collateral movement, posttrade processing, including settlement of securities against cash and asset servicing and provision of (commercial bank) money on the blockchain, enabling payments, settlement, and asset servicing. Finally, Sony Global Education prototyped a blockchain solution built on Hyperledger Fabric to develop a next-generation credentials platform. The prototype achieved all the needed functionality. Now in phase three, Sony Global Education plans to use the solution to manage the educational data of 250,000 participants in the Global Math Challenge.

Hyperledger brings together a community of organizations and individual developers to develop infrastructural software for blockchain applications. This leads to more effective collaboration, more shared code, and less duplication of effort. Any enterprise or other organization looking to blockchain solutions needs to give it careful consideration.

INTRODUCING THE MENOME: IDENTITY ON THE BLOCKCHAIN

On the subject of identity, it seems we really struck a nerve. In chapter 1, we wrote about enforcing the rights of all individuals to establish their own identities and to capture and control their own data. This is a much bigger idea than the word count allotted to it; the idea of a self-sovereign and inalienable digital identity, an identity that is neither bestowed nor revocable by any central administrator and is enforceable in any context—in person and online—anywhere in the world. It builds on the citizen scientist movement to quantify our selves—the quantified self—by lifelogging our physical activity through a Fitbit or other instrument as well as our virtual activity through our Internet browser or mobile app to learn about our health, our habits, and ourselves.[37] It's a real positive when people become interested enough in their own data to take control of it and use it for the greater good.

We have a greater sense of urgency about developing a distributed self-sovereign identity system. Here we need to distinguish between *identity*, which is a social, cultural, and psychological construct, and *identifiers* in a namespace (a 128-bit IP address, a DD Form 214), needed both to participate in and to manage large centralized systems (Google email, Veterans Benefits Administration).[38] Many of us accumulate quite a number of such identifiers in our lifetime, some of them more enduring (a social security number) than others (an employee ID), all generating personal data as we use them. Some of them are inherent (biometric), some are selected (passwords), and some are assigned (resident ID cards).

These are not our identities, which we experience and reveal to others over time as we deem appropriate, and we can do so because we have what developer Moxie Marlinspike calls a *sovereign source authority*, "the actual default design parameter of Human identity, prior to the 'registration' process used to inaugurate participation in Society."[39] Identity is not simply endowed at birth; it is endowed *by* birth. Until now, we haven't had the means to assert this authority.

An identifier, on the other hand, is only one of many attributes of a person's identity. There are five problems with identifiers, which several identity projects in the blockchain space are working to solve. We'll start with the biggest. Before dispensing one of them, most issuers require us first to have some über-identifier, often a birth certificate, recognized as authentic by a government. But getting a birth certificate is actually no small feat. According to UNICEF, "the births of around one fourth of children under the age of five worldwide have never been recorded."[40] Not getting a birth certificate can have life-shattering consequences: these children may have trouble receiving an education or health care. Worse, they may be married off, indentured into labor, or conscripted into the military before they reach the legal age.[41] (Is it a coincidence that "children remain the second most commonly detected group of victims of [human] trafficking globally after women, ranging from 25 to 30 percent of the total over the 2012–2014 period"?[42]) As adults, they may not be able to inherit property, vote in elections, or get jobs or passports, let alone bank accounts.[43] The World Bank estimates that 1.5 billion people on the planet lack a legal identifier.[44] The Syrian refugee crisis has underscored the problem of state-based identification.[45] We need to take action now.

So important is seeding identity by providing a trusted form of identification that the United Nations has made it a Sustainable Development Goal (16.9): all participating countries have committed to provide every person with legal identification by 2030.[46] The World Bank's Identification for Development initiative is designed to support this goal so that more people can participate in the global economy.[47] India has made considerable progress, documenting 99 percent of adults.[48] In a report on peer-to-peer markets, the Blockchain Research Institute highlighted the economic importance of India's efforts. It started with the Aadhaar Act of 2009, which authorized the Unique Identification Authority of India (UIDAI) to create a twelve-digit ID, called an *aadhaar* (meaning "foundation" in Hindi) for every resident.[49] Enrollment agents fanned out to collect demographic and biometric data and upload them to a repository designed to verify ID instantaneously anytime, anywhere. In April 2016, the National Payments Corporation of India unveiled a unified payment interface that would accept *aadhaar* for payment verification. Anyone with *aadhaar* could use it to complete any transaction, conduct peer-to-peer commerce, and receive government benefits. The plan worked: in November 2017, the UIDAI determined that "*aadhaar* data [are] fully safe and secure, and there has been no data leak or breach at UIDAI."[50]

The UIDAI spoke too soon. In early January 2018, Rachna Khaira, a journalist from *The Tribune* in Jalandhar, received an anonymous offer of unrestricted access to the data behind more than one billion *aadhaar* for only 500 rupees. By typing in a twelve-digit number, she was able to see the personal details associated with it—photo, name, address, phone number, and e-mail. For another 300 rupees, she could create an official-looking *aadhaar* card for that person. The system had been hacked, and one billion records exposed.

So the reality of a government-sourced and -sanctioned identity is a big problem—both administratively and philosophically. Why should any government get to rubber-stamp who we are? We should be establishing our own identities and, as Joe Lubin says, bootstrapping ourselves into economic enfranchisement![51] For those of you who think this is a crazy or ill-advised idea, please allow us to underscore the four other major problems with our current identity regime. One, they are system-centric, system-controlled, and vulnerable to cancellation, forgery, and theft. We're dependent on a system administrator who can freeze access, alter terms of access and usage, or delete our student IDs, health care insurance IDs, or land titles altogether.

Two, all the personal data we create and associate with each identifier (biometrics, college transcript, medical history) reside with and belong to the central system administrator, who may entrust it to untrustworthy vendors or sell some of it to unacceptable third parties without our knowledge. Such a system is opaque. If we want to switch colleges or countries, we bear the responsibility of porting our data from one system to the next—sometimes for a fee—and the rules for doing so are often complex and mercurial. Remember, we're going to be generating more of these data, not less.

Three, nothing about this identifier-centric system is user-friendly. Individuals—or, as noted, government or NGO representatives working on their behalf—have to repeat the registration process to obtain nearly every identifier, provide the same forms of über-identification, and maintain a portfolio of ID numbers, usernames, passwords, and the answers to personal questions. It is a system for the über-organized. It asks us whether we're robots and excludes robots from having their own identification—not good for all those robots that want to buy electricity.

Four, we bear most of the risk and responsibility for cleanup, should hackers break into these central systems and steal our identifiers and our data—but we enjoy none of the rewards of third-party data usage. Consider the legendary breach of Anthem, the largest U.S. health insurer. It agreed "to settle litigation over hacking in 2015 that compromised about 79 million people's personal information for $115 million, which lawyers said would be the largest settlement ever for a data breach."[52] Then it was breached again in October 2017 through one of its vendors, exposing the Medicare and health plan IDs of some 18,000 members.[53] Fool me once.

This is not identity management. This is identifier whack-a-mole. Our identities should be informing our selection and management of identifiers. Instead, these identifiers are deforming our identities. If we don't have them, we get the message that we aren't equal, we don't belong. If we do have them, we get the message to watch our backs, violation is a risk of participation, or privacy is overrated. They become a means of manipulation, conforming us to authoritarian rule. It's a lose-lose-lose-lose-lose situation, multiplied by all the identifiers we need to manage in an increasingly data-rich world.

To bootstrap our identity, we need a model that is distributed among and maintained by the people whose identities it protects so that everyone's incentives align—an identity commons—with clear rights for users to steward their own identity, access (and allow others to access) and monetize their own data, and participate in rule making around the preservation and usage of the commons.[54] It must exist independent of any corporate, government, or other third party, not subject to the agency risk of executives or political parties. It must interoperate with these institutions even as it outlasts them. It must outlive its users and enforce their right to be forgotten, which would mean separating data rights from the actual data so that the rights holder could delete them. And, to be inclusive, it must be user-friendly with a low-tech mobile interface and low-cost dispute resolution.

Here's where blockchain technology comes in. In chapter 2, we wrote about the technical and theoretical groundwork that has been laid—such as the deployment of public key infrastructure and the separation of identification and verification layers from the transaction layer, but we focused more on the principles of privacy design, which is the flip side of this identity coin. We alluded to the challenges to using Pretty Good Privacy and why it wasn't widely adopted. To that discussion, we add the promising work of Zooko Wilcox-O'Hearn and his associates at the Zerocoin Electric Coin Company. They launched Zcash, a public blockchain that enables users to conduct transactions while masking their identifiers and the amounts exchanged, compared with the bitcoin blockchain where those data are viewable. Zcash uses what it calls a zero-knowledge proof construction—specifically, a zero-knowledge succinct noninteractive argument of knowledge, or "zk-SNARK" (to which we say, "Gesundheit!")—in which participants can validate transactions and assemble them into a block with zero knowledge about them. Vitalik Buterin told *Fortune*, "I think zk-SNARKs are a hugely important, absolutely game-changing technology. . . . They are the single most under-hyped thing in cryptography right now."[55] We agree: it is an important innovation in privacy.

Let's return to the technology of identity. In *CoinDesk*, veteran developer Christopher Allen wrote a superb overview of the technological "Path to Self-Sovereign Identity," from the centralized IANA and ICANN, to the federated Microsoft Passport and Sun Microsystems Liberty Alliance, and then to the user-centric but registry-controlled OpenID.[56] With the emergence of blockchain, numerous identity projects have sprung up around the logging, storage, and accessibility of personal data. In chapter 2, we looked at a big one, MIT Media Lab's Enigma and its use of homomorphic encryption and secure multiparty computation, both critical to the identity principles of data minimization and algorithm transparency: the data user gets access to only the data needed for a computation but without seeing it, and the data owner can see the algorithms used in processing it. MIT Human Dynamics Group's OpenPDS/SA (for personal data store/safe answers) is a platform for organizing all our personal data streams and allowing people to query our data and get answers but not details.[57]

Blockchain users can already obtain identifiers through such start-ups as Civic, ShoCard, and uPort. We counted at least twenty such companies in the space.[58] The uPort identifier, for example, is a unique and persistent twenty-byte hexadecimal string that is core to uPort's identity system: it serves as the address of a specific type of smart contract known as a *proxy contract*, a piece of special purpose code that executes a complex set of instructions involving identity on the blockchain. The proxy contract is the ultimate mechanism through which a user can digitally sign and verify a transaction, an action, or a claim; manage cryptocurrencies or other tokenized assets; interact with other smart contracts on the Ethereum blockchain; link to the user's off-chain data stored in, say, the distributed InterPlanetary File System; and grant others temporary permission to read or write specific data files in exchange for value.

The uPort system would also work for devices such as driverless cars or 3D printers, virtual entities such as IBM's Watson, or institutions such as the Blockchain Research Institute. For user-friendliness and security, uPort provides a mobile app that holds the user's cryptographic keys. Separating these keys from the proxy contract is another type of smart contract, the *controller contract*, which contains logic for identity recovery: if the device is lost or stolen, the user can replace the private key without having to replace the proxy identifier and all the assets associated with it.[59] So amazing.

Many of these start-ups are collaborating in the Decentralized Identity Foundation (DIF), a consortium consisting of Hyperledger, R3, and Sovrin and incumbents such as Accenture, Microsoft, and IBM. DIF has formed to combine "decentralized identities, blockchain IDs, and zero-trust data stores that are universally discoverable" along the lines of the model of the identity commons we described above.[60] Its working groups are focusing on three big areas—identifiers and discovery, storage and computation of data, and attestation and reputation—with an eye to developing use cases and standards.[61] Separately, Fabian Vogelsteller has put forth "Ethereum request for comment—Issue 725" (ERC-725), a standard that specifies an interface for self-sovereign identity (just as his ERC-20 did for initial coin offerings), where identity resides in a smart contract as it does in uPort.[62] If the standard takes off as the ICO standard did, then we'll make real progress toward realizing this new identity management system, core to the blockchain revolution and the rebalancing of power.

The transition will take time. We expect organizations to take at least two actions to rebuild the trust of those whose data they hold. The first involves data governance. Many large corporations and government agencies have strong governance mechanisms for their hard assets. However, according to Dr. Elizabeth M. Pierce, a program chair for the International Conference on Information Quality hosted by MIT in 2015, "Information assets are often the worst governed, least understood, and most poorly utilized key asset in most firms because [information] is increasingly easy to collect and digitize, has increasing importance in products and services, is very difficult to price, has a decreasing half-life, has increasing security and privacy risk exposure, and is a significant expense in most enterprises."[63]

Dr. Pierce advocates for strong data governance—we couldn't agree more—which she defines as "specifying the decision rights and accountability framework to encourage desirable behaviors in the use of data." She makes an important distinction between data governance and data management: "Governance is about determining who inputs and makes the decisions and how. Management is the process of making and implementing the decisions."[64]

The second involves the discontinuation of practices that collect and store customer data and either destroying these massive customer databases altogether (after returning files and records to customers) or migrating these data to distributed storage systems such as the IPFS and then transferring control to customers.

Consider what Dr. David A. Jaffray is looking to do with patient information. Dr. Jaffray, executive vice president of technology and innovation at the University Health Network and the director of the Techna Institute for the Advancement of Technology for Health, has been involved in the implementation of the patient portal for Toronto's University Health Network. The portal gives patients complete access to their test results, imaging and pathology reports, diagnoses, health care provider's notes from in-person and telephone conversations, health management plans, referrals, and discharge summaries—all of which patients can share as they see fit. The results have been so positive for both patients and medical providers alike that Dr. Jaffray is keen to take the experience to the next level: total patient ownership over this information.

He is working with IBM and Hyperledger to design a blockchain-based pilot project because, for him, it represents a major pivot in thinking about patient data: by putting ownership in the hands of the patients themselves, it obviates a costly complexity of rules, regulations, and contracts among different institutions across jurisdictions—research hospitals, insurance providers, pharmacies, testing sites and laboratories, medical suppliers, drug companies, and National Institutes of Health, to name a few—required to protect patient privacy and security. That it simultaneously solves the data portability problem is a real plus for patients, and it frees them to form communities of interest—member-owned and -governed "health cooperatives," says Dr. Jaffray—around health or medical issues. Through these cooperatives, members could collectively bargain for better prices on specialty drugs in exchange for time-limited access (Dr. Jaffray uses the term "Snapchart") to their collective data on a particular disease.

What interests Dr. Jaffray in particular is the blockchain's ability to support a legal framework for consent at a large scale: through blockchain, patients can not only verify their agreement to share data at a byte level but also track their behavior at a granular level. With these two capabilities, patients can generate unprecedented phenomic data that they can donate or license under very specific terms to medical science, along with their genetic data. Let's call this biodata stream the human *menome*. Human genomics research has been quite useful, of course, but mapping the relationship between genotype and phenotype (e.g., a person's height, weight, health, disease, and fitness over time) will transform our understanding of diet, exercise, occupation, and environment, if not revolutionize medicine altogether. Imagine the health care supply chain of one, tailored to you, and funded in part by your menome.

"It's far more expensive to live a life than to do genetic testing," Dr. Jaffray said. "We need a way for individuals and their heirs to make use of these phenomic data captured over a lifetime."[65] Think of the immortal Henrietta Lacks—both her genes and her phenotypes—except that she now has control over these data, she can decide whether to approve the cultivation of her cells into a cell line, and she can will these data to her family, generation to generation. What an inheritance, potentially a means of transforming how we think of disease to begin with—as an asset, not a liability.

The ability to access and perform data analytics on large sets of (relatively) free data is currently a core competence that bestows competitive advantage; but, as individuals take back control over their data and form their own avatar of data—a *davatar*, if you will—the ability to secure those data sets in a distributed and trust-minimized manner and to help individuals manage and monetize their own data will replace big data analytics as the corporate capability that investors will value. It will remove data as a toxic asset from the corporate balance sheet and make it a fundamental human asset from birth. It will flip the data analytics business model on its head and reward corporations for serving as data brokers on behalf of individuals. We'll see the end of the large centralized data frackers that scrape, hoard, and rent but don't protect these data.

We'll also have a potential solution to the growing fear of mind hacking described by historian Yuval Noah Harari and evidenced by the effectiveness of Cambridge Analytica's psychographic profiling and the Russian manipulation of social media to influence the outcome of the 2016 U.S. presidential election.[66] Harari writes, "Just as divine authority was legitimized by religious mythologies, and human authority was legitimized by humanist ideologies, so high-tech gurus and Silicon Valley prophets are creating a new universal narrative that legitimizes the authority of algorithms and Big Data."[67] Indeed, Cambridge Analytica received a 2017 David Ogilvy Award for its big data practices from the Advertising Research Foundation.[68] Harari refers to this worldview as "Dataism," whose adherents "perceive the entire universe as a flow of data, see organisms as little more than biochemical algorithms, and believe that humanity's cosmic vocation is to create an all-encompassing data-processing system—and then merge into it."[69] In such a data-bio-mind-meld, we risk the loss of free will.

Others think such fears of human menome hacking are unfounded. "This sort of extremely precise and complete mapping of all human metabolic functions and brain activity is a dream," says Marcelo Gleiser, a theoretical physicist, professor, and director of the Institute for Cross-Disciplinary Engagement at Dartmouth College. "There are limits to what technology can do. Every machine has a precision range and is blind to what goes on beyond what it can probe. To monitor the activity of about eighty-five billion neurons and the flowing of neurotransmitters through trillions of synapses seems highly implausible, even if I wear my science-fiction nerd hat."[70]

We prefer to err on the side of caution and to advocate strongly for self-sovereign identities and ownership of all our data through blockchain technology and identity commons.

SMART CONTRACTS COME OF AGE

The use of smart contracts is a big theme of the book—after all, contracts are the building blocks of our identity, economy, and society—and so every chapter highlights potential use cases. In chapter 2, we explain what smart contracts are and how they work. Like traditional contracts, they include incentives—rewards and penalties—for performance in the form of mutually agreeable rules that spell out what happens to assets if certain conditions are met, except that smart contracts can sometimes automate performance (as with the cryptoassets described previously) and call on algorithms and sensors to determine objectively whether those conditions have indeed been met.

Nick Szabo, the father of smart contracts, has come up with the best visual for newcomers to the concept: think of an old-fashioned vending machine as a smart contract, where the terms of a very simple business relationship are programmed into the machine: if the machine has an acceptable type of beverage at an acceptable price, then the buyer selects the beverage and inserts enough coins to cover the price, and the machine verifies the amount, dispenses the chosen beverage, and makes change, if due, for the buyer.[71] In this sense, smart contracts have existed since the first century AD, when the Greek mathematician Hero of Alexandria invented a means of meting out exactly the amount of holy water that worshippers had paid for.[72] Today there are nearly seven million vending machines in the United States alone.[73] We're surrounded by smart contracts.

So they must be legally binding, right? Perhaps. There is no definitive answer yet. Under U.S. common law, for example, parties can express or imply an agreement: they needn't draft or sign a paper contract for the terms to be binding. According legal scholars Primavera De Filippi and Aaron Wright, "Smart contracts memorializing legal agreements are likely to be deemed enforceable under U.S. law. Parties can memorialize their intent using code just as they can with paper; and, to the extent that they set forth recurring performance obligations, smart contracts could even establish a course of performance or dealing."[74] Only time—and courts around the world—will tell.[75]

In chapter 4, we explored how smart contracts could alter the architecture of the firm in Coasian terms of reducing the transaction costs, both shrinking the number of essential employees at the core and expanding the number of gig workers at the edges. Those at the core will work more on retainer-like relationships in ever-changing roles doing whatever the organization needs. Those at the edge will work on more routine tasks that are easy to specify and to verify completion. This transformation will require executives to do what they have most likely never done before. It is what Szabo calls "the most valuable step, but the one traditional scientific management has failed to recognize and take," which is the restructuring of the firm's contractual relationships. We can see why start-ups have an advantage here. Rarely do incumbent firms think about their contract strategy in terms of all the work that needs to be done.

Through smart contracts, we could apply technology strategically to do a greater variety of deals, not just take-it-or-leave-it ones. We could coordinate a greater number of both things and people from a greater diversity of backgrounds across distant legal jurisdictions. We could use cryptocurrencies as a global payment system. In every phase, we could reduce our costs, minimize the need for third-party platforms, and improve productivity, security, and privacy.[76] Szabo points out the scalability of Uber's approach: "Uber substitutes employment with algorithmically negotiated and verified gig work. . . . Since many more people have much more of their labor expended in employment relationships than in spot market relationships, Uber and its similar successors in other logistics industries may be an even bigger deal than eBay and Amazon—and those have been pretty big deals."[77]

Scalability has its downside, if user data aren't secure. Hackers accessed the personal information of 57 million Uber accounts in 2016, but the company didn't disclose the breach until November 2017. Even then, it didn't notify the affected account holders.[78]

Give a blockchain-enabled vending machine wheels, a seat for humans, a trunk for their luggage, mapping software, a global positioning system, and algorithmic pricing, and you've got yourself a driverless and Uber-less ride service (which we call SUber in chapter 6). Blockchain as a Ledger of Things could run each thing's smart contracts—warranties, provenance, registration, insurance, inspection certification, and even operating software written to meet regulatory standards for, say, vehicle emissions. Those contracts could control the operation of that thing. If a machine, a driverless car, or a piece of heavy equipment failed a safety inspection or its liability insurance expired, then the machine could not turn on.

In chapter 5, we talked in theory about a distributed autonomous enterprise that has neither management nor employees; instead, it is a portfolio of smart contracts in the form not of vending machines but of *decentralized applications* that run "on a secure consensus protocol across a network of computers"—in other words, on a blockchain—"rather than on an individual remote computer or centralized server" and that will run properly even if we don't trust the owners of those computers.[79]

Shortly after the book came out, we witnessed the launch of the first such enterprise called the DAO (for decentralized autonomous organization), which crowdfunded a record-breaking $160 million from tens of thousands of global investors. What distinguished the DAO from all other start-ups was the absence of management in the traditional sense. Created by boutique blockchain development firm Slock.it, the DAO was a smart contract for a token with built-in voting rights. Its stakeholders—human beings—could review and vote on proposals curated by a smaller group of stakeholders to determine how the DAO allocated its funds.

Think about that for a moment. There were no agency costs, no information asymmetry between management and stakeholders, because there were no managers. Nor was there moral hazard, where managers could have behaved contrary to stakeholder interests, perhaps taking outsized risks for personal gain in the absence of personal consequences.

Like any corporation, the DAO could invest in new businesses and hire lobbyists or lawyers to represent its interests and advocate on its behalf. Using smart contracts, the DAO could do pretty much what any organization could do, with one important exception: on the blockchain, its agents could not override agreements, mission statements, corporate values, or operating principles without broad stakeholder debate and consensus. That's huge.

Problems with its contract code ultimately caused it to fail: a hacker exploited flaws in its use of recursion, a Turing-complete feature of Ethereum and not found in Bitcoin Script language.

According to Szabo, "Ethereum has a much larger attack surface than Bitcoin because of its Turing-complete smart contracts language and the relative abundance of applications enabled by high-level languages." Still, the DAO's mere existence demonstrated that autonomous entities could raise huge sums of money—peer to peer, without traditional intermediaries. According to Primavera De Filippi and Aaron Wright, "Hundreds of thousands of smart contracts have been deployed since Ethereum's launch."[80]

Still, companies are proceeding with caution. They are identifying pilot projects and experimenting in controlled environments. In chapter 6, we outlined new business models appropriate for experimentation. For example, Slock.it worked with MotionWerk to create a peer-to-peer Share&Charge service in Germany. Owners of electric vehicles can share their charging stations with other EV owners through Ethereum-based smart contracts, all of which are 100 percent updatable so that MotionWerk can adjust or respond quickly in an emergency.[81] Users download a Share&Charge app to handle the blockchain-driven control of the charging station and accounting.[82] The system is also fully backed by a digital euro to facilitate transaction settlement.[83] Finally, it runs on a public blockchain and so it is transparent and open; anyone can engage directly with it through smart contracts.

In chapter 7, we talk about the use of smart contracts to hold the dispensers of humanitarian aid accountable and to level the playing field for entrepreneurs. For example, Siemens AG is working with Slock.it to implement a blockchain-based DAO that will allow for voting on projects with a social purpose. UNICEF Ventures is testing a multisignature smart contract to make its asset transfers transparent and trackable, since traditional international transactions are often difficult to track.[84]

Knowledge networks will be critical. Creating a smart contract is more difficult than writing a traditional contract because we don't have hundreds of years of experience and many well-worn legal templates yet. According to Alan Majer, founder of Good Robot, "Solidity, a language for coding smart contracts, seems easy to use but is actually quite complex. Users must understand the nitty-gritty *anti-patterns*, that is, software design patterns that might be commonly used but can cause code to execute in unintended ways."[85] Competent smart contract developers are rare, with as few as five hundred in the world.[86] Szabo suggests that organizations hire lawyers with computer science backgrounds and software engineers with legal backgrounds. Universities, law schools, and continuing education programs should also be developing coursework and training modules to meet the need for this expertise.

ASSET CHAINS: WHEN BLOCKCHAIN MEETS SUPPLY AND PROCUREMENT

We often get asked, "What is the next big killer app for blockchain?" There is no better candidate than the global supply chain, an industry that runs two thirds of the global economy. Everything we consume is a product of a supply chain. Assets all over the world are extracted, designed, combined, transported, and sold every day through the supply chains that underpin global commerce. While technologies are increasingly disrupting traditional industries, this flow of goods has not been overhauled in years.

In the book, we argue that blockchain holds the potential of decentralizing traditional supply chains and combining them with artificial intelligence, additive manufacturing, and the growing Internet of Things to produce new value networks that scale to the demand of both machines and human beings.

When we were conceiving the Blockchain Research Institute, we began searching for the brightest minds and expert practitioners in this space. At the 2016 TED Summit in Banff, Don met Bettina Warburg and Tom Serres. Bettina was brilliant—her talk has well over three million views. She quoted the work of Nobel Prize–winning economist Douglass North on institutional economics and described blockchain as a new technological institution that would transform the economy and change how we exchange value.

This got us thinking about the unprecedented volume of data that blockchains would be throwing off, enabling us to study large-scale economic systems as never before. Surely a major category of those systems—the global supply chains that manage most of our global trade—would soon be up for change.

We decided to partner with Bettina, Tom, and their company, Animal Ventures, to lead this research. The results have been spectacular.[87] One of our conversations revved up my formulation engine and out popped the phrase *asset chains*, in response to their description of the blockchains that would support the autonomous and distributed management of supply chains.

Their research is foundational and provocative for anyone who deals with assets of some kind—because *every asset has a supply chain.* These new supply chains are autonomous, distributed, and cognitive in the sense that they are learning and bundling what they learn into opportunities for systemic self-improvement in efficiency and responsiveness. Cognitive supply chains require "a network state" function that provides a singular universal truth as the basis for what Bettina and Tom call *machine trust*.[88]

Which is where asset chains come it. This new way of thinking provides a framework for machines to participate autonomously in supply chains and the markets they serve. They allow us to unlock the trading capability of machines without human intermediaries.

The work uncovered some extraordinary initiatives using cryptoeconomics and blockchain to bring about this transformation. Consider Sweetbridge, a company that allows any enterprise to do four things that would be impossible without blockchain: pay suppliers and get paid early while decreasing the amount of money tied up in inventory and receivables to zero; obtain low to zero interest rate loans on their own assets without credit checks or loan applications; share underutilized capacity in supply chain assets with other organizations turning unused capacity into a new source of revenue; and incentivize supply chain experts to help optimize supply chains paying for services based on the outcomes that are measured, such as increase in sales and decreases in expense.

It gets better. The Sweetbridge protocols replace the need for letters of credit, trade financing, and working capital in supply chains. Here's how it works. Sweetbridge uses smart contracts to mint a cryptocurrency that is stable and pegged to the fiat currency of the user's choice based on the collateral value of an asset. The protocol acts like a loan that a company grants itself and must pay off in the same cryptocurrency it borrowed. The Sweetbridge protocols convert the value of any asset into a cash equivalent that a company can hold on its balance sheet as cash, trade with other companies as cash, and convert to cash when fiat currency is needed.

Sweetbridge is also the creator of the discount token discussed earlier—a new idea for funding anything from government infrastructure to supply chain assets. Its customers can buy and use these discount tokens themselves or sell them to other customers. According to Scott Nelson, "The more you buy and lock, the greater your discount. The more the network grows, the greater your discount." In essence, customers are "rewarded for growing the network and using the product or service."[89] It incentivizes network loyalty.

The platform also provides a settlement process that can eliminate counterparty risk in supply chains. The process does not require a bank or credit card network for payments because parties can transfer value in minutes on a blockchain. Nor do parties need an intermediary to hold the original asset until the loan is paid off. To sustain liquidity in the network, Sweetbridge has formed an alliance of projects. Members are working together to build supply chains that can identify faster, less costly, and more secure means of getting products to market and then reconfigure themselves to deliver accordingly. Hence, they are cognitive.

Bettina and Tom also explain what leaders should be doing now to prepare their organizations for this inevitable decentralized future. First is to get comfortable with transparency, an integral component of corporate social responsibility and a source of competitive advantage. According to Bettina and Tom, the winners in the decentralized economy will be those who "drive supply chain transparency toward the most accurate network state possible." Second is to cultivate talent, not just lawyer-coders who can program smart contracts but also artist-engineers like Leonardo da Vinci who imbue their designs with humanity. Third is to form coalitions around common goals, one of which is the shared governance of asset chains and the development of standards and best practices.[90] Now is the time to begin.

BLOCKCHAIN AND THE C-SUITE

For the last century, academics and business leaders have shaped the practice of modern management. The main theories, tenets, and behaviors of managers have worked well overall in building corporations—largely hierarchical, insular, and horizontally or vertically integrated. Until now.

In chapters 3 and 4, we discuss how blockchain will bring about profound changes, not just in the nature of firms, but in how they are funded and managed, how they create value, and how they perform basic functions like marketing and accounting. In some cases, algorithms will replace management altogether.

Because blockchain changes the deep structures and architecture of the firm, it will thereby transform our models of management and the roles of the C-suite. Vertical integration may make sense in some situations, but overall networks will become better structures for creating products, services, and value for stakeholders.

Dr. Irving Wladawsky-Berger, a visiting lecturer at MIT's Sloan School of Management, says, "Navigating this balance between hype and promise is a key responsibility of a company's senior management team."[91] So what should the C-suite prepare for? "Executives must decide whether their companies should adopt blockchain early and start experimenting now or wait until the technology matures and risk lagging behind more aggressive competitors."[92]

Chief Executive Officer

Wladawsky-Berger calls blockchain "the Internet of Transactions, a secure system of record for every transaction that has ever occurred" since its inception. He advises CEOs to "figure out how to best communicate, in the simplest way possible, why every company should embrace" blockchain technology. Business strategy becomes a means not only of proving that you "get it," that your organization gets it, but also of associating your brand with the future. This requires consistently telling your blockchain "stories over a variety of communication channels, including press interviews, conferences around the world, IT and financial analyst meetings, Web articles, and lots of client engagements."[93] The CEO sets the tone.

Wladawsky-Berger provides a word of caution about what he calls "a Wild West mentality" in this second era of the Internet, "where leaders send unproductive messages to the effect of 'the rules don't apply to us,' whether they're talking about the principles of economics—'it's all about eyeballs, not revenues'—or the codes of conduct in a civil society, such that sexual harassment, for example, becomes normalized." He sends a strong message to leaders: "*These rules do apply*," no matter how organizational structure changes.[94]

Since most blockchain initiatives are in the alpha or beta stage, CEOs need to manage expectations, promising only to learn from their experiments and their participation in consortia so that they can anticipate how the future of blockchain will unfold and affect their business.

Chief Information Officer/Chief Technology Officer

CIOs and CTOs have always had to ensure that their organizations incorporated and deployed the right technology at the right time. The Fourth Industrial Revolution centers not only on blockchain but also on machine learning, robotics, the Internet of Things, and even biotechnology.[95] The demands on CIOs and CTOs within an organization will expand from implementing business strategy to formulating it so that it leverages a range of technologies. CIOs and CTOs will need to wear the hats of the visionary and the great communicator so that they can help their peers in the C-suite understand the potential impact of these technologies and move them to action by sharing relevant use cases and suggesting pilot projects.

They will also need to orchestrate innovation across the enterprise. According to Oliver Bussmann, an award-winning CIO and CTO, "Blockchain technology will have a profound impact not only on processes external to the enterprise but also on the architecture stack within the enterprise—generally by moving business logic and processes out of enterprise silos and onto shared blockchains and broader-based ecosystems."[96]

To prepare their organizations, CIOs and CTOs can begin cultivating the necessary skills, talent, and relationships, be they in-house or in the network. There is already a shortage of accomplished blockchain developers and a lack of expertise in smart contracts and blockchain integration. Attending blockchain meetups and participating in relevant consortia can help to make connections.

They should also keep an eye on quantum computing, which uses quantum bits (or "qubits") rather than conventional bits to solve extremely difficult math problems vastly faster than our computers today. We touch on the quantum threat in chapter 10. We have since discovered that it's closer than we originally thought. According to experts at the Institute for Quantum Computing at the University of Waterloo, there's a one-in-seven chance that a quantum computer will be commercially available by 2026. That's less than a decade away! By 2031, the odds become one in two. "The arrival of phenomenally powerful quantum computing will shatter currently deployed public key cryptography and weaken symmetric-key cryptography, thereby undermining the cybersecurity that protects our infrastructure and systems," says Michele Mosca's team at Waterloo. "We cannot assume that blockchains, with their strong reliance on public key cryptography, are immune from this existential threat."[97] CIOs and CTOs should make sure that any blockchain deployed under their watch is quantum-proof.

Chief Human Resources Officer: A Better Way to Engage Talent

Human resources is an area that, when treated properly, can be a strategic asset, not a cost center. As firms move toward contingent labor and operate outside traditional organizational boundaries, the HR function grows more challenging. Perhaps the most immediate concern is diversity. As many have observed, the blockchain movement is overpopulated with men (though many of the best technical minds in the industry are women). In technology, compared with other sectors of the workforce, people of color are underrepresented by 16 to 18 percent, and women hold only 25 percent of all computing jobs.[98] "Everyone in Silicon Valley complains of the gender bias, and perhaps in the blockchain ecosystem even more so," said Pindar Wong, chairman of VeriFi (Hong Kong), former vice-chair of ICANN, and trustee of the Internet Society. "That's unhealthy. We're not getting enough diverse views. Going back to cybernetics' first principles, Ashby's Law of Requisite Variety, we need a variety of viewpoints, be it male, female, gay, straight, old, young, whatever you want to perceive it to be."[99] When problem solving has deadlocked, a key question to ask is, "Do we have enough variety in the room or online?" The goal is to maintain requisite variety to avoid thinking errors, said Wong. "You avoid thinking errors by having a wide variety of views that get equal treatment."[100]

The process of assembling and dispersing a diversity of talent can be far more effective and profitable than the traditional hiring-and-retaining model if HR professionals learn how to leverage blockchain in finding the right people, negotiating all their contracts, implementing the terms, and coordinating their contributions. Andy Spence, founder of Glass Bead Consulting, expects blockchain to transform the HR function in three waves. The first will resolve fundamental issues with recruitment, namely identity management and verification of credentials by querying prospective candidates' black boxes, and payment in nearly real time for output or time worked. The second will provide benefits in the broader talent ecosystem and reduce the number of full-time employment contracts. In the third, he envisages "technology sourcing and executing work projects by bringing in workers and services autonomously."[101]

Spence advises CHROs to "think more in terms of tasks that need to be sourced rather than jobs that need to be filled" and to celebrate portfolio careers, verified career profiles, and the pursuit of digital credentials (aka open badges). He suggests that "HR can be a pioneer in the new technology—not just blockchain but also artificial intelligence, robotics, and the Internet of Things, all of which could eliminate some jobs and create new ones." Ultimately, HR professionals will need to reimagine their function, since the firm "may no longer require many current HR activities in payroll, corporate learning and development, recruitment, performance management, and benefits administration." Instead, they should focus on enabling self-organizing teams, quantifying and predicting team performance, and safeguarding talent systems so that they remain effective, fair, and inclusive.[102]

Chief Marketing Officer: A Better Way to Engage Customers

Since companies will no longer be able to profile customers online by tracking and capturing their behavior, marketing and sales staff will also need to query prospective consumers' black boxes. Some consumers may allow access to their data in exchange for freebies; others will charge companies a fee to license their data. But the quality of results will increase because companies will find their target audience with greater precision. The nature of the blockchain would prevent the Wells Fargo style of customer abuse.

The upside is an end to intermediary fees and institutional bias. Jeremy Epstein, CEO of Never Stop Marketing, thinks smart contracts will improve SEO performance and price negotiation. Advertisers will know exactly which elements of their ad budgets delivered results: "We will have the opportunity to know the exact cost of attention and subsequent acquisition of an individual customer, eventually at scale," he wrote. "As we move to blockchain-based identity systems, we will witness the arrival of a *pay for attention* model."[103] Epstein described Brave Software's approach: it introduced what it calls a "basic attention token" (BAT) and launched a free Web browser that blocks ads and cookies. Epstein explained: "The token is the mechanism through which an advertiser pays for an individual's attention-based effort. With Brave and the BAT, we will pay end users directly for their attention, instead of the 73 percent of all ad dollars going to Facebook and Google."[104] As noted, retailers and manufacturers could eliminate the cost of warehousing and protecting consumer data.

General Counsel: A New Role for Lawyers in Developing Code

Ronald Coase and his successors argued that a firm was essentially a vehicle for creating long-term contracts when short-term contracts were too much effort to negotiate and enforce. Blockchain facilitates contracting, short- or long-term. Through smart contracts, companies can automate terms and use agents known as *oracles* to refer to external data fields, such as commodity prices and foreign exchange rates. They can trigger alerts and ensure payments.

Because smart contracts are self-enforcing, corporations will not want to enter them lightly. Lawyers and other managers will need to learn how to audit legal templates and make sure the contract software supports what its parties agreed to do. The watch phrase is "Don't roll your own crypto," meaning don't create some new and unproven cryptographic means of securing your smart contracts without publishing it for peer review and outside testing. That's apparently what IOTA did: it kept its new Curl hash function to itself.[105] (IOTA is a blockchain protocol, developed on the concept of a tangle—known also as a directed acyclic graph, as with Hashgraph—to disrupt the Internet of Things. Its team named its hash function Curl.[106]) Cryptographers from Boston-based Commonwealth Crypto and MIT's Digital Currency Initiative both identified security problems with Curl and found IOTA's response to their findings equally worrisome.[107] Stick with well-tested methods for creating and running smart contracts, and make sure you have someone on staff who can audit the code of a proposed new blockchain or a DApp behind an ICO.

Lawyers will also need to stay on top of cases involving blockchain, smart contracts, ICOs, and patents, particularly across jurisdictions and in heavily regulated and patent-rich domains such as health care, financial markets, pharmaceuticals, and medical appliances. General counsels will want to understand the patent strategies in the space. According to Thomas

M. Isaacson, a lawyer and shareholder of Polsinelli PC, blockchain innovators file patent applications to profit from an invention, to prevent others from using an invention, or to make an invention available to collaborators. The bar is high for receiving a grant. Isaacson writes, "A blockchain patent application must meet three criteria—eligibility, novelty, and nonobviousness. The question of obviousness is rich and deep. Whether a new useful process based on blockchain technology is patent-eligible is not clear-cut. The courts do not favor existing business practices being implemented on generic computers. The more narrow and focused the claims are, the better the chances of getting claims allowed."[108]

A Fair Deal for Corporate Executives

The year 2017 brought another round of high-profile scandals: bribery charges against Samsung Electronics vice chairman Lee Jae-yong; Uber's alleged use of "Greyball" software for dodging regulators where it was operating unlawfully; the preventable breach of Equifax, the theft of 145 million records, and the alleged insider trading by the Equifax executive who "dumped his stock before the news [of the breach] went public," according to the SEC's Atlanta Regional Office; the falsification of data by Kobe Steel and Mitsubishi Materials on products sold to clients, and Wells Fargo's admission that it had opened another 1.4 million phony accounts (on top of the 2.1 million already disclosed) and billed some 570,000 consumers for car insurance they didn't need, causing some to default on their car loans.[109] Year after year, executives don't always act with integrity or incentivize employees to act with integrity.

Through smart contracts, owners can hold these executives accountable—they must abide by their commitments as enforced and settled by software. Companies can program relationships and parameters of outcomes so that everyone has a better understanding about what each party has signed up to do and can see whether that party is doing it. With multi-signature contracts, shareholders can even vote on high-stakes managerial decisions such as particularly risky investments.

On the blockchain, executives will no longer need to swear that their books are in order once a year; their books will be in order *every ten minutes*, whether executives like it or not. No need for public auditors. The blockchain eliminates human error and prevents fraud in accounting. Shareholders and regulatory agencies alike will be able to examine the books at any point in time. Investors can create their own creditworthy ratings dashboards based on the facts. No more subjective ratings agencies. At last, stakeholders can reward executives for achieving actual results.

GOVERNANCE AND LEADERSHIP FOR THE NEW ERA

Stewarding the Blockchain Revolution

How is this whole blockchain revolution going to play out? As we said in the book, we're of the school that "the future is not something to be predicted: the future is something to be achieved." We argued that, like the first era of the Internet, this blockchain era should *not* be governed by nation-states, state-based institutions, or corporations.

Yes, there is a role for regulation. The first era of the Internet was initially unregulated, but today there are laws in various countries concerning topics ranging from spam and privacy to so-called net neutrality. The second era will require even greater government involvement, because unlike information, it's all about assets—for which there is a clearer public interest.

Still, how we govern the Internet of information as a global resource serves as a model for how to govern this new resource. Rather than relying on governments, blockchain must be primarily self-governed through the bottom-up, multistakeholder approach using what we called "global governance networks"—a concept developed in our previous multimillion-dollar program investigating multistakeholder networks for global problem solving and described in chapter 11.[110]

Since then, we have dived deeper into this issue. We hosted a meeting of key players of the blockchain ecosystem at our family lake house in Muskoka, Ontario, Canada—resulting in the creation of the Muskoka Group Manifesto on stewardship of blockchain.[111] We also wrote a white paper commissioned by the World Economic Forum.[112] Faculty members of the Blockchain Research Institute have explored it and produced some helpful material.

We came to some important conclusions for anyone who cares about making blockchain happen. The Internet of information is a network of similar networks. Blockchain is not—it is balkanized at the basic platform or protocol level. Therefore, unlike the Internet of information, which is a vast network of similar networks, this Internet of value requires stewardship at not just one level but three:

1. Each *platform* needs to govern itself—to develop an ecosystem, standards, and use cases and ensure a robust rollout of its technology. In the last two years, there have been important improvements in this regard, with the bitcoin community forking and implementing different solutions for scalability, the rapid expansion and development of the Lightning Network, Ethereum's crisis management by consensus and planned implementation of proof of stake, and Hyperledger's call for both urgency and moderation around standards.

2. At the *application* level, there have been all kinds of consortia established where companies like FedEx or Pepsi join with industry partners and even competitors to develop standards and common applications. Platforms themselves are encouraging such application-level partnerships as reflected by the Enterprise Ethereum Alliance, which works to build application-level standards for companies using Ethereum.

3. At the overall *ecosystem* level, there are networks like the Blockchain Research Institute conducting research and disseminating knowledge, and advocacy groups like the Global Blockchain Business Council or the Chamber of Digital Commerce.

We can apply our "global solution networks" framework to each of these levels.[113] We urge stakeholders in the space to codify their common ground through standards networks; welcome stakeholders with radically diverse views of what needs to be done through networked institutions; respect members' interests and constraints through advocacy networks; ensure that no one does any harm through watchdog networks; participate in policy debates and coordinate regulation through policy networks; get up to speed through knowledge networks; and keep incentives for mass collaboration in mind through delivery networks.

This is critical work. Whether you are a technology provider or user of this technology, you should care. Think not only about the needs of your own organization but also about the overall challenges of stewarding the blockchain revolution through the maze and even onslaught of difficulties, from technical challenges to bad legislation.

Profile of a Blockchain Hotbed: Seven Conditions for Success

Silicon Valley has been a modern engine of digital innovation, finance, incubation, and transformation of business models. It's the center of venture capital and entrepreneurship, and it produced FANG (Facebook, Apple, Netflix, and Google). Not so for the second era. ICOs and STOs are replacing venture capital, and leaders of the old have difficulty embracing the new.

It's also unclear as to whether a single global hub for blockchain is feasible. As a technology, blockchain is decentralized by design. Early protocols were developed through cross-border collaboration by creative trailblazers in regions largely outside of Silicon Valley. In the process, hubs of blockchain-based innovation have emerged the world over.

If not the Valley, then where? BRI researchers Hilary Carter and Jill Rundle found that not all blockchain hubs are equal, but leading national ecosystems have many of the following in common.[114]

Incubators and Entrepreneurship

Innovation is nurtured in environments established exclusively for this purpose. In Toronto, incubators such as MaRS, OneEleven, and Ryerson University's DMZ have provided a favorable climate in which blockchain entrepreneurship can flourish. Regions with incubators have an automatic advantage over those without.

Corporate Leadership

Centers of blockchain innovation very often enjoy close ties with established business communities. In areas where corporate entities manifest a culture of curiosity and market positioning as innovators, blockchain developments can especially thrive.

Educational Institutions

Great computer science schools can lead to great blockchain innovation ecosystems. Tip the hat to MIT, the National University of Singapore, ETH Zurich, Stanford University, Middlesex University, the University of Toronto, the University of Waterloo, and York University. The University of California at Berkeley may save Silicon Valley yet with its "Blockchain at Berkeley" program.

Investment Climate

Angel capital, venture capital, and strong financial services industries must have a risk tolerance for this kind of innovation. Through ICOs, entrepreneurs have harnessed blockchain as a distributed financing mechanism to overcome the traditional financial barriers that had prevented many ventures from getting off the ground, though regulatory hurdles must first be overcome.

Government Support

One of the most important things that governments can do is to be model users of the technology itself. Government community-based initiatives can also fund innovation directly. In Hangzhou and Guangzhou, China's government is pouring billions into blockchain development. Canada's supercluster initiative has put a billion dollars into projects with a blockchain component. Israel has produced more unicorns per capita than anywhere.

Regulatory Environment

As we have explained, governments can help or impede innovation. ICOs are a new source of funding. Neither overregulation nor no-regulation is sensible. Free-for-all jurisdictions like Belarus or Ukraine will run into big problems. But banning bitcoin, ICOs, STOs, and cryptocurrency exchanges as many countries are considering will hurt innovation for decades. This is just one of many regulatory issues, not the least of which is taxation.

Communities of Talent

Highly educated populations are an important factor for innovation to take root in any given jurisdiction. How do you initiate a national brain *gain* rather than brain drain? Canada reversed this trend, thanks not only to Donald Trump. In November 2017, Mayor John Tory of the city of Toronto spoke to the Blockchain Research Institute about making Toronto a global technology leader, ensuring that the city attracts talent pools from other countries, and persuading Canadians who have flown south to return home.

Leadership of Nations: The Ten Ahead

The opportunity to lead the blockchain revolution is still an open playing field. Whichever country wins will have an innovation economy for decades ahead. Here are the contenders, listed alphabetically.[115]

Australia

Blockchain innovation is thriving in the land Down Under. There's a burgeoning community of collaborators and advocacy groups, including the Blockchain Association of Australia and the Australian Digital Currency Commerce Association (ADCCA), and innovative apps such as ChronoBank are well ahead of competitors in terms of market positioning. Recently, Australia removed taxes from transactions and trades that are made using bitcoin, and in a historic move, the Australian Securities Exchange (ASX) has announced adoption of blockchain following a two-year test of the technology.

Canada

We could argue that Canada has the biggest ecosystem in the world. Vitalik Buterin left the University of Waterloo to create Ethereum. Some of the world's biggest incubators are headquartered in Toronto. There are five innovative banks working to rethink the financial industry. Some of the world's most promising start-ups, such as Nuco/Aion, Paycase, Tendermint/Cosmos, and Decentral, have strong roots in the Toronto area. Vanbex, Axiom Zen, and Frontier Foundry are based in Vancouver. Quebec, with its cool climate and plentiful energy, is quickly becoming the go-to region for cryptocurrency mining operations. There is strong national government support, and the Bank of Canada is an innovator. We'd argue that Toronto is the global center of thought leadership with the Blockchain Research Institute.

China

China's relationship with blockchain and digital currencies could best be described as complex. The country is able to mobilize vast resources to implement any technology that its leaders view as most beneficial or, conversely, to control tightly those it deems potential threats. Massive government initiatives—including banning ICOs, cryptocurrency exchanges, and the mining of bitcoin itself—have simply impeded entrepreneurship. At the same time, China has openly nurtured other aspects of blockchain innovation, keeping the door open to a blockchain-based fiat currency and other innovations that will drive economic growth.

Dubai (United Arab Emirates)

The Dubai blockchain strategy, led by blockchain innovator Vinay Gupta, was launched in 2016 as an exploration and evaluation of technology innovations that could provide simple and secure transactions to make Dubai a leading city in efficiency. The crown prince has targeted 2020 for all government documentation to be entered on a blockchain. As a gateway to Asia and Africa, Dubai's potential impact on supply chains, transportation initiatives, and government services is enormous.

Estonia

This tiny Baltic nation has demonstrated incredible initiative in blockchain transformation. Nearly all of Estonia's public services have been digitized, including identity. The Estonian data aren't central, and so major breaches are unlikely; and the underlying blockchain asserts its legitimacy, making medical, school, financial, and government information readily verifiable across platforms—even allowing emergency personnel to access medical records before they reach victims and preregister them en route to the hospital. With the ability to execute smart contracts for remote e-residents, Estonia is displacing the local governments of its subscribers.

Singapore

Singapore was recently identified as the third-largest ICO market (after the United States and Switzerland) and the leading ICO hub in Asia.[116] Like Dubai, the city-state is moving toward its own version of a "Smart Nation," creating policies and infrastructure to streamline the adoption of new technology in areas such as health care, fintech, education, autonomous vehicles, and public transit.

Sweden (Stockholm and the "Node" Pole)

In 2016, a project to use blockchain technology for the Swedish Land Registry was proposed after it was determined that the time it takes to conclude a land transaction could be reduced from four months to just a few days. Stockholm is a digital leader among European cities, second only to London as a leading fintech hub. The bitcoin mining community was also quick to take advantage of Sweden's northern climate to cool their data farms naturally. Boden, a municipality in northern Sweden, calls itself "The Node Pole."

Switzerland (Zurich and Zug)

Switzerland represents one of the world's most decentralized political systems, which may be why the decentralized ledger system is seen as an opportunity. Nicknamed "Crypto Valley," Zug—perhaps in part because of the Swiss tradition of financial privacy and also low corporate tax rates—has attracted bitcoin asset managers, brokers, and currency exchanges, becoming a center for digital money. As the presence of the industry players grew, Zug leaders responded by embracing the technology. City cashiers now take utility payments in bitcoin, and the bitcoin exchange, Bitcoin Suisse, has facilitated $635 million in ICOs. The Crypto Valley Association facilitates working groups on policy, regulatory development, start-ups, and investments.

United Kingdom (London)

The United Kingdom trails only the United States with 16.7 percent of blockchain start-up activity. Its fintech sector is both vibrant and competitive. The London Blockchain Week, which grew out of an annual conference series, recently focused on government initiatives. Given the high number of identity thefts in the United Kingdom, the government has made new models of identity management a priority. Governor of the Bank of England Mark Carney's Mansion House speech of 2016 also signaled a motivation to advance blockchain applications in financial services as a priority.

United States (New York City and Silicon Valley)

Not surprisingly, the largest market for blockchain development is the United States, with nearly 40 percent of the start-ups in the field. The head start provided by U.S. Internet initiatives is slowly eroding as smaller jurisdictions with fewer regulatory or legacy-system impediments start to gain ground. The legacy power players who have grown up in Silicon Valley, specifically the innovators and venture capitalists who have embraced blockchain technology such as Mark Andreessen and Twitter's Jack Dorsey through his Squarecoin initiative, will likely remain the dominant force in the field.

The ten countries we've listed are showing real leadership. They have, thus far, managed to navigate regulatory and governance uncertainties, giving them an excellent head start in the adoption of blockchain initiatives, in some cases more so than others. There may be no equivalent to Silicon Valley in the second era of the Internet. If economic value can be geographically predistributed by blockchain, and a new economic order realized through sensible regulation, perhaps then, global prosperity centers will simply follow in the technology's decentralized footsteps.

Any country that wants to lead in the second era of the Internet needs to cultivate the seven conditions for success. Consider creating a National Task Force on the Digital Economy, chaired by a well-respected nongovernment leader and consisting of thoughtful and well-respected leaders from business, government, and civil society. Give it six months to hold a national dialogue that engages not only key stakeholders but the population as a whole. Require it to develop an action plan where everyone has a role to play.

What Leaders Can Do

The most exciting development in the past two years has been the emergence of a new generation of leaders for this revolution: the entrepreneurs we've met in visits to thirty countries; the innovative CEOs of major corporations like Fred Smith at FedEx, Jim Smith at Thomson Reuters, and Iain Conn of Centrica; the CFOs, CIOs, and chief legal officers who were curious about the implications for their roles and stepped up to lead; the professionals and managers from every nook of large companies. We've been inspired by central bankers, government regulators, and policy makers who have met the challenge of balancing the innovation, growth, and opportunity that blockchain affords with their ongoing responsibility to steward financial markets. Journalists, academics, and pundits stand on all sides of this issue; and the ones who have discarded their cynicism are inspirations. We've also noted a new generation of "crypto natives" who are growing up with this technology and not only investing but working to develop a career in blockchain.

These are exciting and perilous times. If you are a business leader, we do hope that you will use *Blockchain Revolution* as your playbook, sure, but realize that the rules of the game themselves are changing. You have many other resources at your disposal. Set up Google Alerts for "blockchain," "bitcoin," and other key terms to keep up if you haven't already. Check out Andreessen Horowitz's "Crypto Canon" (a16z.com), which the firm updates constantly.[117] Read *The Truth Machine*, *Cryptoassets*, and *Blockchain and the Law*. Immerse yourself in "crypto Twitter." Find out who at your company is interested in the technology or already using cryptocurrency. Talk to your chief officers and IT department about the technology's implications. Exchange fiat currency for some bitcoin and then use it to buy something just to see how it works—hundreds of thousands of merchants accept bitcoin.[118] Identify nearby blockchain start-ups, take a field trip to see their operations, and talk with their founders. Invite an expert to speak with your team. Most important, act now. This is your chance to reimagine how you create value. If you don't, someone else will.

Read on and join the revolution!

PART I

SAY YOU WANT A REVOLUTION

CHAPTER 1

THE TRUST PROTOCOL

It appears that once again, the technological genie has been unleashed from its bottle. Summoned by an unknown person or persons with unclear motives, at an uncertain time in history, the genie is now at our service for another kick at the can—to transform the economic power grid and the old order of human affairs for the better. If we will it.

Let us explain.

The first four decades of the Internet brought us e-mail, the World Wide Web, dot-coms, social media, the mobile Web, big data, cloud computing, and the early days of the Internet of Things. It has been great for reducing the costs of searching, collaborating, and exchanging information. It has lowered the barriers to entry for new media and entertainment, new forms of retailing and organizing work, and unprecedented digital ventures. Through sensor technology, it has infused intelligence into our wallets, our clothing, our automobiles, our buildings, our cities, and even our biology. It is saturating our environment so completely that soon we will no longer "log on" but rather go about our business and our lives immersed in pervasive technology.

Overall, the Internet has enabled many positive changes—for those with access to it—but it has serious limitations for business and economic activity. *The New Yorker* could rerun Peter Steiner's 1993 cartoon of one dog talking to another without revision: "On the Internet, nobody knows you're a dog." Online, we still can't reliably establish one another's identities or trust one another to transact and exchange money without validation from a third party like a bank or a government. These same intermediaries collect our data and invade our privacy for commercial gain and national security. Even with the Internet, their cost structure excludes some 2.5 billion people from the global financial system. Despite the promise of a peer-to-peer empowered world, the economic and political benefits have proven to be asymmetrical—with power and prosperity channeled to those who already have it, even if they're no longer earning it. Money is making more money than many people do.

Technology doesn't create prosperity any more than it destroys privacy. However, in this digital age, technology is at the heart of just about everything—good and bad. It enables humans to value and to violate one another's rights in profound new ways. The explosion in online communication and commerce is creating more opportunities for cybercrime. Moore's law of the annual doubling of processing power doubles the power of fraudsters and thieves—"Moore's Outlaws"[1]—not to mention spammers, identity thieves, phishers, spies, zombie farmers, hackers, cyberbullies, and datanappers—criminals who unleash ransomware to hold data hostage—the list goes on.

IN SEARCH OF THE TRUST PROTOCOL

As early as 1981, inventors were attempting to solve the Internet's problems of privacy, security, and inclusion with cryptography. No matter how they reengineered the process, there were always leaks because third parties were involved. Paying with credit cards over the Internet was insecure because users had to divulge too much personal data, and the transaction fees were too high for small payments.

In 1993, a brilliant mathematician named David Chaum came up with eCash, a digital payment system that was "a technically perfect product which made it possible to safely and anonymously pay over the Internet. . . . It was perfectly suited to sending electronic pennies, nickels, and dimes over the Internet."[2] It was so perfect that Microsoft and others were interested in including eCash as a feature in their software.[3] The trouble was, online shoppers didn't care about privacy and security online then. Chaum's Dutch company DigiCash went bankrupt in 1998.

Around that time, one of Chaum's associates, Nick Szabo, wrote a short paper entitled "The God Protocol," a twist on Nobel laureate Leon Lederman's phrase "the God particle," referring to the importance of the Higgs boson to modern physics. In his paper, Szabo mused about the creation of a be-all end-all technology protocol, one that designated God the trusted third party in the middle of all transactions: "All the parties would send their inputs to God. God would reliably determine the results and return the outputs. God being the ultimate in confessional discretion, no party would learn anything more about the other parties' inputs than they could learn from their own inputs and the output."[4] His point was powerful: Doing business on the Internet requires a leap of faith. Because the infrastructure lacks the much-needed security, we often have little choice but to treat the middlemen as if they were deities.

A decade later in 2008, the global financial industry crashed. Perhaps propitiously, a pseudonymous person or persons named Satoshi Nakamoto outlined a new protocol for a peer-to-peer electronic cash system using a cryptocurrency called bitcoin. Cryptocurrencies (digital currencies) are different from traditional fiat currencies because they are not created or controlled by countries. This protocol established a set of rules—in the form of distributed computations—that ensured the *integrity* of the data exchanged among these billions of devices *without going through a trusted third party*. This seemingly subtle act set off a spark that has excited, terrified, or otherwise captured the imagination of the computing world and has spread like wildfire to businesses, governments, privacy advocates, social development activists, media theorists, and journalists, to name a few, everywhere.

"They're like, 'Oh my god, this is it. This is the big breakthrough. This is the thing we've been waiting for,'" said Marc Andreessen, the cocreator of the first commercial Web browser, Netscape, and a big investor in technology ventures. "'He solved all the problems. Whoever he is should get the Nobel Prize—he's a genius.' This is the thing! This is the distributed trust network that the Internet always needed and never had."[5]

Today thoughtful people everywhere are trying to understand the implications of a protocol that enables mere mortals to manufacture trust through clever code. This has never happened before—trusted transactions directly between two or more parties, authenticated by mass collaboration and powered by collective self-interests, rather than by large corporations motivated by profit.

It may not be the Almighty, but a trustworthy global platform for our transactions is something very big. We're calling it the Trust Protocol.

This protocol is the foundation of a growing number of global distributed ledgers called blockchains—of which the bitcoin blockchain is the largest. While the technology is complicated and the word *blockchain* isn't exactly sonorous, the main idea is simple. Blockchains enable us to send money directly and safely from me to you, without going through a bank, a credit card company, or PayPal.

Rather than the Internet of Information, it's the Internet of Value or of Money. It's also a platform for everyone to know what is true—at least with regard to structured recorded information. At its most basic, it is an open source code: anyone can download it for free, run it, and use it to develop new tools for managing transactions online. As such, it holds the potential for unleashing countless new applications and as yet unrealized capabilities that have the potential to transform many things.

HOW THIS WORLDWIDE LEDGER WORKS

Big banks and some governments are implementing blockchains as distributed ledgers to revolutionize the way information is stored and transactions occur. Their goals are laudable—speed, lower cost, security, fewer errors, and the elimination of central points of attack and failure. These models don't necessarily involve a cryptocurrency for payments.

However, the most important and far-reaching blockchains are based on Satoshi's bitcoin model. Here's how they work.

Bitcoin or other digital currency isn't saved in a file somewhere; it's represented by transactions recorded in a blockchain—kind of like a global spreadsheet or ledger, which leverages the resources of a large peer-to-peer bitcoin network to verify and approve each bitcoin transaction. Each blockchain, like the one that uses bitcoin, is *distributed*: it runs on computers provided by volunteers around the world; there is no central database to hack. The blockchain is *public*: anyone can view it at any time because it resides on the network, not within a single institution charged with auditing transactions and keeping records. And the blockchain is *encrypted*: it uses heavy-duty encryption involving public and private keys (rather like the two-key system to access a safety deposit box) to maintain virtual security. You needn't worry about the weak firewalls of Target or Home Depot or a thieving staffer of Morgan Stanley or the U.S. federal government.

Every ten minutes, like the heartbeat of the bitcoin network, all the transactions conducted are verified, cleared, and stored in a block which is linked to the preceding block, thereby creating a chain. Each block must refer to the preceding block to be valid. This structure permanently time-stamps and stores exchanges of value, preventing anyone from altering the ledger. If you wanted to steal a bitcoin, you'd have to rewrite the coin's entire history on the blockchain in broad daylight. That's practically impossible. So the blockchain is a distributed ledger representing a network consensus of every transaction that has ever occurred. Like the World Wide Web of information, it's the World Wide Ledger of value—a distributed ledger that everyone can download and run on their personal computer.

Some scholars have argued that the invention of double-entry bookkeeping enabled the rise of capitalism and the nation-state. This new digital ledger of economic transactions can be programmed to record virtually everything of value and importance to humankind: birth and death certificates, marriage licenses, deeds and titles of ownership, educational degrees, financial accounts, medical procedures, insurance claims, votes, provenance of food, and anything else that can be expressed in code.

The new platform enables a reconciliation of digital records regarding just about everything in real time. In fact, soon billions of smart things in the physical world will be sensing, responding, communicating, buying their own electricity and sharing important data, doing everything from protecting our environment to managing our health. This Internet of Everything needs a Ledger of Everything. Business, commerce, and the economy need a Digital Reckoning.

So why should you care? We believe the truth *can* set us free and distributed trust will profoundly affect people in all walks of life. Maybe you're a music lover who wants artists to make a living off their art. Or a consumer who wants to know where that hamburger meat really came from. Perhaps you're an immigrant who's sick of paying big fees to send money home to loved ones in your ancestral land. Or a Saudi woman who wants to publish her own fashion magazine. Maybe you're an aid worker who needs to identify land titles of landowners so you can rebuild their homes after an earthquake. Or a citizen fed up with the lack of transparency and accountability of political leaders. Or a user of social media who values your privacy and thinks all the data you generate might be worth something—to you. Even as we write, innovators are building blockchain-based applications that serve these ends. And they are just the beginning.

A RATIONAL EXUBERANCE FOR THE BLOCKCHAIN

For sure, blockchain technology has profound implications for many institutions. Which helps explain all the excitement from many smart and influential people. Ben Lawsky quit his job as the superintendent of financial services for New York State to build an advisory company in this space. He told us, "In five to ten years, the financial system may be unrecognizable . . . and I want to be part of the change."[6] Blythe Masters, formerly chief financial officer and head of Global Commodities at JPMorgan's investment bank, launched a blockchain-focused technology start-up to transform the industry. The cover of the October 2015 *Bloomberg Markets* featured Masters with the headline "It's All About the Blockchain." Likewise, *The Economist* ran an October 2015 cover story, "The Trust Machine," which argued that "the technology behind bitcoin could change how the economy works."[7] To *The Economist*, blockchain technology is "the great chain of being sure about things." Banks everywhere are scrambling top-level teams to investigate opportunities, some of these with dozens of crackerjack technologists. Bankers love the idea of secure, frictionless, and instant transactions, but some flinch at the idea of openness, decentralization, and new forms of currency. The financial services industry has already rebranded and privatized blockchain technology, referring to it as *distributed ledger technology*, in an attempt to reconcile the best of bitcoin—security, speed, and cost—with an entirely closed system that requires a bank or financial institution's permission to use. To them, blockchains are more reliable databases than what they already have, databases that enable key stakeholders—buyers, sellers, custodians, and regulators—to keep shared, indelible records, thereby reducing cost, mitigating settlement risk, and eliminating central points of failure.

Investing in blockchain start-ups is taking off, as did investing in dot-coms in the 1990s. Venture capitalists are showing enthusiasm at a level that would make a 1990s dot-com investor blush. In 2014 and 2015 alone, more than $1 billion of venture capital flooded into the emerging blockchain ecosystem, and the rate of investment is almost doubling annually.[8] "We're quite confident," said Marc Andreessen in an interview with *The Washington Post*, "that when we're sitting here in 20 years, we'll be talking about [blockchain technology] the way we talk about the Internet today."[9]

Regulators have also snapped to attention, establishing task forces to explore what kind of legislation, if any, makes sense. Authoritarian governments like Russia's have banned or severely limited the use of bitcoin, as have democratic states that should know better, like Argentina, given its history of currency crises. More thoughtful governments in the West are investing considerably in understanding how the new technology could transform not only central banking and the nature of money, but also government operations and the nature of democracy. Carolyn Wilkins, the senior deputy governor of the Bank of Canada, believes it's time for central banks everywhere to seriously study the implications of moving entire national currency systems to digital money. The Bank of England's top economist, Andrew Haldane, has proposed a national digital currency for the United Kingdom.[10]

These are heady times. To be sure, the growing throng of enthusiasts has its share of opportunists, speculators, and criminals. The first tale most people hear about digital currencies is the bankruptcy of the Mt. Gox exchange or the conviction of Ross William Ulbricht, founder of the Silk Road darknet market seized by the Federal Bureau of Investigation for trafficking illegal drugs, child pornography, and weapons using the bitcoin blockchain as a payment system. Bitcoin's price has fluctuated drastically, and the ownership of bitcoins is still concentrated. A 2013 study showed that 937 people owned half of all bitcoin, although that is changing today.[11]

How do we get from porn and Ponzi schemes to prosperity? To begin, it's not bitcoin, the still speculative asset, that should interest you, unless you're a trader. This book is about something bigger than the asset. It's about the power and potential of the underlying technological platform.

This is not to say that bitcoin or cryptocurrencies per se are unimportant, as some people have suggested as they scramble to disassociate their projects from the scandalous ventures of the past. These currencies are critical to the blockchain revolution, which is first and foremost about the peer-to-peer exchange of value, especially money.

ACHIEVING TRUST IN THE DIGITAL AGE

Trust in business is the expectation that the other party will behave according to the four principles of integrity: honesty, consideration, accountability, and transparency.[12]

Honesty is not just an ethical issue; it has become an economic one. To establish trusting relationships with employees, partners, customers, shareholders, and the public, organizations must be truthful, accurate, and complete in communications. No lying through omission, no obfuscation through complexity.

Consideration in business often means a fair exchange of benefits or detriments that parties will operate in good faith. But trust requires a genuine respect for the interests, desires, or feelings of others, and that parties can operate with goodwill toward one another.

Accountability means making clear commitments to stakeholders and abiding by them. Individuals and institutions alike must demonstrate that they have honored their commitments and owned their broken promises, preferably with the verification of the stakeholders themselves or independent outside experts. No passing the buck, no playing the blame game.

Transparency means operating out in the open, in the light of day. "What are they hiding?" is a sign of poor transparency that leads to distrust. Of course, companies have legitimate rights to trade secrets and other kinds of proprietary information. But when it comes to pertinent information for customers, shareholders, employees, and other stakeholders, active openness is central to earning trust. Rather than dressing for success, corporations can undress for success.

Trust in business and other institutions is mostly at an all-time low. The public relations company Edelman's 2015 "Trust Barometer" indicates that trust in institutions, especially corporations, has fallen back to levels from the dismally low period of the 2008 great recession. Edelman noted that even the once impregnable technology industry, still the most trusted business sector, saw declines in the majority of countries for the first time. Globally, CEOs and government officials continue to be the least credible information sources, lagging far behind academic or industry experts.[13] Similarly, Gallup reported in its 2015 survey of American confidence in institutions that "business" ranked second lowest among the fifteen institutions measured; fewer than 20 percent of respondents indicated they had considerable or high levels of trust. Only the U.S. Congress had a lower score.[14]

In the preblockchain world, trust in transactions derived from individuals, intermediaries, or other organizations acting with integrity. Because we often can't know our counterparties, let alone whether they have integrity, we've come to rely on third parties not only to vouch for strangers, but also to maintain transaction records and perform the business logic and transaction logic that powers commerce online. These powerful intermediaries—banks, governments, PayPal, Visa, Uber, Apple, Google, and other digital conglomerates—harvest much of the value.

In the emerging blockchain world, trust derives from the network and even from objects on the network. Carlos Moreira of the cryptographic security company WISeKey said that the new technologies effectively delegate trust—even to physical things. "If an object, whether it be a sensor on a communications tower, a light bulb, or a heart monitor, is not trusted to perform well or pay for services it will be rejected by the other objects automatically."[15] The ledger itself is the foundation of trust.[16]

To be clear, "trust" refers to buying and selling goods and services and to the integrity and protection of information, not trust in all business affairs. However, you will read throughout this book how a global ledger of truthful information can help build integrity into all our institutions and create a more secure and trustworthy world. In our view, companies that conduct some or all of their transactions on the blockchain will enjoy a trust bump in share price. Shareholders and citizens will come to expect all publicly traded firms and taxpayer-funded organizations to run their treasuries, at minimum, on the blockchain. Because of increased transparency, investors will be able to see whether a CEO really deserved that fat bonus. Smart contracts enabled by blockchains will require counterparties to abide by their commitments and voters will be able to see whether their representatives are being honest or acting with fiscal integrity.

RETURN OF THE INTERNET

The first era of the Internet started with the energy and spirit of a young Luke Skywalker—with the belief that any kid from a harsh desert planet could bring down an evil empire and start a new civilization by launching a dot-com. Naïve to be sure, but many people, present company included, hoped the Internet, as embodied in the World Wide Web, would disrupt the industrial world where power was gripped by the few and power structures were hard to climb and harder to topple. Unlike the old media that were centralized and controlled by powerful forces, and where the users were inert, the new media were distributed and neutral, and everyone was an active participant rather than a passive recipient. Low cost and massive peer-to-peer communication on the Internet would help undermine traditional hierarchies and help with the inclusion of developing world citizens in the global economy. Value and reputation would derive from quality of contribution, not status. If you were smart and hardworking in India, your merit would bring you reputation. The world would be flatter, more meritocratic, more flexible, and more fluid. Most important, technology would contribute to prosperity for everyone, not just wealth for the few.

Some of this has come to pass. There have been mass collaborations like Wikipedia, Linux, and Galaxy Zoo. Outsourcing and networked business models have enabled people in the developing world to participate in the global economy better. Today two billion people collaborate as peers socially. We all have access to information in unprecedented ways.

However, the Empire struck back. It has become clear that concentrated powers in business and government have bent the original democratic architecture of the Internet to their will.

Huge institutions now control and own this new means of production and social interaction—its underlying infrastructure; massive and growing treasure troves of data; the algorithms that increasingly govern business and daily life; the world of apps; and extraordinary emerging capabilities, machine learning, and autonomous vehicles. From Silicon Valley and Wall Street to Shanghai and Seoul, this new aristocracy uses its insider advantage to exploit the most extraordinary technology ever devised to empower people as economic actors, to build spectacular fortunes and strengthen its power and influence over economies and societies.

Many of the dark side concerns raised by early digital pioneers have pretty much materialized.[17] We have growth in gross domestic product but not commensurate job growth in most developed countries. We have growing wealth creation and growing social inequality. Powerful technology companies have shifted much activity from the open, distributed, egalitarian, and empowering Web to closed online walled gardens or proprietary, read-only applications that among other things kill the conversation. Corporate forces have captured many of these wonderful peer-to-peer, democratic, and open technologies and are using them to extract an inordinate share of value.

The upshot is that, if anything, economic power has gotten spikier, more concentrated, and more entrenched. Rather than data being more widely and democratically distributed, it is being hoarded and exploited by fewer entities that often use it to control more and acquire more power. If you accumulate data and the power that comes with it, you can further fortify your position by producing proprietary knowledge. This privilege trumps merit, regardless of its origin.

Further, powerful "digital conglomerates" such as Amazon, Google, Apple, and Facebook—all Internet start-ups at one time—are capturing the treasure troves of data that citizens and institutions generate often in private data silos rather than on the Web. While they create great value for consumers, one upshot is that data is becoming a new asset class—one that may trump previous asset classes. Another is the undermining of our traditional concepts of privacy and the autonomy of the individual.

Governments of all kinds use the Internet to improve operations and services, but they now also deploy technology to monitor and even manipulate citizens. In many democratic countries, governments use information and communications technologies to spy on citizens, change public opinion, further their parochial interests, undermine rights and freedoms, and overall to stay in power. Repressive governments like those of China and Iran enclose the Internet, exploiting it to crack down on dissent and mobilize citizens around their objectives.

This is not to say that the Web is dead, as some have suggested. The Web is critical to the future of the digital world and all of us should support efforts under way to defend it, such as those of the World Wide Web Foundation, who are fighting to keep it open, neutral, and constantly evolving.

Now, with blockchain technology, a world of new possibilities has opened up to reverse all these trends. We now have a true peer-to-peer platform that enables the many exciting things we've discussed in this book. We can each own our identities and our personal data. We can do transactions, creating and exchanging value without powerful intermediaries acting as the arbiters of money and information. Billions of excluded people can soon enter the global economy. We can protect our privacy and monetize our own information. We can ensure that creators are compensated for their intellectual property. Rather than trying to solve the problem of growing social inequality through the redistribution of wealth only, we can start to change the way wealth is *distributed*—how it is created in the first place, as people everywhere from farmers to musicians can share more fully, a priori, in the wealth they create. The sky does seem to be the limit.

It's more Yoda than God. But this new protocol, if not divine, does enable trusted collaboration to occur in a world that needs it, and that's a lot. Excited, we are.

YOUR PERSONAL AVATAR AND THE BLACK BOX OF IDENTITY

Throughout history, each new form of media has enabled mankind to transcend time, space, and mortality. That—dare we say—divine ability inevitably raises anew the existential question of identity: Who are we? What does it mean to be human? How do we conceptualize ourselves? As Marshall McLuhan observed, the medium becomes the message over time. People shape and are shaped by media. Our brains adapt. Our institutions adapt. Society adapts.

"Today you need an organization with endowed rights to provide you with an identity, like a bank card, a frequent flyer card, or a credit card,"[18] said Carlos Moreira of WISeKey. Your parents gave you a name, the state-licensed obstetrician or midwife who delivered you took your footprint and vouched for your weight and length, and both parties attested to the time, date, and place of your arrival by signing your birth certificate. Now they can record this certificate on the blockchain and link birth announcements and a college fund to it. Friends and family can contribute bitcoin to your higher education. There, your data flow begins.

In the early days of the Internet, Tom Peters wrote, "You are your projects."[19] He meant that our corporate affiliations and job titles no longer defined us. What is equally true now: "You are your data." Trouble is, Moreira said, "That identity is now yours, but the data that comes from its interaction in the world is owned by someone else."[20] That's how most corporations and institutions view you, by your data contrail across the Internet. They aggregate your data into a virtual representation of you, and they provide this "virtual you" with extraordinary new benefits beyond your parents' happiest dreams.[21] But convenience comes with a price: privacy. Those who say "privacy is dead—get over it" are wrong.[22] Privacy is the foundation of free societies.

"People have a very simplistic view of identity,"[23] said blockchain theorist Andreas Antonopoulos. We use the word *identity* to describe the self, the projection of that self to the world, and all these attributes that we associate with that self or one of its projections. These may come from nature, from the state, from private organizations. We may have one or more roles and a series of metrics attached to those roles, and the roles may change. Consider your last job. Did your role change organically because of changes in the work that needed to be done or because of revisions to your job description?

What if "the virtual you" was in fact owned by you—your personal avatar—and "lived" in the black box of your identity so that you could monetize your data stream and reveal only what you needed to, when asserting a particular right. Why does your driver's license contain more information than the fact that you have passed your driving test and demonstrated your ability to drive? Imagine a new era of the Internet where your personal avatar manages and protects the contents of your black box. This trusty software servant could release only the required detail or amount for each situation and at the same time whisk up your data crumbs as you navigate the digital world.

This may sound like the stuff of science fiction as portrayed in films like *The Matrix* or *Avatar*. But today blockchain technologies make it possible. Joe Lubin, CEO of Consensus Systems, refers to this concept as a "persistent digital ID and persona" on a blockchain. "I show a different aspect of myself to my college friends compared to when I am speaking at the Chicago Fed," he said. "In the online digital economy, I will represent my various aspects and interact in that world from the platform of different personas." Lubin expects to have a "canonical persona," the version of him that pays taxes, obtains loans, and gets insurance. "I will have perhaps a business persona and a family persona to separate the concerns that I choose to link to my canonical persona. I may have a gamer persona that I don't want linked to my business persona. I might even have a dark web persona that is never linkable to the others."[24]

Your black box may include information such as a government-issued ID, Social Security number, medical information, service accounts, financial accounts, diplomas, practice licenses, birth certificate, various other credentials, and information so personal you don't want to reveal it but do want to monetize its value, such as sexual preference or medical condition, for a poll or a research study. You could license these data for specific purposes to specific entities for specific periods of time. You could send a subset of your attributes to your eye doctor and a different subset to the hedge fund that you would like to invest in. Your avatar could answer yes-no questions without disclosing who you are: "Are you twenty-one years or older? Did you earn more than $100,000 in each of the last three years? Do you have a body mass index in the normal range?"[25]

In the physical world, your reputation is local—your local shopkeeper, your employer, your friends at a dinner party all have a certain opinion about you. In the digital economy, the reputations of various personas in your avatar will be portable. Portability will help bring people everywhere into the digital economy. People with a digital wallet and avatar in Africa could establish the reputation required to, say, borrow money to start a business. "See, all these people know me and have vouched for me. I am financially trustworthy. I am an enfranchised citizen of the global digital economy."

Identity is only a small part of it. The rest is a cloud—an identity cloud—of particulates loosely or tightly linked to your identity. If we try to record all these into the blockchain, an immutable ledger, we lose not only the nuance of social interaction but also the gift of forgetting. People ought never be defined by their worst day.

A PLAN FOR PROSPERITY

In this book, you'll read dozens of stories about initiatives enabled by this trust protocol that create new opportunities for a more prosperous world. Prosperity first and foremost is about one's standard of living. To achieve it, people must have the means, tools, and opportunities to create material wealth and thrive economically. But for us it includes more—security of the person, safety, health, education, environmental sustainability, opportunities to shape and control one's destiny and to participate in an economy and society. In order to achieve prosperity, an individual must possess, at minimum, access to some form of basic financial services to reliably store and move value, communication, and transactional tools to connect to the global economy, and security, protection, and enforcement of the title to land and other assets they possess legally.[26] This and more is the promise of the blockchain. The stories you will read should give you a sense of a future where there is prosperity for everyone, not just more wealth and power for the wealthy and powerful. Perhaps even a world where we own our data and can protect our privacy and personal security. An open world where everyone can contribute to our technology infrastructure, rather than a world of walled gardens where big companies offer proprietary apps. A world where billions of excluded people can now participate in the global economy and share in its largesse. Here's a preview.

Creating a True Peer-to-Peer Sharing Economy

Pundits often refer to Airbnb, Uber, Lyft, TaskRabbit, and others as platforms for the "sharing economy." It's a nice notion—that peers create and share in value. But these businesses have little to do with sharing. In fact, they are successful precisely because they do not share—they aggregate. It is an aggregating economy. Uber is a $65 billion corporation that aggregates driving services. Airbnb, the $25 billion Silicon Valley darling, aggregates vacant rooms. Others aggregate equipment and handymen through their centralized, proprietary platforms and then resell them. In the process, they collect data for commercial exploitation. None of these companies existed a decade ago because the technological preconditions were not there: ubiquitous smart phones, full GPS, and sophisticated payment systems. Now with blockchains, the technology exists to reinvent these industries again. Today's big disrupters are about to get disrupted.

Imagine instead of the centralized company Airbnb, a distributed application—call it blockchain Airbnb or bAirbnb—essentially a cooperative owned by its members. When a renter wants to find a listing, the bAirbnb software scans the blockchain for all the listings and filters and displays those that meet her criteria. Because the network creates a record of the transaction on the blockchain, a positive user review improves their respective reputations and establishes their identities—now without an intermediary. Says Vitalik Buterin, founder of the Ethereum blockchain: "Whereas most technologies tend to automate workers on the periphery doing menial tasks, blockchains automate away the center. Instead of putting the taxi driver out of a job, blockchain puts Uber out of a job and lets the taxi drivers work with the customer directly."[27]

Rewiring the Financial System for Speed and Inclusion

The financial services industry makes our global economy hum, but the system today is fraught with problems. For one, it is arguably the most centralized industry in the world and the last industry to feel the transformational effect of the technological revolution. Bastions of the old financial order such as banks go to great lengths to defend monopolies and often stymie disruptive innovation. The financial system also runs on outmoded technology and is governed by regulations dating back to the nineteenth century. It is rife with contradictions and uneven developments, making it sometimes slow, oftentimes insecure, and largely opaque to many stakeholders.

Distributed ledger technology can liberate many financial services from the confines of old institutions, fostering competition and innovation. That's good for the end user. Even when connected to the old Internet, billions of people are excluded from the economy for the simple reason that financial institutions don't provide services like banking to them because they would be unprofitable and risky customers. With the blockchain these people can not only become connected, but more important become included in financial activity, able to purchase, borrow, sell, and otherwise have a chance at building a prosperous life.

Similarly incumbent institutions can transform themselves around blockchain technology, if they can find the leadership to do it. The technology holds great promise to revolutionize the industry for the good—from banks to stock exchanges, insurance companies to accounting firms, brokerages, microlenders, credit card networks, real estate agents, and everything in between. When everyone shares the same distributed ledger, settlements don't take days, they occur instantly for all to see. Billions will benefit, and this shift could liberate and empower entrepreneurs everywhere.

Protecting Economic Rights Globally

Property rights are so inexorably tied to our system of capitalist democracy that Jefferson's first draft of the Declaration of Independence listed the inalienable rights of man as life, liberty, and the pursuit of *property*, not happiness.[28] While those aspirational tenets laid the groundwork for the modern economy and society we enjoy in much of the developed world, to this day much of the world's population does not reap their benefits. Even though some progress has been made in the departments of life and liberty, a majority of the world's property holders can have their homes or their bit of land seized arbitrarily by corrupt government functionaries, with the flick of a software switch in their centralized government property database. Without proof of property ownership, landowners can't secure a loan, get a building permit, or sell the property and they can be expropriated—all serious impediments to prosperity.

Peruvian economist and president of the Institute for Liberty and Democracy Hernando de Soto, one of the world's foremost economic minds, suggests that as many as five billion people in the world are barred from participating fully in the value created through globalization because they have a tenuous right to their land. Blockchain, he argues, could change all that. "The central idea to blockchain is that the rights to goods can be transacted, whether they be financial, hard assets or ideas. The goal is not merely to record the plot of land but rather to record the rights involved so that the rights holder cannot be violated."[29] Universal property rights could lay the groundwork for a new agenda of global justice, economic growth, prosperity, and peace. In this new paradigm, rights are protected, not by guns or militias or minutemen, but by technology. "Blockchain is for a world that's governed by real things instead of fictitious things. And I think that's good,"[30] said de Soto. And it's decentralized. No central authority controls it, everybody knows what's happening, and it remembers forever.

Ending the Remittance Rip-off

Just about every report, article, or book reviewing the benefits of cryptocurrencies discusses the opportunity of remittances. And for good reason. The largest flow of funds into the developing world is not foreign aid or direct foreign investment. Rather, it is remittance money repatriated to poor countries from their diasporas living abroad. The process takes time, patience, and sometimes courage to travel each week to the same wire transfer office's seedy neighborhood, fill out the same paperwork each time, and pay the same 7 percent fee. There is a better way.

Abra and other companies are building payment networks using the blockchain. Abra's goal is to turn every one of its users into a teller. The whole process—from the funds leaving one country to their arriving in another—takes an hour rather than a week and costs 2 percent versus 7 percent or higher. Abra wants its payment network to outnumber all physical ATMs in the world. It took Western Union 150 years to get to 500,000 agents worldwide. Abra will have that many tellers in its first year.

Cutting Out Bureaucracy and Corruption in Foreign Aid

Could blockchain solve problems with foreign aid? The 2010 Haiti earthquake was one of the deadliest natural disasters in recorded history. Somewhere between 100,000 and 300,000 people perished. The government in Haiti proved itself a liability in the aftermath. The global community donated more than $500 million to the Red Cross, a known brand. An after-action investigation revealed that funds were misspent or went missing altogether.

The blockchain can improve the delivery of foreign aid by eliminating the middlemen who take the aid before it reaches its destination. Second, as an immutable ledger of the flow of funds, blockchain holds institutions more accountable for their actions. Imagine if you could track each dollar you gave to the Red Cross from its starting point on your smart phone to the person it benefited. You could park your funds in escrow, releasing amounts after the Red Cross reached each milestone.

Feeding the Creators of Value First

Under the first generation of the Internet, many creators of intellectual property did not receive proper compensation for it. Exhibit A was musicians and composers who had signed with record labels whose leaders failed to imagine how the Internet would affect their industry. They failed to embrace the digital age and reinvent their own business models, slowly ceding control to innovative online distributors.

Consider the major labels' reaction to Napster, the peer-to-peer music file-sharing platform launched in 1999. Incumbents in the music industry teamed up to sue the new venture, its founders, *and eighteen thousand of its users*, dismantling the platform by July 2001. Alex Winter, director of a documentary on Napster, told *The Guardian*, "I have a problem with black-and-white thinking when it comes to big cultural changes. . . . With Napster, there was an enormous amount of grey" between the 'I can share everything I've paid for' position and the 'You're a criminal even if you share only one of the files you've purchased' point of view."[31]

We agree. Cocreating with consumers is usually a more sustainable business model than suing them. The whole incident turned a huge hot spotlight on the music industry, exposing its outdated marketing practices, gross distribution inefficiencies, and what some interpreted as antimusician policies.

Very little has changed since then. Until now. We look at the new music ecosystem emerging on the blockchain, led by British singer-songwriter Imogen Heap, cellist Zoë Keating, and blockchain developers and entrepreneurs. Every cultural industry is up for disruption, and the promise is that creators get fully compensated for the value they create.

Reconfiguring the Corporation as the Engine of Capitalism

With the rise of a global peer-to-peer platform for identity, trust, reputation, and transactions, we will finally be able to re-architect the deep structures of the firm for innovation, shared-value creation, and perhaps even prosperity for the many, rather than just wealth for the few. This doesn't mean smaller firms in terms of revenue or impact. To the contrary, we're talking about building twenty-first-century companies, some that may be massive wealth creators and powerful in their respective markets. We do think enterprises will look more like networks rather than the vertically integrated hierarchies of the industrial age. As such there is an opportunity to distribute (not redistribute) wealth more democratically.

We'll also take you on a stroll through the mind-boggling world of smart contracts, new autonomous economic agents, and what we call distributed autonomous enterprises where intelligent software takes over the management and organization of many resources and capabilities, perhaps displacing corporations. Smart contracts enable the creation of what we call open networked enterprises based on a new set of business models, or old business models with a blockchain twist.

Animating Objects and Putting Them to Work

Technologists and science fiction writers have long envisioned a world where a seamless global network of Internet-connected sensors could capture every event, action, and change on earth. Blockchain technology will enable things to collaborate, exchange units of value—energy, time, and money—and reconfigure supply chains and production processes according to shared information on demand and capacity. We can attach metadata to smart devices and program them to recognize other objects by their metadata and to act or react to defined circumstances without risk of error or tampering.

As the physical world comes to life, everyone can prosper—from small farmers in the Australian outback who need electrical power for their businesses to home owners everywhere who can become part of a distributed blockchain power grid.

Cultivating the Blockchain Entrepreneur

Entrepreneurship is essential to a thriving economy and a prosperous society. The Internet was supposed to liberate entrepreneurs, giving them the tools and capabilities of big companies without many of the liabilities, such as legacy culture, ossified processes, and dead weight. However, the high-flying success of dot-com billionaires obfuscates an unsettling truth: entrepreneurship and new business starts have been steadily declining for thirty years in many developed economies.[32] In the developing world, the Internet has done little to lower the barriers of would-be entrepreneurs who must suffer deadening government bureaucracies. The Internet has also not liberated the financial tools essential to starting a business available to billions of people. Not everyone is destined to be an entrepreneur, of course, but even for the average person trying to earn a decent wage, the lack of financial tools and the prevalence of government red tape make doing so challenging.

This is a complex issue, but blockchain can help supercharge entrepreneurship and therefore prosperity in many important ways. For the average person living in the developing world to have a reliable store of value and a way to conduct business beyond his community, all he needs now is an Internet-enabled device. Access to the global economy means greater access to new sources of credit, funding, suppliers, partners, and investment opportunities. No talent or resource is too small to monetize on the blockchain.

Realizing Governments by the People for the People

Buckle up for big changes in government and governance too. Blockchain technology is already revolutionizing the machinery of government and how we can make it high performance—better and cheaper. It's also creating new opportunities to change democracy itself—how governments can be more open and free from lobbyist control, and behave with the four values of integrity. We look at how blockchain technologies can change what it means to be a citizen and participate in the political process, from voting and accessing social services to solving some of society's big hairy problems and holding elected representatives accountable for the promises that got them elected.

PROMISE AND PERIL OF THE NEW PLATFORM

If there are six million people in the naked city,[33] then there are six million obstacles to this technology fulfilling its potential. Further, there are some worrisome downsides. Some say the technology is not ready for prime time; that it's still hard to use, and that the killer applications are nascent. Other critics point to the massive amount of energy consumed to reach consensus in just the bitcoin network: What happens when thousands or perhaps millions of interconnected blockchains are each processing billions of transactions a day? Are the incentives great enough for people to participate and behave safely over time, and not try to overpower the network? Is blockchain technology the worst job killer ever?

These are questions of leadership and governance, not of technology. The first era of the Internet took off because of the vision and common interests of its key stakeholders—governments, civil society organizations, developers, and everyday people like you. Blockchain requires similar leadership. We discuss at greater length in the book why leaders of this new distributed paradigm will need to stake their claim and unleash a wave of economic and institutional innovation, to ensure this time that the promise is fulfilled. We invite you to be one of these.

This book grew out of the $4 million Global Solution Networks program at the Rotman School of Management at the University of Toronto. Funded primarily by large technology corporations along with the Rockefeller and Skoll foundations, the U.S. State Department, and Industry Canada, the initiative explored new approaches to global problem solving and governance. We were both involved in running the program. (Don founded it; Alex led the project on cryptocurrencies.) In 2014, we launched a one-year initiative on the blockchain revolution and its implications for business and society, culminating in this book. In it, we have attempted to put the promise and the peril of the new platform into perspective.

If business, government, and civil society innovators get this right, we will move from an Internet driven primarily by the falling costs of search, coordination, data collection, and decision making—where the name of the game was monitoring, mediating, and monetizing information and transactions on the Web—to one driven by the falling costs of bargaining, policing, and enforcing social and commercial agreements, where the name of the game will be integrity, security, collaboration, the privacy of all transactions, and the creation and distribution of value. That's a 180-degree turn in strategy. The result can be an economy of peers with institutions that are truly distributed, inclusive, and empowering—and thereby legitimate. By fundamentally changing what we can do online, how we do it, and who can participate, the new platform may even create the technological preconditions to reconciling some of our most vexing social and economic challenges.

If we get this wrong, blockchain technology, which holds so much promise, will be constrained or even crushed. Worse, it could become a tool powerful institutions use to entrench their wealth or, if hacked by governments, a platform for some kind of new surveillance society. The tightly related technologies of distributed software, cryptography, autonomous agents, and even artificial intelligence could get out of control and turn against their human progenitors.

It is possible that this new technology may be delayed, stalled, underutilized, or worse. The blockchain and cryptocurrencies, particularly bitcoin, already have massive momentum, but we're not predicting whether or not all this will succeed, and if it does, how fast it will occur.[34] Prediction is always a risky business. Says technology theorist David Ticoll: "Many of us did a bad job of predicting the full impact of the Internet. ISIS type bad phenomena are among what we missed, and some big optimistic predictions turned out wrong." He says, "If the blockchain is as big and universal as the Net, we are likely to do a comparably bad job of predicting both its upsides and downsides."[35]

So rather than predicting a blockchain future, we're advocating for it. We're arguing that it should succeed, because it could help us usher in a new era of prosperity. We believe that the economy works best when it works for everyone, and this new platform is an engine of inclusion. It drastically lowers the cost of transmitting such funds as remittances. It significantly lowers the barrier to having a bank account, obtaining credit, and investing. And it supports entrepreneurship and participation in global trade. It catalyzes distributed capitalism, not just a redistributed capitalism.

Everyone should stop fighting it and take the right steps to get on board. Let's harness this force not for the immediate benefit of the few but for the lasting benefit of the many.

Today, both of us are excited about the potential of this next round of the Internet. We're enthusiastic about the massive wave of innovation that is being unleashed and its potential for prosperity and a better world. This book is our case to you to become interested, understand this next wave, and take action to ensure that the promise is fulfilled.

So hang on to your seat and read on! We're at one of those critical junctures in human history.

CHAPTER 2

BOOTSTRAPPING THE FUTURE: SEVEN DESIGN PRINCIPLES OF THE BLOCKCHAIN ECONOMY

Freedom is predicated on privacy," said Ann Cavoukian, executive director of the Privacy and Big Data Institute at Ryerson University. "I first learned that thirty years ago when I started going to conferences in Germany. It is no accident that Germany is the leading privacy and data protection country in the world. They had to endure the abuses of the Third Reich and the complete cessation of all of their freedoms, which started with the complete removal of their privacy. When that ended, they said, 'Never again.'"[1]

And so it is ironic—or totally fitting—that one of the first decentralized peer-to-peer computational platforms to guarantee user privacy is called Enigma, also the name given to the machine developed by German engineer Arthur Scherbius to transcribe coded information. Scherbius designed Enigma for commercial use: through his device, global companies could quickly and safely communicate their trade secrets, stock tips, and other insider information. Within a few years, Germany's military forces were manufacturing their own versions of Enigma to broadcast coded messages over radio to troops. During the war, the Nazis used Enigma to disseminate strategic plans, details of targets, and the timing of attacks. It was a tool of suffering and oppression.

Our contemporary Enigma is a tool of freedom and prosperity. Designed at MIT Media Lab by Guy Zyskind and Oz Nathan, the new Enigma combines the virtues of blockchain's public ledger, the transparency of which "provides strong incentives for honest behavior," with something known as *homomorphic encryption* and *secure multiparty computation.*[2] More simply put, "Enigma takes your information—any information—breaks it up, and encrypts it into pieces of data that are randomly distributed to nodes in the network. It doesn't exist in one spot," said Cavoukian. "Enigma uses blockchain technology to embed the data and track all the pieces of information."[3] You can share it with third parties and those parties can use it in computations without ever decrypting it.[4] If it works, it could reshape how we approach our own identity online. Imagine having a black box of your personal information that you alone control and can access.

No matter how cool it may sound, there are reasons to tread cautiously on the cryptographic frontier. First, it needs to bootstrap a large network of participants. Second, "cryptography is an area where you never want to be using the newest and greatest, because there is an entire history of an algorithm that everyone believes is secure, that's out there for four or five years, and some very inspired scientist will come out and say, there's a flaw, and the entire thing tumbles," said Austin Hill of Blockstream. "That's why we generally prefer conservative, very well-established, long-standing algorithms. This stuff is very, very well future-proofed, and bitcoin was designed with that in mind."[5]

Still, the concept is worth taking very seriously, as it has profound implications for privacy, security, and sustainability. "Enigma is offering what they say guarantees privacy," Cavoukian said. "That is a big claim, but that's the kind of thing we increasingly need in this connected, interconnected world."[6]

In our research, we came across a number of projects initiated on blockchain technologies whose developers had similar aspirations for enabling basic human rights—not only the rights to privacy and security, but also the rights to property, recognition as a person under the law, and participation in government, culture, and the economy. Imagine a technology that could preserve our freedom to choose for ourselves and our families, to express these choices in the world, and to control our own destiny, no matter where we lived or were born. What new tools and new jobs could we create with those capabilities? What new businesses and services? How should we think about the opportunities? The answers were right in front of us, compliments of Satoshi Nakamoto.

THE SEVEN DESIGN PRINCIPLES

We believe that this next era could be inspired by Satoshi Nakamoto's vision, designed around a set of implicit principles, and realized by the collaborative spirit of many passionate and equally talented leaders in the community.

His grand vision was limited to money, not to some greater goal of creating a second generation of the Internet. There was no discussion of reinventing the firm, changing our institutions, or transforming civilization for the better. Still, Satoshi's vision was stunning in its simplicity, originality, and insight into humankind. It became clear to those who read the 2008 paper that a new era of the digital economy was about to begin. Where the first era of the digital economy was sparked by a convergence of computing and communications technologies, this second era would be powered by a clever combination of computer engineering, mathematics, cryptography, and behavioral economics.

Folksinger Gordon Lightfoot crooned, "If you could read my mind, love, what a tale my thoughts could tell." Satoshi has been incommunicado since 2011 (though the name pops up on discussion boards from time to time), but we think the trust protocol he bootstrapped lends itself to principles for reconfiguring our institutions and economy.

Everyone we talked to has been eager to share insights into blockchain technology with us. Each conversation, each white paper, each forum thread has surfaced a number of themes that we've reverse-engineered into design principles—principles for creating software, services, business models, markets, organizations, and even governments on the blockchain. Satoshi never wrote about these principles, but they are implicit in the technology platform he unleashed. We see them as principles for shaping the next era of the digital economy, and an era of renewed trust.

If you're new to this space, we hope these principles will help you understand the basics of the blockchain revolution. If you're a die-hard skeptic of the bitcoin blockchain, they should still serve you as you contemplate your future as an entrepreneur, inventor, engineer, or artist who seeks creative collaborations with like-minded people; as an owner or investor in assets of all kinds; or as a manager who wants to reimagine your role in this nascent blockchain economy.

1. Networked Integrity

Principle: Trust is intrinsic, not extrinsic. Integrity is encoded in every step of the process and distributed, not vested in any single member. Participants can exchange value directly with the expectation that the other party will act with integrity. That means that the values of integrity—honesty in one's words and deeds, consideration for others' interests, accountability for the consequences of one's decisions and actions, and transparency in decision making and action taking—are coded in decision rights, incentive structures, and operations so that acting without integrity either is impossible or costs a lot more time, money, energy, and reputation.

Problem to Be Solved: On the Internet, people haven't been able to transact or do business directly for the simple reason that money isn't like other information goods and intellectual property per se. You can send the same selfie to all your friends, but you ought not give your friend a dollar that you've already given to someone else. The money must leave your account and go into your friend's. It can't exist in both places, let alone multiple places. And so there's a risk of your spending a unit of digital currency in two places and having one of them bounce like a bad check. That's called the *double-spend problem*. That's good for fraudsters who want to spend their money twice. It's bad for the recipient of the bounced amount and bad for your reputation online. Traditionally, when making online payments, we solve the double-spend problem by clearing every transaction through the central databases of one or many third parties, such as a money transfer service (like Western Union), a commercial bank (Citicorp), a government body (Commonwealth Bank of Australia), a credit card company (Visa), or an online payment platform (PayPal). Settlement can take days or even weeks in some parts of the world.

Breakthrough: Satoshi leveraged an existing distributed peer-to-peer network and a bit of clever cryptography to create a *consensus mechanism* that could solve the double-spend problem as well as, if not better than, a trusted third party. On the bitcoin blockchain, the network time-stamps the first transaction where the owner spends a particular coin and rejects subsequent spends of the coin, thus eliminating a double spend. Network participants who run fully operating bitcoin nodes—called *miners*—gather up recent transactions, settle them in the form of a block of data, and repeat the process every ten minutes. Each block must refer to the preceding block to be valid. The protocols also include a method for reclaiming disk space so that all nodes can efficiently store the full blockchain. Finally, the blockchain is public. Anyone can see transactions taking place. No one can hide a transaction, and that makes bitcoin more traceable than cash.

Satoshi sought not only to disintermediate the central banking powers but also to eliminate the ambiguity and conflicting interpretations of what happened. Let the code speak for itself. Let the network reach consensus algorithmically on what happened and record it cryptographically on the blockchain. The mechanism for reaching consensus is critical. "Consensus is a social process," blogged Vitalik Buterin, pioneer of the Ethereum blockchain. "Human beings are fairly good at engaging in consensus . . . without any help from algorithms." He explained that, once a system scales beyond an individual's ability to do the math, people turn to software agents. In peer-to-peer networks, the consensus algorithm divvies up the right to update the status of the network, that is, to vote on the truth. The algorithm doles out this right to a group of peers who constitute an economic set, a set that has skin in the game, so to speak. According to Buterin, what's important about this economic set is that its members are securely distributed: no single member or cartel should be able to overtake a majority, even if they had the means and incentive to do so.[7]

To achieve consensus, the bitcoin network uses what's called a *proof of work* (PoW) mechanism. This may sound complicated but the idea is a simple one. Because we can't rely on the identity of the miners to select who creates the next block, we instead create a puzzle that is hard to solve (i.e., it takes a lot of *work*), but easy to verify (i.e., everyone else can check the answer very quickly). Participants agree that whoever solves the problem first gets to create the next block. Miners have to expend resources (computing hardware and electricity) to solve the puzzle by finding the right hash, a kind of unique fingerprint for a text or a data file. For each block they find, miners receive bitcoin as a reward. The puzzle is mathematically set up to make it impossible to find a shortcut to solve it. That's why, when the rest of the network sees the answer, everyone trusts that a lot of work went into producing it. Also, this puzzle solving is continuous "to the tune of 500,000 trillion hashes per second," according to Dino Mark Angaritis. Miners are "looking for a hash that meets the target. It is statistically bound to occur every ten minutes. It's a Poisson process, so that sometimes it takes one minute and sometimes one hour, but on average, it's ten minutes." Angaritis explained how it works: "Miners gather all the pending transactions that they find on the network and run the data through a cryptographic digest function called the secure hash algorithm (SHA-256), which outputs a 32-byte *hash value*. If the hash value is below a certain target (set by the network and adjusted every 2,016 blocks), then the miner has found the answer to the puzzle and has 'solved' the block. Unfortunately for the miner, finding the right hash value is very difficult. If the hash value is wrong, the miner adjusts the input data slightly and tries again. Each attempt results in an *entirely different* hash value. Miners have to try many times to find the right answer. As of November 2015, the number of hash attempts is on average 350 million trillion. That's a lot of work!"[8]

You may hear about other consensus mechanisms. The first version of the Ethereum blockchain—Frontier—also uses proof of work, but the developers of Ethereum 1.1 expect to replace it with a *proof of stake* mechanism. Proof of stake requires miners to invest in and hang on to some store of value (i.e., the native token of the blockchain such as Peercoin, NXT, etc.). They needn't spend energy to vote. Other blockchains, such as Ripple and Stellar, rely on social networks for consensus and may recommend that new participants (i.e., new nodes) generate a *unique node list* of at least one hundred nodes they can trust in voting on the state of affairs. This type of proof is biased: newcomers need social intelligence and reputation to participate. *Proof of activity* is another mechanism; it combines proof of work and proof of stake, where a random number of miners must sign off on the block using a cryptokey before the block becomes official.[9] *Proof of capacity* requires miners to allot a sizable volume of their hard drive to mining. A similar concept, *proof of storage*, requires miners to allocate and share disk space in a distributed cloud.

Storage does matter. Data on blockchains are different from data on the Internet in one important way. On the Internet, most of the information is malleable and fleeting, and the exact date and time of its publication isn't critical to past or future information. On the blockchain, bitcoin movement across the network is permanently stamped, from the moment of its coinage. For a bitcoin to be valid, it must reference its own history as well as the history of the blockchain. Therefore, the blockchain must be preserved in its entirety.

So important are the processes of mining—assembling a block of transactions, spending some resource, solving the problem, reaching consensus, maintaining a copy of the full ledger—that some have called the bitcoin blockchain a public utility like the Internet, a utility that requires public support. Paul Brody of Ernst & Young thinks that all our appliances should donate their processing power to the upkeep of a blockchain: "Your lawnmower or dishwasher is going to come with a CPU that is probably a thousand times more powerful than it actually needs, and so why not have it mine? Not for the purpose of making you money, but to maintain your share of the blockchain,"[10] he said. Regardless of the consensus mechanism, the blockchain ensures integrity through clever code rather than through human beings who choose to do the right thing.

Implications for the Blockchain Economy: Rather than trusting big companies and governments to verify people's identities and vouch for their reputations, we can trust the network. *For the first time ever, we have a platform that ensures trust in transactions and much recorded information no matter how the other party acts.*

The implications for most social, political, and economic activity are staggering. It's not just about who married whom, who voted for whom, who paid whom, it's about any endeavor that requires trusted records and assured transactions. Who owns what? Who holds which rights to this intellectual property? Who graduated from medical school? Who bought guns? Who made these Nike shoes, this Apple device, or this baby formula? Where did these diamonds come from? Trust is the sine qua non of the digital economy, and a platform for secure and reliable mass collaboration holds many possibilities for a new kind of organization and society.

2. Distributed Power

Principle: The system distributes power across a peer-to-peer network with no single point of control. No single party can shut the system down. If a central authority manages to black out or cut off an individual or group, the system will still survive. If over half the network attempts to overwhelm the whole, everyone will see what's happening.

Problem to Be Solved: In the first era of the Internet, any large institution with a large established base of users, be they employees, citizens, customers, or other organizations, thought little of their social contract. Time and time again, central powers have proven that they're willing and able to override users, warehouse and analyze user data, respond to government requests for data without users' knowledge, and implement large-scale changes without users' consent.

Breakthrough: The energy costs of overpowering the bitcoin blockchain would outweigh the financial benefits. Satoshi deployed a proof-of-work method that requires users to expend a lot of computing power (which requires a lot of electricity) to defend the network and mint new coins. He was inspired by cryptographer Adam Back's solution, Hashcash, to mitigate spam and denial-of-service attacks. Back's method required e-mailers to provide proof of work when sending the message. It in effect stamped "special delivery" on an e-mail to signal the message's importance to its sender. "This message is so critical that I've spent all this energy in sending it to you." It increases the costs of sending spam, malware, and ransomware.

Anyone can download the bitcoin protocol for free and maintain a copy of the blockchain. It leverages *bootstrapping*, a technique for uploading the program onto a volunteer's computer or mobile device through a few simple instructions that set the rest of the program in motion. It's fully distributed across a volunteer network like BitTorrent, a shared database of intellectual property that resides on tens of thousands of computers worldwide.

To be sure, this shields the network from the hands of the state, which could be good or bad depending on the situation—say a dissident in a totalitarian country fighting for women's rights versus a criminal in a democratic country conducting extortion. Totalitarian regimes could not freeze bank accounts or seize funds of political activists. States could not arbitrarily seize assets on the blockchain as Franklin Delano Roosevelt's administration did through FDR's Executive Order 6102, which required citizens to turn their "gold coin, gold bullion, and gold certificates" over to the government or risk fines or imprisonment.[11] Josh Fairfield of Washington and Lee University put it bluntly: "There's no middleman to go after anymore."[12] The blockchain resides everywhere. Volunteers maintain it by keeping their copy of the blockchain up to date and lending their spare computer processing units for mining. No backdoor dealing. Every action or transaction is broadcast across the network for subsequent verification and validation. Nothing passes through a central third party; nothing is stored on a central server.

Satoshi also distributed the *mint* by linking the issuance of bitcoins to the creation of a new block in the ledger, putting the power to mint into all the hands of the peer network. Whichever miner solved the puzzle and submitted proof of work first could receive a number of new bitcoins. There is no Federal Reserve, central bank, or treasury with control over the money supply. Moreover, each bitcoin contains direct links to its genesis block and all subsequent transactions.

So no intermediaries are required. The functioning of the blockchain is mass collaboration at its best. You have power over your data, your property, and your level of participation. It's distributed computing power enabling distributed and collective human power.

Implications for the Blockchain Economy: Perhaps such a platform could enable new distributed models of wealth creation. Perhaps new kinds of peer-to-peer collaborations could target humanity's most vexing social problems. Perhaps we could solve the crisis of confidence and even legitimacy in today's institutions by shifting real power toward citizens, equipping them with real opportunities for prosperity and participation in society, rather than through PR trickery.

3. Value as Incentive
Principle: The system aligns the incentives of all stakeholders. Bitcoin or some token of value is integral to this alignment and correlative of reputation. Satoshi programmed the software to reward those who work on it and belong to those who hold and use its tokens, so that they all take care of it. Sort of the ultimate Tamagotchi, the blockchain is a globally distributed nest egg.[13]

Problem to Be Solved: In the first era of the Internet, the concentration of power in corporations, combined with their sheer size, complexity, and opacity, enabled them to extract disproportionate value from the very networks that endowed them with rights. Large banks exploited the financial system to its breaking point because "incentive structures for most of the top executives and many of the lending officers of these banks [were] designed to encourage short-sighted behavior and excessive risk-taking," according to economist Joseph Stiglitz. That included "preying on the poorest Americans." He summed up the problem: "If you give people bad incentives, they behave badly, and they behaved just as one would have expected."[14]

Large dot-coms dangled free services in retail, search, and social media in exchange for user data. According to an Ernst & Young survey, nearly two thirds of managers polled said they collected consumer data to drive business, and nearly 80 percent claimed to have increased revenues from this data mining. But when these firms get hacked, it's the consumers who have to clean up the mess of stolen credit card and bank account information. It's not surprising that, in the same survey, nearly half of consumers said they'd be cutting off access to their data in the next five years, and over half said they were already providing less data, including censoring themselves on social media, than in the previous five years.[15]

Breakthrough: Satoshi expected participants to act in their own self-interests. He understood game theory. He knew that networks without gatekeepers have been vulnerable to Sybil attacks, where nodes forge multiple identities, dilute rights, and depreciate the value of reputation.[16] The integrity of the peer-to-peer network and the reputation of its peers both diminish if you don't know whether you're dealing with three parties or one party using three identities. So Satoshi programmed the source code so that, no matter how selfishly people acted, their actions would benefit the system overall and accrue to their reputations, however they chose to identify themselves. The resource requirements of the consensus mechanism, combined with bitcoins as reward, could compel participants to do the right thing, making them trustworthy in the sense that they were predictable. Sybil attacks would be economically unviable.

Satoshi wrote, "By convention, the first transaction in a block is a special transaction that starts a new coin owned by the creator of the block. This adds an incentive for nodes to support the network."[17] Bitcoin is an incentive for miners to participate in creating a block and linking it to the previous block. Those who complete a block first get a quantity of bitcoins for their efforts. Satoshi's protocol rewarded early adopters handsomely with bitcoin: for the first four years, miners received 50 bitcoins (BTC) for each block. Every four years, the reward per block would halve: 25 BTC, 12.5 BTC, and so on. Because they now own bitcoin, they have an incentive to ensure the platform's long-term success, buying the best equipment to run mining operations, spending energy as efficiently as possible, and maintaining the ledger. Bitcoin is also a claim on the blockchain, not just as an incentive to participate in mining and transacting with others but through ownership in the platform itself. Distributed user accounts are the most basic element of the cryptographic network infrastructure. By owning and using bitcoin, one is financing the blockchain's development.

Satoshi chose as the economic set the *owners of computing power*. This requires these miners to consume a resource external to the network, namely electricity, if they want to participate in the reward system. Every so often, different miners find two equally valid blocks of equal height, and the rest of the miners must choose which block to build on next. They generally pick whichever they think will win rather than building on both, because they'd otherwise have to split their processing power between the forks, and that's a strategy for losing value. The longest chain represents the greatest amount of work and therefore participants choose it as the canonical state of the blockchain. In contrast, Ethereum chose *owners of coin* as its economic set. Ripple and Stellar chose the social network.

The paradox of these consensus schemes is that by acting in one's self-interest, one is serving the peer-to-peer (P2P) network, and that in turn affects one's reputation as a member of the economic set. Before blockchain technologies, people couldn't easily leverage the value of their reputation online. It wasn't only because of Sybil attacks, where a computer could inhabit multiple roles. Identity is multifaceted, nuanced, and transient. Few people see all sides, let alone the subtleties and the arc of our identity. For different contexts, we have to produce some document or other to attest to some detail of our identity. People "without papers" are confined to collaborating with their social circle. On blockchains like Stellar, that's an excellent start, a means of creating a persistent digital presence and establishing reputation that is portable well beyond one's geographic community.

Another breakthrough to preserve value is the *monetary policy* programmed into the software. "All money mankind has ever used has been insecure in one way or another," said Nick Szabo. "This insecurity has been manifested in a wide variety of ways, from counterfeiting to theft, but the most pernicious of which has probably been inflation."[18] Satoshi capped the supply of bitcoins at 21 million to be issued over time to prevent arbitrary inflation. Given the halving every four years of bitcoins mined in a block and the current rate of mining—six blocks per hour—those 21 million BTC should be in circulation around the year 2140. No hyperinflation or currency devaluation caused by incompetent or corrupt bureaucracies.

Currencies are not the only assets that we can trade on the blockchain. "We've only begun to scratch the surface on what's possible," said Hill of Blockstream. "We're still at that 1994 point in terms of applications and protocols that really take advantage of the network and show the world, 'Here's what you can do that is totally groundbreaking.'"[19] Hill expects to see different financial instruments, from proof-of-asset authenticity to proof-of-property ownership. He also expects to see bitcoin applications in the Metaverse (a virtual world) where you can convert bitcoin into Kongbucks and hire Hiro Protagonist to hack you some data.[20] Or jack yourself into the OASIS (a world of multiple virtual utopias) where you actually do discover the Easter egg, win Halliday's estate, license OASIS's virtual positioning rights to Google, and buy a self-driving car to navigate Toronto.[21]

And, of course, there's the Internet of Things, where we register our devices, assign them an identity (Intel is already doing this), and coordinate payment among them using bitcoin rather than multiple fiat currencies. "You can define all these new business cases that you want to do, and have it interoperate within the network, and use the network infrastructure without having to bootstrap a new blockchain, just for yourself," said Hill. [22]

Unlike fiat currency, each bitcoin is divisible to eight decimal places. It enables users to combine and split value over time in a single transaction, meaning that an input can have multiple outputs over multiple periods of time, which is far more efficient than a series of transactions. Users can set up smart contracts to meter usage of a service and make tiny fractions of payments at regular intervals.

Implications for the Blockchain Economy: The first era of the Internet missed all this. Now we have a platform where people and even things have proper financial incentives to collaborate effectively and create just about anything. Imagine online discussion groups where participants have reputations worth enhancing, in part because bad behavior will cost them financially. Trolls need not apply. Imagine a peer-to-peer network of solar panels where home owners receive real-time compensation on the blockchain for generating sustainable energy. Imagine an open source software project where a community of developers compensates supercontributors for acceptable code. Imagine there's no countries. It isn't hard to do.[23]

4. Security

Principle: Safety measures are embedded in the network with no single point of failure, and they provide not only confidentiality, but also authenticity and nonrepudiation to all activity. Anyone who wants to participate must use cryptography—opting out is not an option—and the consequences of reckless behavior are isolated to the person who behaved recklessly.

Problem to Be Solved: Hacking, identity theft, fraud, cyberbullying, phishing, spam, malware, ransomware—all of these undermine the security of the individual in society. The first era of the Internet, rather than bringing transparency and impairing violations, seems to have done little to increase security of persons, institutions, and economic activity. The average Internet user often has to rely on flimsy passwords to protect e-mail and online accounts because service providers or employers insist on nothing stronger. Consider the typical financial intermediary: it doesn't specialize in developing secure technology; it specializes in financial innovation. In the year that Satoshi published his white paper, data breaches at such financial firms as BNY Mellon, Countrywide, and GE Money accounted for over 50 percent of all identity thefts reported that year, according to the Identity Theft Resource Center.[24] By 2014, that figure had fallen to 5.5 percent for the financial sector, but breaches in medical and health care jumped to 42 percent of the year's total. IBM reported that the average cost of a data breach is $3.8 million, which means that data breaches have cost at least $1.5 billion over the last two years.[25] The average cost to an individual of medical identity fraud is close to $13,500, and offenses are on the rise. Consumers don't know which aspect of their life will be hacked next.[26] If the next stage of the digital revolution involves communicating money directly between parties, then communication needs to be hackproof.

Breakthrough: Satoshi required participants to use public key infrastructure (PKI) for establishing a secure platform. PKI is an advanced form of "asymmetric" cryptography, where users get two keys that don't perform the same function: one is for encryption and one for decryption. Hence, they are asymmetric. The bitcoin blockchain is now the largest civilian deployment of PKI in the world, and second overall to the U.S. Department of Defense common access system.[27]

Pioneered in the 1970s,[28] asymmetric cryptography gained some traction in the 1990s in the form of e-mail encryption freeware such as Pretty Good Privacy. PGP is pretty secure, and pretty much a hassle to use because everyone in your network needs to be using it, and you have to keep track of your two keys and everyone's public keys. There's no password-reset function. If you forget yours, you have to start all over. According to the Virtru Corporation, "the use of email encryption is on the rise. Still, only 50 percent of emails are encrypted in transit, and end-to-end email encryption is rarer still."[29] Some people use *digital certificates*, pieces of code that protect messages without the encrypt-decrypt operations, but users must apply (and pay an annual fee) for their individual certificates, and the most common e-mail services—Google, Outlook, and Yahoo!—don't support them.

"Past schemes failed because they lacked incentive, and people never appreciated privacy as incentive enough to secure those systems,"[30] Andreas Antonopoulos said. The bitcoin blockchain solves nearly all these problems by providing the incentive for wide adoption of PKI for all transactions of value, not only through the use of bitcoin but also in the shared bitcoin protocols. We needn't worry about weak firewalls, thieving employees, or insurance hackers. If we're both using bitcoin, if we can store and exchange bitcoin securely, then we can store and exchange highly confidential information and digital assets securely on the blockchain.

Here's how it works. Digital currency isn't stored in a file per se. It's represented by transactions indicated by a cryptographic *hash*. Users hold the cryptokeys to their own money and transact directly with one another. With this security comes the responsibility of keeping one's private keys private.

Security standards matter. The bitcoin blockchain runs on the very well-known and established SHA-256 published by the U.S. National Institute of Standards and Technology and accepted as a U.S. Federal Information Processing Standard. The difficulty of the many repetitions of this mathematical calculation required to find a block solution forces the computational device to consume substantial electricity in order to solve a puzzle and earn new bitcoin. Other algorithms such as proof of stake burn much less energy.

Remember what Austin Hill said at the start of this chapter about never using the newest and greatest in algorithms. Hill, who works with cryptographer Adam Back at Blockstream, expressed concern over cryptocurrencies that don't use proof of work. "I don't think proof of stake ultimately works. To me, it's a system where the rich get richer, where people who have tokens get to decide what the consensus is, whereas proof of work ultimately is a system rooted in physics. I really like that because it's very similar to the system for gold."[31]

Finally, the longest chain is generally the safest chain. The security of Satoshi's blockchain benefits greatly from its relative maturity and its established base of bitcoin users and miners. Hacking it would require more computing power than attacking short chains. Hill said, "Whenever one of these new networks start up with an all new chain, there's a bunch of people who direct their latent computer power, all the computers and CPUs that they took offline from mining bitcoin, they point at these new networks to manipulate them and to essentially attack the networks."[32]

Implications for the Blockchain Economy: In the digital age, technological security is obviously the precondition to security of a person in society. Today bits can pass through our firewalls and wallets. Thieves can pick our pockets or hijack our cars from the other side of the world. As each of us relies more on digital tools and platforms, such threats have multiplied in ways that most of us do not understand. With the bitcoin blockchain, with its more secure design and its transparency, we can make transactions of value and protect what happens to our data.

5. Privacy

Principle: People should control their own data. Period. People ought to have the right to decide what, when, how, and how much about their identities to share with anybody else. Respecting one's right to privacy is not the same as actually respecting one's privacy. We need to do both. By eliminating the need to trust others, Satoshi eliminated the need to know the true identities of those others in order to interact with them. "I've spoken to many engineers and computer scientists, and they all tell me—every single one—'Of course, we can embed privacy into data architecture, into the design of the programs. Of course we can,'"[33] said Ann Cavoukian.

Problem to Be Solved: Privacy is a basic human right and the foundation of free societies. In the last twenty years of the Internet, central databases in both public and private sectors have accumulated all sorts of confidential information about individuals and institutions, sometimes without their knowledge. Everywhere people worry that corporations are creating what we could call *cyberclones* of them by fracking the digital world for their data. Even democratic governments are creating surveillance nations, evidenced by the recent U.S. National Security Agency's overextending its surveillance rights by conducting warrantless spying over the Internet. These are double privacy offenses, first collecting and using our data without our understanding or our permission, then not protecting the honeypot from hackers. "It's all about abandoning zero-sum pursuits, either-or propositions, win-lose, you can have one interest or the other. That, to me, is so dated, so yesterday, and so counterproductive," said Cavoukian. "We substitute a positive-sum model which is, essentially, you can have privacy and—fill in the blank."[34]

Breakthrough: Satoshi installed no identity requirement for the network layer itself, meaning that no one had to provide a name, e-mail address, or any other personal data in order to download and use the bitcoin software. The blockchain doesn't need to know who anybody is. (And Satoshi didn't need to capture anybody's data to market other products. His open source software was the ultimate in thought leadership marketing.) That's how the Society for Worldwide Interbank Financial Telecommunication works—if you pay in cash, then SWIFT doesn't generally ask for identification—but we're guessing that many SWIFT offices have cameras, and financial institutions must comply with anti–money laundering/know your customer (AML/KYC) requirements to join and use SWIFT.

Additionally, the identification and verification layers are separate from the transaction layer, meaning that Party A broadcasts the transfer of bitcoins from Party A's address to Party B's address. There's no reference to anyone's identity in that transaction. Then the network confirms that Party A not only controlled the amount of bitcoin specified but also authorized the transaction before recognizing Party A's message as "unspent transaction output" associated with Party B's address. Only when Party B goes to spend that amount does the network verify that Party B now controls that bitcoin.

Compare that with using credit cards, a very identity-centric model. That's why millions of people's addresses and phone numbers are stolen every time a database gets breached. Consider the number of records attached to a few of the more recent data breaches: T-Mobile, 15 million records; JPMorgan Chase, 76 million; Anthem Blue Cross Blue Shield, 80 million; eBay, 145 million; Office of Personnel Management, 37 million; Home Depot, 56 million; Target, 70 million; and Sony, 77 million; and there were smaller breaches of airlines, universities, gas and electric utilities, and hospital facilities, some of our most precious infrastructure assets.[35]

On the blockchain, participants can choose to maintain a degree of personal anonymity in the sense that they needn't attach any other details to their identity or store those details in a central database. We can't underscore how huge this is. *There are no honeypots of personal data on the blockchain.* The blockchain protocols allow us to choose the level of privacy we're comfortable with in any given transaction or environment. It helps us to better manage our identities and our interaction with the world.

A start-up called Personal BlackBox Company, LLC, is aiming to help large corporations transform their relationship to consumer data. PBB's chief marketing officer, Haluk Kulin, told us, "Companies such as Unilever or Prudential are coming to us and saying, 'We're very interested in building better data relationships. Can we leverage your platform? We're very interested in reducing our data liability.' They're seeing that data is increasingly a toxic asset inside of corporations."[36] Its platform gives clients access to anonymous data—much like a clinical trial, where pharmaceuticals know only the relevant aspects of patients' health—without taking on any data security risk. Some consumers may give away more information in exchange for bitcoins or other corporate benefits. On the back end, PBB's platform deploys PKI so that only consumers have access to their data through their private keys. Not even PBB has access to consumer data.

The blockchain offers a platform for doing some very flexible forms of selective and anonymous attestation. Austin Hill likened it to the Internet. "A TCP/IP address is not identified to a public ID. The network layer itself doesn't know. Anyone can join the Internet, get an IP address, and start sending and receiving packets freely around the world. As a society, we've seen an incredible benefit allowing that level of pseudonymity. . . . Bitcoin operates almost exactly like this. The network itself does not enforce identity. That's a good thing for society and for proper network design."[37]

So while the blockchain is public—anyone can view it at any time because it resides on the network, not within a centralized institution charged with auditing transactions and keeping records—users' identities are pseudonymous. This means that you have to do a considerable amount of triangulating of data to figure out who or what owns a particular public key. The sender can provide only the metadata that the recipient needs to know. Moreover, anyone can own multiple public/private key sets, just as anyone can have multiple devices or access points to the Internet and multiple e-mail addresses under various pseudonyms.

That said, Internet service providers like Time Warner that assign IP addresses do keep records linking identities to accounts. Likewise, if you get a bitcoin wallet from a licensed online exchange such as Coinbase, that exchange is required to do its due diligence under AML/KYC requirements. For example, here is Coinbase's privacy policy: "We collect information sent to us through your computer, mobile phone, or other access device. This information may include your IP address, device information including, but not limited to, identifier, device name and type, operating system, location, mobile network information and standard web log information, such as your browser type, traffic to and from our site and the pages you accessed on our website."[38] So governments can subpoena ISPs and exchanges for this type of user data. But they can't subpoena the blockchain.

It's also important to know that we can design higher levels of transparency into any set of transactions, application, or business model, should all the stakeholders agree to do so. In varying situations we will see new capabilities where radical transparency makes a lot of sense. When companies tell the truth to customers, shareholders, or business partners, they build trust.[39] That is, privacy for individuals, transparency for organizations, institutions, and public officials.

Implications for the Blockchain Economy: To be sure, the blockchain provides opportunities to stop the stampede to a surveillance society. Now think about the problem of corporate big data for each of us. What does it mean for a corporation to have perfect information about you? We are some twenty years into the global Internet era, and only at the beginning of corporate access to the most intimate details of our personal lives. Coming up fast are personal health and fitness data, our daily comings and goings, the inner lives of our homes, and, well, you name it. Many people are simply unaware of the many micro-Faustian deals they make online every day. By simply using Web sites, consumers authorize their owners to convert trails of digital crumbs into detailed road maps for private commercial benefit.

Unless we shift to the new paradigm, it's not science fiction to foresee hundreds of millions of avatars humming away in tomorrow's data centers. With blockchain technology, you could own your personal avatars as you do in the Second Life virtual world, but with real-world implications. The Virtual You could protect your personal information, giving away only the information required in any social or economic exchange work under your command and make sure you receive compensation for any of your data that has value to another party. It's a shift from big data to private data. Call it "little data."

6. Rights Preserved

Principle: Ownership rights are transparent and enforceable. Individual freedoms are recognized and respected. We hold this truth to be self-evident—that all of us are born with certain inalienable rights that should and can be protected.

Problem to Be Solved: The first era of the digital economy was about finding ways to exercise these rights more efficiently. The Internet became a medium for new forms of art, news, and entertainment, for establishing copyright of poems, songs, stories, photographs, and audio and video recordings. We could apply the Uniform Commercial Code further to do online what the code had already expedited in physical space, which was to eliminate the need to negotiate and create contracts for every single item, like a tube of toothpaste, no matter how small its price. Even so, we had to trust middlemen to manage transactions, and they had the power to deny the transaction, delay it, and hold the money in their own account (bankers call this "float"), or clear it only to reverse it later. They expected a percentage of people to cheat and accepted a certain level of fraud as unavoidable.

In this great burst of efficiency, legitimate rights got trampled, the rights not only to privacy and security but also free speech, reputation, and equal participation. People could anonymously censor us, defame us, and block us at little cost or risk to themselves. Filmmakers who depended on revenues from syndication, video on demand, enhanced DVD sales, and cable rights to films released decades earlier found their revenue stream drying up to a trickle as their fans uploaded digital files for others to download for free.

Breakthrough: The proof of work required to mint coins also timestamps transactions, so that only the first spend of a coin would clear and settle. Combined with PKI, the blockchain not only prevents a double spend but also confirms ownership of every coin in circulation, and each transaction is immutable and irrevocable. In other words, we can't trade what isn't ours on the blockchain, whether it's real property, intellectual property, or rights of personhood. Nor can we trade what we aren't authorized to trade on somebody else's behalf in an agency role, perhaps as a lawyer or a company manager. And we can't stifle people's freedom of expression, assembly, and religion.

Haluk Kulin of Personal BlackBox said it best: "In the thousands of years of human social interaction, every time we've taken the right of participation from the people, they have come back and broken the system. We're discovering that, even in digital, stealing their consent is not sustainable."[40] As the Ledger of Everything, the blockchain can serve as a public registry through such tools as Proof of Existence (PoE), a site that creates and registers cryptographic digests of deeds, titles, receipts, or licenses on the blockchain. Proof of Existence doesn't maintain a copy of any original document; the hash of the document is calculated on the user's machine, not on the PoE site, thus ensuring confidentiality of content. Even if a central authority shuts down Proof of Existence, the proof remains on the blockchain.[41] So the blockchain provides means of proving ownership and preserving records without censorship.

On the Internet, we couldn't necessarily enforce contractual rights or oversee implementation. And so, for more complex transactions involving bundles of rights and multiple parties, we now have the *smart contract*, a piece of special purpose code that executes a complex set of instructions on the blockchain. "That intersection of legal descriptions and software is fundamental, and the smart contracts are the first step in that direction," said Steve Omohundro, president of think tank Self-Aware Systems. "Once the principles of how you codify law digitally become more understood, then I think every country will start doing it. . . . Each jurisdiction would encode its laws, precisely and digitally, and there would be translation programs between them. . . . Getting rid of the friction of all legal stuff is going to be a huge economic gain."[42]

A smart contract provides a means for assigning usage rights to another party, as a composer might assign a completed song to a music publisher. The code of the contract could include the term or duration of the assignment, the magnitude of royalties that would flow from the publisher's to the composer's bitcoin account during the term, and some triggers for terminating the contract. For example, if the composer's account received less than a quarter of a bitcoin in a consecutive thirty-day period, then all rights would automatically revert to the composer, and the publisher would no longer have access to the composer's work registered on the blockchain. To set this smart contract in motion, both the composer and the publisher—and perhaps representatives of the publisher's finance and legal teams—would sign using their private keys.

A smart contract also provides a means for owners of assets to pool their resources and create a corporation on the blockchain, where the articles of incorporation are coded into the contract, clearly spelling out and enforcing the rights of those owners. Associated agency-employment contracts could define the decision rights of managers by coding what they could and couldn't do with corporate resources without ownership permission.

Smart contracts are unprecedented methods of ensuring contractual compliance, including social contracts. "If you have a big transaction with a specific control structure, you can predict the outcome at any period in time," said Antonopoulos. "If I have a fully verified signed transaction with a number of signatures in a multisignature account, I can predict whether that transaction will be verifiable by the network. And if it is verifiable by the network, then that transaction can be redeemed and irrevocably so. No central authority or third party can revoke it, no one can override the consensus of the network. That's a new concept in both law and finance. The bitcoin system provides a very high degree of certainty as to the outcome of a contract."[43]

The contract couldn't be seized, stopped, or redirected to a different bitcoin address. You need only to transmit the signed transaction to any of the bitcoin network nodes from anywhere using any medium. Said

Antonopoulos, "People could shut down the Internet, and I could still transmit that transaction over shortwave radio with Morse code. A government agency could try to censor my communication, and I could still transmit that transaction as a series of smiley emoticons over Skype. As long as someone on the other end could decode the transaction and record it in the blockchain, I could effect the [smart contract]. So we've converted something that, in law, is almost impossible to guarantee into something that has verifiable mathematical certainty."[44]

Consider property rights, both real and intellectual: "Ownership is just a recognition by a government or an agency that you own something and they will defend your claims on that ownership," said Stephen Pair, CEO of BitPay. "That's just a contract that can be signed by whatever authority that will defend your rights for you and they sign it over to your identity, and then once you have that, and that ownership is recorded, you then can transfer it to other people. That's very straightforward."[45] Communities with shared resources could consider a spectrum of rights, borrowing from Nobel Prize–winning economist Elinor Ostrom's pyramid of rights, a pecking order of sorts. At the lowest level, there are authorized users who may only access and withdraw resources; claimants who have those rights but can also exclude others from access; proprietors who hold management rights beyond access and exclusion; and owners who can access, use, exclude others, manage, and sell the resource (i.e., right of alienation).[46]

Now consider the rights to privacy and publicity: "Our model is really rights applied to the market," said Kulin of Personal BlackBox. His company uses blockchain technology to represent and enforce the rights of individuals to extract value from their personal data. "The blockchain provides us a whole group of people who are both mission-aligned and technology-aligned to create different ways that enterprises can leverage these unique data sets rather than protect their data silos."[47] Simply put, people create better data than what a company can frack from them, and consumers are much better at emotionally aligning with brands and influencing their peers than companies are.

Implications for the Blockchain Economy: As an economic design principle, enforcing rights must start with clarifying rights. In the field of management science, the holacracy movement is an interesting, if not controversial, example of how members of organizations are defining the work that needs to be done and then assigning rights and the responsibility to do this work as part of a whole.[48] Who did we agree should have this set of decisions and activities at our company? The answer to that question can be codified in a smart contract and placed on the blockchain so that the decisions, progress toward the goal, and incentives are all transparent and reached by consensus.

To be sure, this is not simply about technology. It's much bigger than physical assets, intellectual property, or Personal BlackBox's privacy tool with a publicity rights module for the Kardashians. We need greater education about rights and the development of new understandings about rights management systems. We'll have voting rights management systems and property rights management systems. Some start-up will create a rights dashboard that will indicate a person's level of civic engagement, where voting is but one of several measures, like donating skills, reputation, time, and bitcoin or providing free access to one's physical or intellectual property. Buckle up.

7. Inclusion

Principle: The economy works best when it works for everyone. That means lowering the barriers to participation. It means creating platforms for ***distributed capitalism***, not just a redistributed capitalism.

Problem to Be Solved: The first era of the Internet created many wonders for many people. But as we have pointed out, a majority of the world's population is still excluded—not just from access to technology but also from access to the financial system and economic opportunity. Moreover, the promise that this new communications medium would bring prosperity to all has rung hollow. Yes, it helped companies in the developed world provide jobs for millions in the emerging economies. It lowered the barriers to entry for entrepreneurs and gave the disadvantaged access to opportunities and basic information.

That's not enough. There are still two billion[49] people without a bank account, and in the developed world, prosperity is actually declining as social inequality continues to grow. In developing economies, mobile is often the only affordable means of connecting. Most financial institutions have mobile payment apps that combine cameras and QR codes. However, the fees needed to support these intermediaries make micropayments impractical. Consumers at the bottom of the pyramid still can't afford the minimum account balances, minimum payment amounts, or transaction fees to use the system. Its infrastructure costs make micropayments and microaccounts unfeasible.

Breakthrough: Satoshi designed the system to work on top of the Internet stack (TCP/IP), but it could run without the Internet if necessary. Satoshi imagined that the typical person would be interacting with the blockchain through what he called "simplified payment verification" (SPV) mode that can work on cell phones to mobilize the blockchain. Now anyone with a flip phone can participate in the economy, or in a market, as a producer or consumer. No bank account required, no proof of citizenship required, no birth certificate required, no home address required, no stable local currency required to use the blockchain technologies. The blockchain drastically lowers the cost of transmitting such funds as remittances. It significantly lowers the barrier to having a bank account, obtaining credit, and investing. And it supports entrepreneurship and participation in global trade.

That was part of Satoshi's vision. He understood that, for people in developing economies, the situation was worse. When corrupt or incompetent bureaucrats in failed states need funding to run the government, their central banks and treasuries simply print more currency and then profit from the difference between the cost of manufacturing and the face value of the currency. That's seigniorage. The increase in the money supply debases the currency. If the local economy really tanked—as it did in Argentina and Uruguay, and more recently in Cyprus and Greece—these central bodies could freeze the bank assets of whoever couldn't afford a bribe. Given such a possibility, the wealthy could store their assets in more trustworthy jurisdictions and more stable currencies.

But not the poor. Whatever money they have becomes worthless. Officials could siphon off inflows of foreign aid and ribbon their borders with red tape, adding friction to every attempt at helping their people, from mothers and children needing food and medicine to victims of war, prolonged drought, and other natural disasters.

The Australia micropayment service mHITs (short for Mobile Handset Initiated Transactions) has launched a new service, BitMoby, that enables consumers in more than one hundred countries to top up their mobile phone credit by texting mHITs an amount of bitcoin.[50] According to bitcoin core developer Gavin Andresen, "You don't see every transaction; you see only the transactions you care about. You're not trusting peers with your money, you're just trusting them to give you the information touring across the network."[51]

"The potential of using the blockchain for property records in the emerging world, where that's a huge issue related to poverty," is significant, said Austin Hill. "There isn't a trusted entity that has governance over land title, and so allowing people to actually say, 'I own this property,' and then use that for collateral to improve them and their family situation is a fascinating use case."[52]

On a technical note, Andresen called on Nielsen's law of Internet bandwidth, where high-end user bandwidth increases by 50 percent each year, whereas the bandwidth of the masses tends to lag by two or three years. Bandwidth lags behind computer processing power, which increases by about 60 percent annually (Moore's law). So bandwidth is the gating factor, according to Jakob Nielsen.[53] Most designs—interfaces, Web sites, digital products, services, organizations, and so forth—will need to accommodate the technology of the masses to leverage network effects. So inclusion means considering the full spectrum of usage—not just the state of the science of high-end users, but the slow tech and sporadic power outages of users in remote regions of the world's poorest countries.

Implications for the Blockchain Economy: Later in the book, we tackle the issue of the prosperity paradox—how the first era of the Internet benefited many, but overall prosperity in the Western world for most people is no longer improving. The foundation for prosperity is inclusion, and blockchains can help. Let's be clear that inclusion has multiple dimensions. It means an end to social, economic, and racial hegemony, an end to discrimination based on health, gender, sexual identification, or sexual preference. It means ending barriers to access because of where a person lives, whether a person spent a night in jail, or how a person voted, but also an end to glass ceilings, and good ol' boys' clubs of countless varieties.

DESIGNING THE FUTURE

Our conversation with Ann Cavoukian inspired us to follow up on Germany's "Never again" promise. We came across the words of German federal president Joachim Gauck on the Day of Remembrance of the Victims of National Socialism, victims of Hitler's regime. "Our moral obligations cannot be fulfilled solely at the level of remembrance. There also exists within us a deep and abiding certainty that remembrance bestows a mission on us. That mission tells us to protect and preserve humanity. It tells us to protect and preserve the rights of every human being."[54] Was he alluding to genocide in Syria, Iraq, Darfur, Srebrenica, Rwanda, and Cambodia, after the German people had vowed, "Never again"?

We believe that blockchain technology could be an important tool for protecting and preserving humanity and the rights of every human being, a means of communicating the truth, distributing prosperity, and—as the network rejects the fraudulent transactions—of rejecting those early cancerous cells from a society that can grow into the unthinkable.

Admittedly, a bold statement. Read on and judge for yourself.

From a more parochial and practical perspective, these seven principles can serve as a guide to designing the next generation of high-performance and innovative companies, organizations, and institutions. If we design for integrity, power, value, privacy, security, rights, and inclusion, then we will be redesigning our economy and social institutions to be worthy of trust. We now turn our attention to how this could roll out and what you should consider doing.

PART II

TRANSFORMATIONS

CHAPTER 3

REINVENTING FINANCIAL SERVICES

The global financial system moves trillions of dollars daily, serves billions of people, and supports a global economy worth more than $100 trillion.[1] It's the world's most powerful industry, the foundation of global capitalism, and its leaders are known as the Masters of the Universe. Closer up, it's a Rube Goldberg contraption of uneven developments and bizarre contradictions. First, the machine hasn't had an upgrade in a while. New technology has been welded onto aging infrastructure helter-skelter. Consider the bank offering Internet banking but still issuing paper checks and running mainframe computers from the 1970s. When one of its customers taps her credit card on a state-of-the-art card reader to buy a Starbucks grande latte, her money passes through no fewer than five different intermediaries before reaching Starbucks's bank account. The transaction takes seconds to clear but days to settle.

Then there are the large multinationals like Apple or GE that have to maintain hundreds of bank accounts in local currencies around the world just to facilitate their operations.[2] When such a corporation needs to move money between two subsidiaries in two different countries, the manager of one subsidiary sends a bank wire from his operation's bank account to the other subsidiary's bank account. These transfers are needlessly complicated and take days, sometimes weeks to settle. During that time, neither subsidiary can use the money to fund operations or investment, but the intermediaries can earn interest on the float. "The advent of technology essentially took paper-based processes and turned them into semiautomated, semielectronic processes but the logic was still paper based," said Vikram Pandit, former CEO of Citigroup.[3]

Around every corner, another bizarre paradox: Traders buy and sell securities on the world's stock exchanges in nanoseconds; their trades clear instantly but take three full days to settle. Local governments use no fewer than ten different agents—advisers, lawyers, insurers, bankers, and more—to facilitate the issuance of a municipal bond.[4] A day laborer in Los Angeles cashes his paycheck at a money mart for a 4 percent fee, and then walks his fistful of dollars over to a convenience store to wire it home to his family in Guatemala, where he gets dinged again on flat fees, exchange rates, and other hidden costs. Once his family has divvied up the sum among its many members, nobody has enough to open a bank account or get credit. They are among the 2.2 billion people who live on less than two dollars a day.[5] The payments they need to make are tiny, too small for conventional payment networks such as debit and credit cards, where minimum fees make so-called micropayments impossible. Banks simply don't view serving these people as a "profitable proposition," according to a recent Harvard Business School study.[6] And so the money machine isn't truly global in scale and scope.

Monetary policy makers and financial regulators often find themselves lacking all the facts, thanks to the planned opacity of many large financial operations and the compartmentalization of oversight. The global financial crisis of 2008 was a case in point. Excess leverage, a lack of transparency, and a sense of complacency driven by skewed incentives prevented anyone from identifying the problem until it was nearly too late. "How can you have anything work, from the police force to a monetary system, if you don't have numbers and locations?" pondered Hernando de Soto.[7] Regulators are still trying to manage this machine with rules devised for the industrial age. In New York State, money transmission laws date back to the Civil War when the primary means of moving money around was horse and buggy.

It's Franken-finance, full of absurd contradictions, incongruities, hot pipes, and pressure pots. Why, for example, does Western Union need 500,000 points of sale around the world, when more than half the world's population has a smart phone?[8] Erik Voorhees, an early bitcoin pioneer and outspoken critic of the banking system, told us, "It is faster to mail an anvil to China than it is to send money through the banking system to China. That's crazy! Money is already digital, it's not like they're shipping pallets of cash when you do a wire!"[9]

Why is it so inefficient? According to Paul David, the economist who coined the term *productivity paradox*, laying new technologies over existing infrastructure is "not unusual during historical transitions from one technological paradigm to the next."[10] For example, manufacturers needed forty years to embrace commercial electrification over steam power, and often the two worked side by side before manufacturers finally switched over for good. During that period of retrofitting, productivity actually decreased. In the financial system, however, the problem is compounded because there has been no clean transition from one technology to the next; there are multiple legacy technologies, some hundreds of years old, never quite living up to their full potential.

Why? In part, because finance is a monopoly business. In his assessment of the financial crisis, Nobel laureate Joseph Stiglitz wrote that banks "were doing everything they could to increase transaction costs in every way possible." He argued that, even at the retail level, payments for basic goods and services "should cost a fraction of a penny." "Yet how much do they charge?" he wondered. "One, two, or three percent of the value of what is sold or more. Capital and sheer scale, combined with a regulatory and social license to operate allows banks to extract as much as they can, in country after country, especially in the United States, making billions of dollars of profits."[11] Historically, the opportunity for large centralized intermediaries has been enormous. Not only traditional banks (e.g., Bank of America), but also charge card companies (Visa), investment banks (Goldman Sachs), stock exchanges (NYSE), clearinghouses (CME), wire/remittance services (Western Union), insurers (Lloyd's), securities law firms (Skadden, Arps), central banks (Federal Reserve), asset managers (BlackRock), accountancies (Deloitte), consultancies (Accenture), and commodities traders (Vitol Group) make up this expansive leviathan. The gears of the financial system—powerful intermediaries that consolidate capital and influence and often impose monopoly economics—make the system work, but also slow it down, add cost, and generate outsized benefits for themselves. Because of their monopoly position, many incumbents have no incentive to improve products, increase efficiency, improve the consumer experience, or appeal to the next generation.

A NEW LOOK FOR THE WORLD'S SECOND-OLDEST PROFESSION

The days of Franken-finance are numbered as blockchain technology promises to make the next decade one of great upheaval and dislocation but also immense opportunity for those who seize it. The global financial services industry today is fraught with problems: It is antiquated, built on decades-old technology that is at odds with our rapidly advancing digital world, making it oftentimes slow and unreliable. It is exclusive, leaving billions of people with no access to basic financial tools. It is centralized, exposing it to data breaches, other attacks, or outright failure. And it is monopolistic, reinforcing the status quo and stifling innovation. Blockchain promises to solve these problems and many more as innovators and entrepreneurs devise new ways to create value on this powerful platform.

There are six key reasons why blockchain technology will bring about profound changes to this industry, busting the finance monopoly, and offering individuals and institutions alike real choice in how they create and manage value. Industry participants the world over should take notice.

Attestation: For the first time in history, two parties who neither know nor trust each other can transact and do business. Verifying identity and establishing trust is no longer the right and privilege of the financial intermediary. Moreover, in the context of financial services, the trust protocol takes on a double meaning. The blockchain can also establish trust when trust is needed by verifying the identity and capacity of any counterparty through a combination of past transaction history (on the blockchain), reputation scores based on aggregate reviews, and other social and economic indicators.

Cost: On the blockchain, the network both clears and settles peer-to-peer value transfers, and it does so continually so that its ledger is always up to date. For starters, if banks harnessed that capability, they could eliminate an estimated $20 billion in back-office expenses *without changing their underlying business model*, according to the Spanish bank Santander, though the actual number is surely much greater.[12] With radically lower costs, banks could offer individuals and businesses greater access to financial services, markets, and capital in underserved communities. This can be a boon not only to incumbents but also to scrappy upstarts and entrepreneurs everywhere. Anyone, anywhere, with a smart phone and an Internet connection could tap into the vast arteries of global finance.

Speed: Today, remittances take three to seven days to settle. Stock trades take two to three days, whereas bank loan trades take on average a staggering twenty-three days to settle.[13] The SWIFT network handles fifteen million payment orders a day between ten thousand financial institutions globally but takes days to clear and settle them.[14] The same is true of the Automated Clearing House (ACH) system, which handles trillions of dollars of U.S. payments annually. The bitcoin network takes an average of ten minutes to clear and settle all transactions conducted during that period. Other blockchain networks are even faster, and new innovations, such as the Bitcoin Lightning Network, aim to dramatically scale the capacity of the bitcoin blockchain while dropping settlement and clearing times to a fraction of a second.[15] "In the corresponding banking world, where you have a sender in one network and a receiver in another, you have to go through multiple ledgers, multiple intermediaries, multiple hops. Things can literally fail in the middle. There's all kinds of capital requirements for that," said Ripple Labs CEO Chris Larsen.[16] Indeed, the shift to instant and frictionless value transfer would free up capital otherwise trapped in transit, bad news for anyone profiting from the float.

Risk Management: Blockchain technology promises to mitigate several forms of financial risk. The first is *settlement risk*, the risk that your trade will bounce back because of some glitch in the settlement process. The second is *counterparty risk*, the risk that your counterparty will default before settling a trade. The most significant is *systemic risk*, the total sum of all outstanding counterparty risk in the system. Vikram Pandit called this *Herstatt risk*, named after a German bank that couldn't meet its liabilities and subsequently went under: "We found through the financial crisis one of the risks was, if I'm trading with somebody, how do I know they're going to settle on the other side?" According to Pandit, instant settlement on the blockchain could eliminate that risk completely. Accountants could look into the inner workings of a company at any point in time and see which transactions were occurring and how the network was recording them. Irrevocability of a transaction and instant reconciliation of financial reporting would eliminate one aspect of *agency risk*—the risk that unscrupulous managers will exploit the cumbersome paper trail and significant time delay to conceal wrongdoing.

Value Innovation: The bitcoin blockchain was designed for moving bitcoins, not for handling other financial assets. However, the technology is open source, inviting experimentation. Some innovators are developing separate blockchains, known as altcoins, built for something other than bitcoin payments. Others are looking to leverage the bitcoin blockchain's size and liquidity to create "spin-off" coins on so-called sidechains that can be "colored" to represent any asset or liability, physical or digital—a corporate stock or bond, a barrel of oil, a bar of gold, a car, a car payment, a receivable or a payable, or of course a currency. *Sidechains* are blockchains that have different features and functions from the bitcoin blockchain but that leverage bitcoin's established network and hardware infrastructure without diminishing its security features. Sidechains interoperate with the blockchain through a *two-way peg*, a cryptographic means of transferring assets off the blockchain and back again without a third party exchange. Others still are trying to remove the coin or token altogether, building trading platforms on private blockchains. Financial institutions are already using blockchain technology to record, exchange, and trade assets and liabilities, and could eventually use it to replace traditional exchanges and centralized markets, upending how we define and trade value.

Open Source: The financial services industry is a technology stack of legacy systems standing twenty miles high and on the verge of teetering over. Changes are difficult to make because each improvement must be backward compatible. As open source technology, blockchain can constantly innovate, iterate, and improve, based on consensus in the network.

These benefits—attestation, dramatically lower costs, lightning speed, lower risks, great innovation of value, adaptability—have the potential to transform not only payments, but also the securities industry, investment banking, accounting and audit, venture capital, insurance, enterprise risk management, retail banking, and other pillars of the industry. Read on.

THE GOLDEN EIGHT: HOW THE FINANCIAL SERVICES SECTOR WILL CHANGE

Here are what we believe to be the eight core functions ripe for disruption. They are also summarized in the table on page 64.

1. Authenticating Identity and Value: Today we rely on powerful intermediaries to establish trust and verify identity in a financial transaction. These intermediaries are the ultimate arbiters for access to basic financial services, such as bank accounts and loans. Blockchain lowers and sometimes eliminates trust altogether in certain transactions. The technology will also enable peers to establish identity that is verifiable, robust, and cryptographically secure and to establish trust when trust is needed.

2. Moving Value: Daily, the financial system moves money around the world, making sure that no dollar is spent twice: from the ninety-nine-cent purchase of a song on iTunes to the transfer of billions of dollars to settle an intracompany fund transfer, purchase an asset, or acquire a company. Blockchain can become the common standard for the movement of anything of value—currencies, stocks, bonds, and titles—in batches big and small, to distances near and far, and to counterparties known and unknown. Thus, blockchain can do for the movement of value what the standard shipping container did for the movement of goods: dramatically lower cost, improve speed, reduce friction, and boost economic growth and prosperity.

3. Storing Value: Financial institutions are the repositories of value for people, institutions, and governments. For the average Joe, a bank stores value in a safety deposit box, a savings account, or a checking account. For large institutions that want ready liquidity with the guarantee of a small return on their cash equivalents, so-called risk-free investments such as money market funds or Treasury bills will do the trick. Individuals need not rely on banks as the primary stores of value or as providers of savings and checking accounts, and institutions will have a more efficient mechanism to buy and hold risk-free financial assets.

4. Lending Value: From household mortgages to T-bills, financial institutions facilitate the issuance of credit such as credit card debt, mortgages, corporate bonds, municipal bonds, government bonds, and asset-backed securities. The lending business has spawned a number of ancillary industries that perform credit checks, credit scores, and credit ratings. For the individual, it's a credit score. For an institution, it's a credit rating—from investment grade to junk. On the blockchain, anyone will be able to issue, trade, and settle traditional debt instruments directly, thereby reducing friction and risk by increasing speed and transparency. Consumers will be able to access loans from peers. This is particularly significant for the world's unbanked and for entrepreneurs everywhere.

5. Exchanging Value: Daily, markets globally facilitate the exchange of trillions of dollars of financial assets. Trading is the buying and selling of assets and financial instruments for the purpose of investing, speculating, hedging, and arbitraging and includes the posttrade life cycle of clearing, settling, and storing value. Blockchain cuts settlement times on all transactions from days and weeks to minutes and seconds. This speed and efficiency creates opportunities for unbanked and underbanked people to participate in wealth creation.

6. Funding and Investing: Investing in an asset, company, or new enterprise gives an individual the opportunity to earn a return, in the form of capital appreciation, dividends, interest, rents, or some combination. The industry makes markets: matching investors with entrepreneurs and business owners at every stage of growth—from angels to IPOs and beyond. Raising money normally requires intermediaries—investment bankers, venture capitalists, and lawyers to name a few. The blockchain automates many of these functions, enables new models for peer-to-peer financing, and could also make recording dividends and paying coupons more efficient, transparent, and secure.

7. Insuring Value and Managing Risk: Risk management, of which insurance is a subset, is intended to protect individuals and companies from uncertain loss or catastrophe. More broadly, risk management in financial markets has spawned myriad derivative products and other financial instruments meant to hedge against unpredictable or uncontrollable events. At last count the notional value of all outstanding over-the-counter derivatives is $600 trillion. Blockchain supports decentralized models for insurance, making the use of derivatives for risk management far more transparent. Using reputational systems based on a person's social and economic capital, their actions, and other reputational attributes, insurers will have a much clearer picture of the actuarial risk and can make more informed decisions.

8. Accounting for Value: Accounting is the measurement, processing, and communication of financial information about economic entities. It is a multibillion-dollar industry controlled by four massive audit firms—Deloitte Touche Tohmatsu, PricewaterhouseCoopers, Ernst & Young, and KPMG. Traditional accounting practices will not survive the velocity and complexity of modern finance. New accounting methods using blockchain's distributed ledger will make audit and financial reporting transparent and occur in real time. It will also dramatically improve the capacity for regulators and other stakeholders to scrutinize financial actions within a corporation.

FROM STOCK EXCHANGES TO BLOCK EXCHANGES

"Wall Street has woken up in a big way,"[17] said Austin Hill of Blockstream. He was speaking of the financial industry's deep interest in blockchain technologies. Consider Blythe Masters, one of the most powerful people on Wall Street. She built JPMorgan's derivatives and commodities desk into a global juggernaut and pioneered the derivatives market. After a brief pseudoretirement, she joined a New York–based start-up, Digital Asset Holdings, as CEO. The decision surprised many. She understood that the

THE GOLDEN EIGHT

Blockchain Transformations of Financial Services

Function	Blockchain Impact	Stakeholder
1. Authenticating Identity and Value	Verifiable and robust identities, cryptographically secured	Rating agencies, consumer data analytics, marketing, retail banking, wholesale banking, payment card networks, regulators
2. Moving Value—make a payment, transfer money, and purchase goods and services	Transfer of value in very large and very small increments without intermediary will dramatically reduce cost and speed of payments	Retail banking, wholesale banking, payment card networks, money transfer services, telecommunications, regulators
3. Storing Value—currencies, commodities, and financial assets are stores of value. Safety deposit box, a savings account, or a checking account. Money market funds or Treasury bills	Payment mechanism combined with a reliable and safe store of value reduces need for typical financial services; bank savings and checking accounts will become obsolete	Retail banking, brokerages, investment banking, asset management, telecommunications, regulators
4. Lending Value—credit card debt, mortgages, corporate bonds, municipal bonds, government bonds, asset-backed securities, and other forms of credit	Debt can be issued, traded, and settled on the blockchain; increases efficiency, reduces friction, improves systemic risk. Consumers can use reputation to access loans from peers; significant for the world's unbanked and for entrepreneurs	Wholesale, commercial, and retail banking, public finance (i.e., government finance), microlending, crowdfunding, regulators, credit rating agencies, credit score software companies
5. Exchanging Value—speculating, hedging, and arbitraging. Matching orders, clearing trades, collateral management and valuation, settlement and custody	Blockchain takes settlement times on all transactions from days and weeks to minutes and seconds. This speed and efficiency also creates opportunities for unbanked and underbanked to participate in wealth creation	Investment, wholesale banking, foreign exchange traders, hedge funds, pension funds, retail brokerage, clearinghouses, stock, futures, commodities exchanges; commodities brokerages, central banks, regulators
6. Funding and Investing in an Asset, Company, Start-up—capital appreciation, dividends, interest, rents, or some combination	New models for peer-to-peer financing, recording of corporate actions such as dividends paid automatically through smart contracts. Titles registry to automate claims to rental income and other forms of yield	Investment banking, venture capital, legal, audit, property management, stock exchanges, crowdfunding, regulators
7. Insuring Value and Managing Risk—protect assets, homes, lives, health, business property, and business practices, derivative products	Using reputational systems, insurers will better estimate actuarial risk, creating decentralized markets for insurance. More transparent derivatives	Insurance, risk management, wholesale banking, brokerage, clearinghouses, regulators
8. Accounting for Value—new corporate governance	Distributed ledger will make audit and financial reporting real time, responsive, and transparent, will dramatically improve capacity of regulators to scrutinize financial actions within a corporation	Audit, asset management, shareholder watchdogs, regulators

blockchain would transform her business as the Internet transformed other industries: "I would take it about as seriously as you should have taken the concept of the Internet in the 1990s. It's a big deal and it is going to change the way our financial world operates."[18]

Masters had dismissed many of the early tales of bitcoin, exploited by drug dealers, harnessed by gamblers, and hailed by libertarians as creating a new world order. That changed in late 2014. Masters told us, "I had an 'aha moment' where I began to appreciate the potential implications of the technology for the world that I knew well. Whilst the cryptocurrency application of the distributed ledgers technology was interesting and had implications for payments, the underlying database technology itself had far broader implications."[19] According to Masters, blockchain could reduce inefficiencies and costs "by allowing multiple parties to rely on the same information rather than duplicating and replicating it and having to reconcile it." As a mechanism for shared, decentralized, replicated transaction records, blockchain is the "golden source," she says.[20]

"Bear in mind that financial services infrastructures have not evolved in decades. The front end has evolved but not the back end," says Masters. "It's been an arms race in technology investment oriented toward speeding up transaction execution so that, nowadays, competitive advantages are measured in fractions of nanoseconds. The irony is that the posttrade infrastructure hasn't really evolved at all." It still takes "days and in some cases weeks of delay to do the posttrade processing that goes into actually settling financial transactions and keeping record of them."[21]

Masters is not alone in her enthusiasm for blockchain technology. NASDAQ CEO Bob Greifeld said, "I am a big believer in the ability of blockchain technology to effect fundamental change in the infrastructure of the financial service industry."[22] Greifeld is integrating blockchain's distributed ledger technology into NASDAQ's private markets platform through a platform called NASDAQ Linq. Exchanges are centralized marketplaces for securities and they are also ripe for disruption. On January 1, 2016, NASDAQ Linq completed its first trade on blockchain. According to Blockstream's Hill, one of the largest asset managers in the world "has more people dedicated to its blockchain innovation group than we have in our entire company." Hill's company has raised over $75 million and employs more than twenty people. "These guys are serious about making sure that they understand how they can use the technology to change how they do business."[23] The NYSE, Goldman Sachs, Santander, Deloitte, RBC, Barclays, UBS, and virtually every major financial firm globally have taken a similar serious interest. In 2015, Wall Street's opinion of blockchain technology became *universally* positive: in one study, 94 percent of respondents said blockchain could play an important role in finance.[24]

Although many other applications pique the interest of Wall Street, what interests financial executives everywhere is the notion of using the blockchain to process any trade securely from beginning to end, which could dramatically lower costs, increase speed and efficiency, and mitigate risk in their businesses. Masters said, "The entire life cycle of a trade including its execution, the netting of multiple trades against each other, the reconciliation of who did what with whom and whether they agree, can occur at the trade entry level, much earlier in the stack of process, than occurs in the mainstream financial market."[25] Greifeld put it this way: "We currently settle trades 'T+3' (that is, three days). Why not settle in five to ten minutes?"[26]

Wall Street trades in risk, and this technology can materially reduce counterparty risk, settlement risk, and thus systemic risk across the system. Jesse McWaters, financial innovation lead at the World Economic Forum, told us, "The most exciting thing about distributed ledger technology is how traceability can improve systemic stability." He believes these "new tools allow regulators to use a lighter touch."[27] The blockchain's public nature—its transparency, its searchability—plus its automated settlement and immutable time stamps, allow regulators to see what's happening, even set up alerts so that they don't miss anything.

DR. FAUST'S BLOCKCHAIN BARGAIN

Banks and transparency rarely go hand in hand. Most financial actors gain competitive advantage from information asymmetries and greater know-how than their counterparties. However, the bitcoin blockchain as constructed is a radically transparent system. For banks, this means opening the kimono, so to speak. So how do we reconcile an open platform with the closed-door policy of banks?

Austin Hill called it Wall Street's "Faustian bargain," an onerous trade-off.[28] "People love the idea of not having to wait three days to settle transactions but having them cleared within minutes and knowing that they're final and that they're true," said Hill. "The counterpart to that is all transactions on the [bitcoin] blockchain are completely public. That terrifies a number of people on Wall Street." The solution? Confidential transactions on so-called permissioned blockchains, also known as private blockchains. Whereas the bitcoin blockchain is entirely open and *permissionless*—that is, anyone can access it and interact with it—*permissioned* blockchains require users to have certain credentials, giving them a license to operate on that particular blockchain. Hill has developed the technology whereby only a few stakeholders see the various components of a transaction and can ensure its integrity.

At first blush, private and permissioned blockchains would appear to have a few clear advantages. For one, its members can easily change the rules of the blockchain if they so desire. Costs can be kept down as transactions need only validation from the members themselves, removing the need for anonymous miners who use lots of electricity. Also, because all parties are trusted, a 51 percent attack is unlikely. Nodes can be trusted to be well connected, as in most use cases they are large financial institutions. Furthermore, they are easier for regulators to monitor. However, these advantages also create weaknesses. The easier it is to change the rules, the more likely a member is to flout them. Private blockchains also prevent the network effects that enable a technology to scale rapidly. Intentionally limiting certain freedoms by creating new rules can inhibit neutrality. Finally, with no open value innovation, the technology is more likely to stagnate and become vulnerable.[29] This is not to say private blockchains won't flourish, but financial services stakeholders must still take these concerns seriously.

Ripple Labs, which has gained traction within banking circles, is developing other clever ways to relieve Faust. "Ripple Labs is aimed at wholesale banking, and we use a consensus method, rather than a proof-of-work system," said CEO Chris Larsen, meaning no miners and no anonymous nodes are validating transactions.[30] The company Chain has its own strategy. With $30 million in funding from Visa, NASDAQ, Citi, Capital One, Fiserv, and Orange, Chain plans on building enterprise-focused blockchain solutions, targeting the financial services industry first, where it already has a deal with NASDAQ. "All assets in the future will be digital bearer instruments running on multiple blockchains," argued Chain CEO Adam Ludwin. But this won't be the siloed world Wall Street is accustomed to, "because everyone is building on the same open specs."[31] Wall Streeters might want to capture this technology, but they will have to contend with the value innovation it enables, something they can't control or predict.

Masters also sees the virtues of permissioned blockchains. For her, only a small coterie of trading partners, some vendors and other counterparties, and regulators need have access. Those select few chosen will be granted blockchain credentials. To Masters, "permissioned ledgers have the advantage of never exposing a regulated financial institution to the risk of either transacting with an unknown party, an unacceptable activity from a regulatory point of view, or creating a dependency upon an unknown service provider such as a transaction processor, also unacceptable from a regulatory point of view."[32] These permissioned blockchains, or *private chains*, appeal to traditional financial institutions wary of bitcoin and everything associated with it.

While Blythe Masters is the CEO of a start-up, her keen interest represents broader involvement of traditional financial actors in this sector. This embrace of new technology reflects a growing concern that tech start-ups can also upend high finance. For Eric Piscini of Deloitte, whose clients have undergone a great awakening over the past year, the "sudden interest in tech was not something that anyone was expecting."[33] The enthusiasm is spreading like a contagion into some of the largest and oldest financial institutions in the world.

Barclays is one of dozens of financial institutions exploring opportunities in blockchain technology. According to Derek White, Barclays's chief design and digital officer, "technologies like the blockchain are going to reshape our industry." White is building an open innovation platform that will allow the bank to engage a wide array of builders and thinkers in this industry. "We're keen to be shapers. But we're also keen to connect with the shapers of the technologies and the translators of those technologies," he said.[34] Barclays is putting its money where its mouth is, cutting tens of thousands of jobs in traditional areas and doubling down on technology, notably by launching the Barclays Accelerator. According to White, "three out of the ten companies in our last cohort were blockchain or bitcoin companies. Blockchain is the greatest evidence of the world moving from closed systems to open systems and has huge potential impact on the future of not just financial services but many industries."[35] Banks talking about open systems—*mon Dieu!*

The Financial Utility

In the autumn of 2015, nine of the world's largest banks—Barclays, JPMorgan, Credit Suisse, Goldman Sachs, State Street, UBS, Royal Bank of Scotland, BBVA, and Commonwealth Bank of Australia—announced a plan to collaborate on common standards for blockchain technology, dubbed the R3 Consortium. Thirty-two more have since joined the effort and every few weeks a new batch of the industry's Who's Who signs up.[36] Questions remain about how seriously these banks are taking the initiative. After all, the barrier to joining the group is a commitment of only $250,000, yet R3's formation marks a clear leap forward for the industry. Setting standards is critical to accelerate adoption and usage of a new technology and so we are optimistic about the initiative. R3 has poached some of the leading visionaries and technical practitioners in the sector to move the ball forward. Mike Hearn joined in November, adding to a team that includes Richard Gendal Brown, formerly the executive architect for banking innovation at IBM, and James Carlyle, now chief engineer of R3 and ex-chief engineer at Barclays.[37]

In December 2015, the Linux Foundation, in collaboration with a huge group of yet more blue-chip corporate partners, launched another blockchain initiative, dubbed the Hyperledger Project. This is not a competitor to R3; indeed, Hyperledger Project counts R3 as a founding member, along with Accenture, Cisco, CLS, Deutsche Börse, Digital Asset Holdings, DTCC, Fujitsu Limited, IC3, IBM, Intel, JPMorgan, the London Stock Exchange Group, Mitsubishi UFJ Financial Group (MUFG), State Street, SWIFT, VMware, and Wells Fargo.[38] Still, it demonstrates how seriously the industry is taking this technology and also how reluctant it is to embrace fully open, decentralized blockchains like bitcoin. Unlike R3, Hyperledger Project is an open source project that has tasked a community to develop a "blockchain for business." This is certainly laudable and may very well work. But don't be mistaken: This is an open source project designed to build *gated* technologies by, for example, limiting the number of nodes in a network or requiring credentials. As with R3, one of Hyperledger's priorities is standard setting. David Treat of Accenture, a founding member of the group, said, "Key to this journey is to have standards and shared platforms that are utilized across industry participants."

Blockchain has also opened up a broader discussion about the role of governments in overseeing the financial services industry. A "utility" conjures images of natural monopolies, highly regulated by the state. However, because blockchain technology promises to reduce risks and increase transparency and responsiveness, some industry players suggest that the technology *itself* functions like a regulation.[39] If regulators can peer into the inner workings of banks and markets, then surely we can simplify some laws and repeal others, right? This is a tricky question to answer. On the one hand, regulators will have to rethink their oversight role, given the breakneck pace of innovation. On the other hand, banks have a track record of acting without integrity when government steps away.

Will the big banks reign supreme by deploying the blockchain without bitcoin, cherry-picking elements of distributed ledger technology and welding them to existing business models? R3 is only one of many signs banks are moving in this direction. On November 19, 2015, Goldman Sachs filed a patent for "methods for settling securities in financial markets using distributed, peer-to-peer and cryptographic techniques," using a proprietary coin called SETLcoin.[40] The irony of a bank *patenting* a technology originally intended as an open source gift to the world is not lost on us, nor should it be on you. Perhaps this is what Andreas Antonopolous feared when he warned an audience that banks would turn bitcoin from "punk rock to smooth jazz"?[41] Or perhaps banks will have to compete with best-in-class products and services amid radically different types of organizations whose leaders oppose everything these companies represent.

The financial utility of the future could be a walled and well-groomed garden, harvested by a cabal of influential stakeholders, or it could be an organic and spacious ecosystem, where people's economic fortunes grow wherever there is light. The debate rages on, but if the experience of the first generation of the Internet has taught us anything, it's that open systems scale more easily than closed ones.

THE BANK APP: WHO WILL WIN IN RETAIL BANKING

The Google of Capital—that's what Jeremy Allaire is building, "a consumer finance company providing products to consumers to hold money, send money, send and receive payments; the fundamental utilities that people expect out of retail banking."[42] He sees it as a powerful, instant, and free utility for anyone with access to an Internet-enabled device. His company, Circle Internet Financial, is one of the largest and best-funded ventures in the space.

Call Circle what you like, just don't call it a bitcoin company. "Amazon was not an HTTP company and Google was not an SMTP company. Circle is not a bitcoin company," said Allaire. "We look at bitcoin as a next generation of fundamental Internet protocols that are used in society and the economy."[43]

Allaire sees financial services as the last holdouts, and perhaps the largest prize, to be fundamentally transformed by technology. "If you look at retail banking, there are three or four things that retail banks do. One is that they provide a place to store value. A second is that they provide some kind of payment utility. Beyond that, they extend credit and provide a place for you to store wealth and generate potential income."[44] His vision: "Within three to five years, a person should be able to download an app, store value digitally in whatever currency they want—dollars, euro, yen, renminbi, as well as digital currency—and be able to make payments instantly or nearly instantly with global interoperability, with a very high level of security and without privacy leakage. Most importantly, it will be free."[45] As the Internet transformed information services, the blockchain will transform financial services, instigating unimagined new categories of capability.

According to Allaire, the benefits of blockchain technology—instant settlement, global interoperability, high levels of security, and nearly no-cost transactions—benefit everyone whether you're a person or a business. And what of his plan to make it all free? Heresy! say the world's bankers. Surely, Goldman Sachs and the Chinese venture firm IDG did not commit $50 million to create a nonprofit or public benefit company![46] "If we're successful in building out a global franchise with tens of millions of users and we're sitting at the center of transaction behavior of users, then we are sitting on some powerful assets." Allaire expects Circle to have "the underlying capabilities to deliver other financial products." Though he wouldn't speak to it specifically, the financial data of millions of customers could become more valuable to the company than their financial assets. "We want to reinvent the consumers' experience and their relationship to money and give them the choice of how their money is used and applied and how they can generate money from their money."[47] Leaders of the old paradigm, take notice.

Companies like Circle are unburdened by legacy and culture. Their fresh approach can be a big advantage. Many of the great innovators of the past were consummate outsiders. Netflix wasn't invented by Blockbuster. iTunes wasn't invented by Tower Records. Amazon wasn't invented by Barnes & Noble—you get the idea.

Stephen Pair, CEO of BitPay, an early mover in the industry, believes newcomers have a distinct advantage. "Issuing fungible assets like equities, bonds, and currencies on the blockchain and building the necessary infrastructure to scale it and make it commercial don't require a banker's CV," he said. For one, "You don't require all the legacy infrastructure or institutions that make up Wall Street today. . . . Not only can you issue these assets on the blockchain, but you can create systems where I can have an instantaneous atomic transaction where I might have Apple stock in my wallet and I want to buy something from you. But you want dollars. With this platform I can enter a single atomic transaction (i.e., all or none) and use my Apple stock to send you dollars."[48]

Is it really that easy? The battle to reinvent the financial services industry differs from the battle for e-commerce in the early days of the Web. For businesses like Allaire's to scale, they must facilitate one of the largest value transfers in human history, moving trillions of dollars from millions of traditional bank accounts to millions of Circle wallets. Not so easy. Banks, despite their enthusiasm for blockchain, have been wary of these companies, arguing blockchain businesses are "high-risk" merchants. Perhaps their reluctance stems from the fear of hastening their own demise. Intermediaries have sprung up between the old and new worlds. Vogogo, a Canadian company, is already working with Coinbase, Kraken, BitPay, Bitstamp, and others to open bank accounts, meet compliance standards, and enable customers to move money *into* bitcoin wallets through traditional payment methods.[49] Oh, the irony. Whereas Amazon could leapfrog incumbent retailers with ease, the leaders of this new paradigm must play nice with the leaders of the old.

Perhaps we need a banker with Silicon Valley's willingness to experiment. Suresh Ramamurthi fits that bill. The Indian-born former Google executive and software engineer surprised many when he decided to buy CBW Bank in Wier, Kansas, population 650. For him, this small local bank was a laboratory for using the blockchain protocol and bitcoin-based payment rails for free cross-border remittance payments. In his view, would-be blockchain entrepreneurs who don't understand the nuances of financial services are doomed to fail. He said, "They are drawing a window on the building. Making it look nice and colorful. But you can't assess the problem from the outside. You need to talk to someone from inside the building, who knows the plumbing."[50] In the past five years, Suresh has served as the bank's CEO, CIO, chief compliance officer, teller, janitor, and, yes, plumber. Suresh now knows the plumbing of banking.

Many Wall Street veterans don't see a battle between old and new. Blythe Masters believes there are "at least as many ways for banks to improve the efficiency and operations of Wall Street as there are opportunities for disruption from new entrants."[51] We can't help feeling the tides turning toward the radically new. That's why the Big Three TV networks didn't come up with YouTube, why the Big Three automakers didn't come up with Uber, why the Big Three hotel chains didn't come up with Airbnb. By the time the C-suites of the Fortune 1000 decide to pursue a new avenue of growth, a new entrant has broadsided them with speed, agility, and a superior offering. Regardless of who lands on top, the collision between the unstoppable force of technological change and the immovable object of financial services, the most entrenched industry in the world, promises to be an intense one.

GOOGLE TRANSLATE FOR BUSINESS: NEW FRAMEWORKS FOR ACCOUNTING AND CORPORATE GOVERNANCE

"Accountants are like mushrooms—they're kept in the dark and fed shit,"[52] said Tom Mornini, CEO of Subledger, a start-up targeting the accounting industry. Accounting has become known as the language of finance, unintelligible to all but a few disciples. If every transaction is available on a shared, globally distributed ledger, then why would we need public accountancies to translate for us?

Modern accounting sprang from the curious mind of Luca Pacioli in Italy during the fifteenth century. His deceptively simple invention was a formula known as double-entry accounting, where every transaction has two effects on each participant, that is, each must enter both a debit and a credit onto the balance sheet, the ledger of corporate assets and liabilities. By codifying these rules, Pacioli provided order to an otherwise ad hoc practice that prevented enterprises from scaling.

Ronald Coase thought accounting was cultlike. While a student at the London School of Economics, Coase saw "aspects of a religion" in the practice. "The books entrusted to the accountants' keeping were apparently sacred books." Accounting students deemed his challenges "sacrilegious."[53] How dare he question their "many methods of calculating depreciation, valuing inventories, allocating on-costs, and so on, all of which gave different results but all of which were perfectly acceptable accounting practices," and other *nearly* identical practices that were nonetheless deemed entirely "unrespectable." So Tom Mornini is by no means the first to criticize the profession.

We see four problems with modern accounting. First, the current regime relies upon managers to swear that their books are in order. Dozens of high-profile cases—Enron, AIG, Lehman Brothers, WorldCom, Tyco, and Toshiba—show that management doesn't always act with integrity. Greed too often gets the best of people. Cronyism, corruption, and false reporting precipitate bankruptcies, job losses, and market crashes, but also high costs of capital and tighter reins on equity.[54]

Second, human error is a leading cause of accounting mistakes, according to AccountingWEB. Often the problems begin when Randy in finance fat-fingers a number into a spreadsheet and, like a butterfly flapping its wings, the small mistake becomes a big problem as it factors into calculations across financial statements.[55] Nearly 28 percent of professionals reported that people plugged incorrect data into their firm's enterprise system.[56]

Third, new rules such as Sarbanes-Oxley have done little to curb accounting fraud. If anything, the growing complexity of companies, more multifaceted transactions, and the speed of modern commerce create new ways to hide wrongdoing.

Fourth, traditional accounting methods cannot reconcile new business models. Take microtransactions. Most audit software allows for two decimal places (i.e., one penny), useless for microtransactions of any kind.

Accounting—the measurement, processing, and communication of financial information—is not the problem. It performs a critical function in today's economy. However, the implementation of accounting methods must catch up with the modern era. Consider that in Pacioli's day, audits were done daily. Today they happen with the cycles of the moon and the seasons. Name another industry where five hundred years of technological advancement have *increased* the time it takes to complete a task by 9,000 percent.

The World Wide Ledger

Today, companies record a debit and credit with each transaction—two entries, hence double-entry accounting. They could easily add a third entry to the World Wide Ledger, instantly accessible to those who need to see it—the company's shareholders, auditors, or regulators. Imagine that when a massive company like Apple sells products, buys raw materials, pays its employees, or accounts for assets and liabilities on its balance sheet, the World Wide Ledger recorded the transaction and published a time-stamped receipt to a blockchain. The financial reports for a company would become a living ledger—auditable, searchable, and verifiable. Generating any up-to-the-minute financial statement should be as simple as a spreadsheet function, where the click of a button gives you an immutable, complete, and searchable financial statement, free of error. Companies might not want everyone seeing these numbers, and so executives might give only regulators, managers, and other key stakeholders permissioned access.

Many in the industry see the inherent implications of this World Wide Ledger for accounting. According to Simon Taylor of Barclays, such a ledger could streamline bank compliance with regulators and reduce risk. "We do a lot of regulatory reporting where we're basically saying, here's everything we've done, because what we've done sits inside a system that nobody else can see."[57] A World Wide Ledger and a transparent record of everything "means that a regulator would have access to the same base layer of data. That would mean less work, less cost, and we could be held to account in near real time. That's really powerful."[58] For Jeremy Allaire of Circle, regulators benefit the most. "Bank examiners have had to rely upon opaque, privately controlled, proprietary ledgers and financial accounting systems to do their work—the 'books and records,'" said Allaire. "With a shared public ledger, auditors and bank examiners could have automated forms of examination to look at the underlying health of a balance sheet and the strength of a corporation—a powerful innovation that could automate meaningful parts of regulation as well as audit and accounting."[59]

It bakes integrity into the system. "All fraud would be much harder. You have to do fraud on an ongoing basis and at no point can you go back and change your records," said Christian Lundkvist, of Balanc3, an Ethereum-based triple-entry accounting start-up.[60] Austin Hill argued, "A public ledger that is constantly audited and verified means you don't have to trust the books of your partner; there is integrity in the statements or the transaction logs, because the network itself is verifying it. It's like a continuous a priori audit that is done cryptographically. You're not relying on PricewaterhouseCoopers or Deloitte. There is no counterparty risk. If the ledger says this is true, then it's true."[61]

Deloitte, one of the world's Big Four accounting firms, has been trying to understand the impact of blockchain. Eric Piscini, who heads up the Deloitte cryptocurrency center, tells clients that the blockchain is "a big risk for your own business model because now the business of banking is to manage risk. If tomorrow that risk disappears, what are you going to do?"[62] Overripe for disruption is the audit business, and audit is a third of Deloitte's revenue.[63] Piscini said, "That's a disruption to our own business model, right? Today we spend a lot of time auditing companies, and we charge fees accordingly. Tomorrow, if that process is completely streamlined because there is a time stamp in the blockchain, that changes the way we audit companies."[64] Or perhaps eliminates the audit firm altogether?

Deloitte has developed a solution called PermaRec (for Permanent Record) whereby "Deloitte would record those transactions into the blockchain and would then be able to audit one of the two partners, or both of them, very quickly, because that transaction is recorded."[65] But if the third entry on the blockchain—time-stamped and ready for all to see—happens automatically, anyone, anywhere, could determine whether the books balanced. Conversely, the fastest area of growth for Deloitte and the other big three audit firms is consulting services. Many clients are already scratching their heads over the blockchain. This bewilderment provides opportunities for migration up the advisory value chain.

Mornini, a plucky entrepreneur and self-described "eternal optimist," likened periodic accounting to "watching a person stand up and dance in front of a strobe light. You know they're dancing, but you can't quite figure out what's going on. And it looks interesting, but it's hard to figure out all the steps in between."[66] Periodic accounting gives a snapshot. Audit is, by definition, a backward-looking process. Creating a complete picture of a company's financial health by looking at periodic financial statements is like turning a hamburger into a cow.

According to Mornini, most large corporations would never want a completely transparent accounting record in the public domain or even readily accessible by people with special privileges, such as auditors or regulators. A company's financials are one of its most guarded secrets. Furthermore, many companies want to ensure that management has a certain degree of flexibility in how it accounts for certain items, such as how to recognize revenue, depreciate an asset, or account for a goodwill charge.

But Mornini believes that companies would benefit from greater transparency—not only in terms of streamlining their finance department or lowering the cost of audit, but in how the market values their company. He said, "The first public company with this system in place will see a significant price per share advantage, or price to earnings ratio advantage, over other companies where investors have to anxiously await the dribble of financial information that they are provided quarterly." After all, he argues, "Who is going to invest in a company that shows you what's going on quarterly, compared to one that shows you what's going on all the time?"[67]

Will investors demand triple-entry accounting to meet corporate governance standards? It's not a far-fetched question. Many institutional investors, such as the California Public Employees' Retirement System, have developed strict corporate governance standards, and will not invest in a company unless those standards are met.[68] Triple-entry accounting could be next.

Triple-Entry Accounting: Privacy Is for Individuals, Not Corporations

Triple-entry accounting is not without skeptics. Izabella Kaminska, a *Financial Times* reporter, believes mandating triple-entry accounting will lead to an increasing number of transactions moving off balance sheets. "There will always be those who refuse to follow the protocol, who abscond and hide secret value in parallel off-grid networks, what we call the black market, off balance sheet, shadow banking."[69]

How does one reconcile non-transaction-based accounting measures, particularly the recognition of intangible assets? How are we going to track intellectual property rights, brand value, or even celebrity status—think Tom Hanks? How many bad films must this Oscar winner make before the blockchain impairs the Hanks brand value?

The argument for triple-entry accounting is not against traditional accounting. There will always be areas where we will need competent auditors. But if triple-entry accounting can vastly increase transparency and responsiveness through real-time accruals, verifiable transaction records, and instant audit, then the blockchain could solve many of accounting's biggest problems. Deloitte will need someone to assess in real time the value of intangibles and perform the other accounting functions that the blockchain cannot, rather than a large task force of auditors.

Finally, is an immutable record of everything truly desirable? In Europe, courts are upholding the "right to be forgotten," enforcing people's petitions to remove their history from the Internet. Shouldn't the same principle apply to corporations? No. Why do Uber drivers get rated on customer satisfaction but corporate executives get a pass? Imagine a mechanism—let's call it a trust app—to record feedback in a public ledger and maintain an independent, searchable score for corporate integrity. Inside the black box of corporations, sunshine is the best disinfectant.

Triple-entry accounting is the first of many blockchain innovations in corporate governance. Like many institutions in society, our corporations are suffering from a crisis of legitimacy. Shareholder activist Robert Monks wrote, "Capitalism has become a kleptocracy, run by and for the enrichment of CEOs, or what I term manager-kings."[70]

The blockchain returns power to shareholders. Imagine that a token representing a claim on an asset, a "bitshare," could come with a vote or many votes, each colored to a particular corporate decision. People could vote their proxies instantly from anywhere, thereby making the voting process for major corporate actions more responsive, more inclusive, and less subject to manipulation. Decisions within companies would require real consensus, multiple signatures on an industrial scale, where each shareholder held a key to the company's future. Once the votes are in, the decision as well as the board meeting minutes would be time-stamped and recorded in an immutable ledger.

Shouldn't corporations have a right to change their history, to be forgotten?[71] No. As artifacts of society, companies have responsibilities that accompany their license to operate. Indeed, corporations have an obligation to society to publish any and all information about their dealings. Sure, corporations have a right and obligation to protect trade secrets and the privacy of their employees, staff, and other stakeholders. But that's different from privacy. Increasing transparency is a huge opportunity for managers everywhere: uphold the highest standards of corporate governance, seize the mantle of trust as corporate leaders, and do it by embracing the blockchain.

REPUTATION: YOU ARE YOUR CREDIT SCORE

Whether you're applying for your first credit card or seeking a loan, the bank will value one number above all else: your credit score. This number is meant to reflect your creditworthiness and therefore your risk of default. It is the amalgamation of a number of inputs, from how long you've borrowed to your payment track record. Most retail credit depends on it. But the calculation is deeply flawed. First, it is incredibly narrow. A young person with no credit history might have a sterling reputation, a track record of fulfilling commitments, or a rich aunt. None factors into a credit score. Second, the score creates perverse incentives for individuals. Increasingly, people use debit cards, that is, the cash in their account. Because they have no credit score, they get penalized. Yet credit card firms encourage individuals without resources to apply for credit cards anyway. Third, scores are very laggy: data inputs can be outdated and have little relevance. A late payment at the age of twenty has little bearing on one's credit risk at fifty.

FICO, an American company originally called Fair, Isaac and Company, dominates the U.S. market for credit scores, yet it doesn't factor most relevant information into its analysis. Marc Andreessen said, "PayPal can do a real-time credit score in milliseconds, based on your eBay purchase history—and it turns out that's a better source of information than the stuff used to generate your FICO score."[72] These factors, combined with transaction and business data and other attributes generated by blockchain technology, can enable a far more robust algorithm for issuing credit and managing risk.

What's your reputation? We all have at least one. Reputation is critical to trust in business and in everyday life. To date, financial intermediaries have not used reputation as the basis for establishing trust between individuals and banks. Consider a small business owner who wants to get a loan. More often than not, the loan officer will base the decision on the person's documentation, a one-point perspective of identity, and their credit score. Of course, a human being is more than the sum of a Social Security number, place of birth, primary residence, and credit history. However, the bank does not know, and does not care, whether you're a reliable employee, an active volunteer, an engaged citizen, or the coach of your kid's soccer team. The loan officer might appreciate your acting with integrity, but the bank's scoring system does not. These components of reputation are simply difficult to formulate, document, and use as social and economic systems are currently constructed. Most of these are ethereal and ephemeral.

So what do the billions of people do who have no reputation beyond their immediate social circle? Where financial services are available to the global poor, many can't meet the requisite identity thresholds, such as ID cards, proof of residence, or financial history. This is a problem in the developed world too. In December 2015, many large U.S. banks rejected the newly formed New York ID cards as a valid credential to open a bank account, despite the fact that more than 670,000 people signed up for them and the banks' federal regulators had approved their use.[73] Blockchain could solve this problem by empowering people to form unique identities with a variety of attributes, previous transaction history among them, and give them new alternatives beyond the traditional banking system.

There are still many use cases—particularly in credit—where the blockchain establishes trust between parties when trust is needed. Blockchain technology not only works to ensure that loan funds move to the borrower, but also assures that the borrower repays with interest. It empowers both parties with their own data, strengthens their privacy, and generates a new kind of persistent economic identity based on factors such as one's past economic history on the blockchain and one's social capital. Patrick Deegan, CTO at identity start-up Personal BlackBox, said that individuals will someday "deploy and manage their own identity, and form trusted connections with other peers and nodes," thanks to blockchain technology.[74] Because the blockchain records and stores all transactions in an immutable record, every transaction can count incrementally toward reputation and creditworthiness. Further, individuals can decide which persona interacts with which institution. Deegan said, "I can create different personas that represent different sides of myself, and I choose the persona that interacts with the company."[75] Banks and other companies on the blockchain ought not ask for and aggregate more information than they need to provide service.

This model has proven to work. BTCjam is a peer-to-peer lending platform that uses reputation as the basis for extending credit. Users can link their profile on BTCjam to Facebook, LinkedIn, eBay, or Coinbase to add more depth and texture. Friends can volunteer recommendations from Facebook. You can even submit your actual credit score as one of many attributes. None of this private information is released. Users start on the platform with a low credit score. But you can quickly build a reputation by showing you are a reliable borrower. The best strategy is to start with a "reputation loan" to prove you're reliable. As a user, you will have to respond to investor questions during the funding process. Ignoring these questions is a red flag; the community will hesitate to fund you. With your first loan, start with a manageable amount, and pay it back on time. Once you have, your quantitative score will improve, and other members in the community might give you a positive review. As of September 2015, BTCjam had funded eighteen thousand loans in excess of $14 million.[76]

Entrepreneur Erik Voorhees called for common sense: "With a reputation-based system, people who are more likely to be able to afford a house should be able to purchase one more easily. Those who are less likely should have a harder time getting a loan." To him, this method "will drive down costs for good actors and drive up costs for our bad actors, which is the proper incentive."[77] In reputation systems, your creditworthiness is derived not from a FICO score, but from an amalgamation of attributes that form your identity and inform your ability to repay a loan. Credit ratings for companies will also change to reflect new information and insight made possible by blockchain. Imagine tools that can aggregate reputation and track different reputational aspects, such as financial trustworthiness, vocational competence, and social consciousness. Imagine getting credit based on shared values, where the people loaning you money appreciate your role in the community and your goals.

THE BLOCKCHAIN IPO

The week of August 17, 2015, was an ugly one: The Chinese stock market crashed, the S&P 500 had its worst performance in four years, and financial pundits everywhere were talking about another global economic slowdown and possible crisis. Traditional IPOs were pulled from the market, mergers were stalled, and Silicon Valley was getting antsy about the overinflated valuation of its cherished unicorns, private companies valued at more than $1 billion.

Amid the carnage, an enterprise called Augur launched one of the most successful crowdfunding campaigns in history. In the first week, more than 3,500 people from the United States, China, Japan, France, Germany, Spain, the United Kingdom, Korea, Brazil, South Africa, Kenya, and Uganda contributed a total of $4 million. There was no brokerage, no investment bank, no stock exchange, no mandatory filings, no regulator, and no lawyers. There wasn't even a Kickstarter or Indiegogo. Ladies and gentlemen, welcome to the blockchain IPO.

Matching investors with entrepreneurs is one of the eight functions of the financial services industry most likely to be disrupted. The process of raising equity capital—through private placements, initial public offerings, secondary offerings, and private investments in public equities (PIPEs)—has not changed significantly since the 1930s.[78]

Thanks to new crowdfunding platforms, small companies can access capital using the Internet. The Oculus Rift and the Pebble Watch were early successes of this model. Still, participants couldn't buy equity directly. Today, the U.S. Jumpstart Our Business Startups Act allows small investors to make direct investments in crowdfunding campaigns, but investors and entrepreneurs still need intermediaries such as Kickstarter or Indiegogo, and a conventional payment method, typically credit cards and PayPal, to participate. The intermediary is the ultimate arbiter of everything, including who owns what.

The blockchain IPO takes the concept further. Now, companies can raise funds "on the blockchain" by issuing tokens, or cryptosecurities, of some value in the company. They can represent equity, bonds, or, in the case of Augur, market-maker seats on the platform, granting owners the right to decide which prediction markets the company will open. Ethereum was an even greater success than Augur, funding the development of a whole new blockchain through a crowd sale of its native token, ether. Today Ethereum is the second-longest and fastest-growing public blockchain. The average investment in the Augur crowdfunding was $750, but one can easily imagine minimum subscriptions of a dollar or even ten cents. Anyone in the world—even the poorest and most remote people—could become stock market investors.

Overstock, the e-retailer, is launching perhaps the most ambitious cryptosecurity initiative yet. Overstock's forward-thinking founder, Patrick Byrne, believes blockchain "can do for the capital markets what the Internet has done for consumers." The project, dubbed Medici, enables companies to issue securities on the blockchain and recently received the support of the Securities and Exchange Commission.[79] The company began issuing its first blockchain-based securities, such as the $5 million cryptobond for an affiliate of FNY Capital, in 2015.[80] Overstock claims many financial services firms and other companies are lining up to use the platform. Surely, the tacit approval of the SEC will give Overstock a head start on what is sure to be a long journey.

Should blockchain IPOs continue to gain traction, they will ultimately disrupt many of the roles in the global financial system—brokers, investment bankers, and securities lawyers—and change the nature of investment. By integrating blockchain IPOs with new platforms for value exchange such as Circle, Coinbase (the most well-funded bitcoin exchange start-up), Smartwallet (a global asset exchange for all forms of value), and other emerging companies, we expect a distributed virtual exchange to emerge. The old guard is taking notice. The NYSE invested in Coinbase and NASDAQ is integrating blockchain technology into its private market. Bob Greifeld, CEO of NASDAQ, is starting small, using blockchain to "streamline financial record keeping while making it cheaper and more accurate,"[81] but evidently NASDAQ and other incumbents have bigger plans.

THE MARKET FOR PREDICTION MARKETS

Augur is building a decentralized prediction market platform that rewards users for correctly predicting future events—sporting events, election results, new product launches, the genders of celebrity babies. How does it work? Augur users can purchase or sell shares in the outcome of a future event, the value of which is an estimate of the probability of an event happening. So if there are even odds (i.e., 50/50), the cost of buying a share would be fifty cents.

Augur relies on "the wisdom of the crowd," the scientific principle that a large group of people can often predict the outcome of a future event with far greater accuracy than one or more experts.[82] In other words, Augur brings the spirit of the market to bear on the accuracy of predictions. There have been a few attempts at centralized prediction markets, such as the Hollywood Stock Exchange, Intrade, and HedgeStreet (now Nadex), but most have been shut down or failed to launch over regulatory and legal concerns. Think assassination contracts and terrorism futures.

Using blockchain technology makes the system more resilient to failure, more accurate, and more resistant to crackdowns, error, coercion, liquidity concerns, and what the Augur team calls euphemistically "dated jurisdictional regulation." The arbiters on the Augur platform are known as referees and their legitimacy derives from their reputation points. For doing the right thing—that is, correctly stating that an event happened, who won a sporting match, or who won an election—they receive more reputation points. Maintaining the integrity of the system has other monetary benefits: the more reputation points you have, the more markets you can make, and thus the more fees you can charge. In Augur's words, "our prediction markets eliminate counterparty risks, centralized servers, and create a global market by employing cryptocurrencies including bitcoin, ether, and stable cryptocurrencies. All funds are stored in smart contracts, and no one can steal the money."[83] Augur resolves the issue of unethical contracts by having a zero-tolerance policy for crime.

To Augur's leadership team, human imagination is the only practical limit to the utility of prediction markets. On Augur, anyone can post a clearly defined prediction about anything with a clear end date—from the trivial, "Will Brad Pitt and Angelina Jolie divorce?" to the vital, "Will the European Union dissolve by June 1, 2017?" The implications for the financial services industry, for investors, economic actors, and entire markets, are huge. Consider the farmer in Nicaragua or Kenya who has no robust tools to hedge against currency risk, political risk, or changes to the weather and climate. Accessing prediction markets would allow that person to mitigate the risk of drought or disaster. For example, he could buy a prediction contract that pays out if a crop yield is below a certain level, or if the country gets less than a predetermined amount of rain.

Prediction markets are useful for investors who want to place bets on the outcome of specific events such as "Will IBM beat its earnings by at least ten cents this quarter?" Today the reported "estimate" for corporate earnings is nothing more than the mean or median of a few so-called expert analysts. By harnessing the wisdom of the crowds, we can form more realistic predictions of the future, leading to more efficient markets. Prediction markets can serve as a hedge against global uncertainty and "black swan" events: "Will Greece's economy shrink by more than 15 percent this year?"[84] Today, we rely on a few talking heads to sound the alarms; a prediction market would act more impartially as an early warning system for investors globally.

Prediction markets could complement and ultimately transform many aspects of the financial system. Consider prediction markets on the outcomes of corporate actions—earnings reports, mergers, acquisitions, and changes in management. Prediction markets would inform the insurance of value and the hedging of risk, potentially even displacing esoteric financial instruments like options, interest rate swaps and credit default swaps.

Of course, not everything needs a prediction market. Enough people need to care to make it liquid enough to attract attention. Still, the potential is vast, the opportunity significant, and access available to all.

ROAD MAP FOR THE GOLDEN EIGHT

Blockchain technologies will impact every form and function of the financial services industry—from retail banking and capital markets to accounting and regulation. They will also force us to rethink the role of banks and financial institutions in society. "Bitcoin cannot have bail-ins, bank holidays, currency controls, balance freezes, withdrawal limits, banking hours,"[85] said Andreas Antonopoulos.

Whereas the old world was hierarchical, slow-moving, reluctant to change, closed and opaque, and controlled by powerful intermediaries, the new order will be flatter, offering a peer-to-peer solution; more private and secure; transparent, inclusive, and innovative. To be sure, there will be dislocation and disruption, but there is also a remarkable opportunity for the industry's leaders to do something about it today. The financial services industry will both shrink and grow over the coming years; fewer intermediaries will be able to offer more products and services at a much lower cost to a much larger population. That's a good thing. Whether permissioned and closed blockchains will find a place in a decentralized world is up for debate. Barry Silbert, who founded SecondMarket and is now CEO of the Digital Currency Group, said, "I have a very cynical view of the objectivities put forward by large financial incumbents. When all you have is a hammer, everything looks like a nail."[86] We believe that the unstoppable force of blockchain technology is barreling down on the entrenched, regulated, and ossified infrastructure of modern finance.[87] Their collision will reshape the landscape of finance for decades to come. We would like it to finally transform from an industrial age money machine into a prosperity platform.

CHAPTER 4

RE-ARCHITECTING THE FIRM: THE CORE AND THE EDGES

BUILDING CONSENSYS

July 30, 2015, was a big day for a global group of coders, investors, entrepreneurs, and corporate strategists who think that Ethereum is the next big thing—not just for business, but possibly for civilization. Ethereum, the blockchain platform eighteen months in the making, went live.

We witnessed the launch firsthand in the Brooklyn office of Consensus Systems (ConsenSys), one of the first Ethereum software development companies. Around 11:45 a.m., there were high fives all around as the Ethereum network created its "genesis block," after which a frenzy of miners raced to win the first block of ether, Ethereum's currency. The day was eerily suspenseful. A massive thunderstorm broke over the East River, triggering loud and random emergency flood warnings on everyone's smart phones.

According to its Web site, Ethereum is a platform that runs decentralized applications, namely smart contracts, "exactly as programmed without any possibility of downtime, censorship, fraud, or third party interference." Ethereum is like bitcoin in that its ether motivates a network of peers to validate transactions, secure the network, and achieve consensus about what exists and what has occurred. But unlike bitcoin it contains some powerful tools to help developers and others create software services ranging from decentralized games to stock exchanges.

Ethereum was conceived in 2013 by then-nineteen-year-old Vitalik Buterin, a Canadian of Russian descent. He had argued to the bitcoin core developers that the platform needed a more robust scripting language for developing applications. When they rejected him, he decided to craft his own platform. ConsenSys was first off the block, so to speak, launched to create Ethereum-based apps. Flash-forward a couple of years and the analogy is clear: Linus Torvalds is to Linux what Vitalik Buterin is to Ethereum.

When discussing the rise of blockchain and Ethereum technology, Joseph Lubin, ConsenSys's cofounder, said, "It became clear to me that instead of people wasting their time walking down the street with posters on sticks, we could all work together to just build the new solutions to this broken economy and society."[1] Don't occupy Wall Street. Invent our own street.

Like many entrepreneurs, Lubin has a bold mission, not just to build a great company but to solve important problems in the world. He deadpans that the company is a "blockchain venture production studio, building decentralized applications, mostly on Ethereum." Pretty low-key. But, if implemented, the applications that ConsenSys is building would shake the windows and rattle the walls of a dozen industries. Projects include a distributed triple-entry accounting system; a decentralized version of the massively popular Reddit discussion forum, plagued of late by controversy over its centralized control; a document formation and management system for self-enforcing contracts (aka smart contracts); prediction markets for business, sports, and entertainment; an open energy market; a distributed music model to compete with Apple and Spotify, though those two firms could use it too;[2] and a suite of business tools for mass collaboration, mass creation, and mass management of a management-less company.

Our story of ConsenSys is not so much about its ambitious blockchain-based products or services. It's about its efforts to cultivate a company of its own, pioneering important new ground in management science along the lines of holacracy, a collaborative rather than hierarchical process for defining and aligning the work to be done. "While I don't want us to implement holacracy as is—it feels way too rigid and structured to me—we are working to incorporate many of its philosophies in our structure and processes," said Lubin. Among those holacratic tenets are "dynamic roles rather than traditional job descriptions; distributed, not delegated authority; transparent rules rather than office politics; and rapid reiterations rather than big reorganizations," all of which describe how blockchain technologies work. How ConsenSys is structured, how it creates value, and how it manages itself differs not only from the industrial corporation but also from the typical dot-com.

Joe Lubin is not an ideologue, and certainly not an anarchist or libertarian as some in the cryptocurrency movement are. But he does think that we need to change capitalism if we want it to survive, specifically to move away from the command-and-control hierarchies inappropriate for a networked world. He notes that today, even though vast networks enmesh the world and enable us all to communicate inexpensively, richly, and immediately, hierarchies prevail. Blockchain technology is the countervalence: "Global human society can now agree on the truth and make decisions in ten minutes, or ten seconds. This surely creates an opportunity to have a more enfranchised society," he said. The greater the engagement, the greater the prosperity.

The End of Managers. Long Live Management

ConsenSys operates according to a plan that all employees ("members") developed, modified, voted on, and adopted. Joe Lubin describes its structure as a "hub" rather than a hierarchy, and each of its projects is a "spoke" in which major contributors hold equity.

For the most part, members of ConsenSys choose what they work on. No top-down assignments. Lubin said, "We share as much as possible, including shared software components. We build small agile teams but there is collaboration among them. We have tons of immediate, open, rich communication." Members choose to work on two to five projects. When someone sees a piece of work that needs to get done, he or she jumps in and pushes it a little or a lot farther in a valuable direction, as appropriate for her role. "We talk about things quite a bit so people are aware of the many things that could be pushed forward," he said. But these many things can and do change constantly. "Part of being agile means that priorities are dynamic."

Lubin is not the boss. His main operational role is advisory: "In many cases, individuals ask me or others what would be good to work on," he said. Through Slack[3] and GitHub,[4] he suggests directions they might pursue "to build all the services and platforms that we want to build, and many that we want to build but don't know it yet."

Member ownership explicitly incentivizes this behavior. Everyone owns a piece of every project directly or indirectly: the Ethereum platform issues tokens that members can exchange for ether and then convert into any other currency. "Our goal is to achieve a nice balance between independence and interdependence," Lubin said. "We view ourselves as a collective of closely collaborating entrepreneurlike agents. At some point, it may prove necessary to suggest that a certain thing really needs to get done and if nobody steps up, to hire someone initially for that role or incentivize internal people to do it," said Lubin. But, overall, "everyone is a self-managed adult. Did I mention we communicate a lot? Then we all make our own decisions."

The watchwords are *agility*, *openness*, and *consensus*: identify the work to be done, distribute the load among the people eager and able to do it, agree on their roles, responsibilities, and compensation, and then codify these rights in "explicit, detailed, unambiguous, self-enforcing agreements that can serve as the glue to hold all of the business aspects of our relationships together," he said. Some agreements pay for performance, others mete out annual salary in ether, and still others are more like "requests for participation" with bounties attached to task completion such as writing a line of code. If the code passes the test, then the bounty is automatically released. "Everything can be surfaced and appropriately transparent. Incentives are explicit and granular," he said. "This leaves us free to communicate, be creative, and adapt based on these expectations."

Dare we coin the neologism *blockcom*, a company formed and functioning on blockchain technologies? That's the goal, to run as much of ConsenSys as possible on Ethereum, from governance and day-to-day operations to project management, software development and testing, hiring and outsourcing, compensation, and funding. The blockchain also enables reputation systems where members can rate one another's performance as collaborators, thereby syndicating trust in the community. Lubin said, "Persistent digital identity or persona and reputation systems will keep us more honest and well behaved toward one another."

These capabilities blur the boundaries of a company. There are no default settings for incorporation. Members of the ConsenSys ecosystem can form spokes by reaching consensus on strategy, architecture, capital, performance, and governance. They may decide to launch a company that competes within an existing market or provides an infrastructure for a new market. Once it is launched, they can adjust those settings.

Decentralizing the Enterprise

The blockchain will reduce friction for companies everywhere. "Lower friction means lower costs as the price of valuable intermediation is determined via the most efficient price discovery mechanism: decentralized free markets. No longer will incumbents be able to leverage legal, regulatory, informational, and power asymmetries to extract far more value from a transaction in their role as intermediary than they add to it," Lubin said.

Could ConsenSys build some kind of truly decentralized autonomous organization owned and controlled by its nonhuman value creators, governed through smart contracts rather than human agency? "All the way!" said Lubin. "Massive intelligence on a decentralized global computational substrate, an underlying layer, should change the architecture of the firm from a large collection of specialized departments run by humans to software agents that can cooperate and compete in free markets." Some agents will organize for longer periods of time to serve ongoing customer needs, such as utility and maintenance. Others will swarm around a short-term problem, solve it, and dissolve just as quickly, having served their purpose.

Is there a risk that radical decentralization and automation removes human agency in decision making (e.g., the risk of rogue algorithms)? "I am not concerned about machine intelligence. We will evolve with it and for a long time it will be in the service of, or an aspect of, *Homo sapiens cybernetica*. It may evolve beyond us but that is fine," Lubin said. "If so, it will occupy a different ecological niche. It will operate at different speeds and different relevant time scales. In that context, artificial intelligence will not distinguish between humans, a rock, or a geological process. We evolved past lots of species, many of which are doing fine (in their present forms)."

ConsenSys is still a tiny company. Its grand experiment may or may not succeed. But its story provides a glimpse into radical changes in corporate architecture that may help unleash innovation and harness the power of human capital for not just wealth creation but prosperity. Blockchain technology is enabling new forms of economic organization and new portfolios of value. There are distributed models of the firm emerging—ownership, structure, operations, rewards, and governance—that go far beyond enhancing innovation, employee motivation, and collective action. They may be the long-awaited precondition for a more prosperous and inclusive economy.

Business leaders have another opportunity to rethink how they organize value creation. They could negotiate, contract, and enforce their agreements on the blockchain; deal seamlessly with suppliers, customers, employees, contractors, and autonomous agents; and maintain a fleet of these agents for others to use, and these agents could rent out or license any excess capacity in their value chain.

CHANGING THE BOUNDARIES OF THE FIRM

Throughout the first era of the Internet, management thinkers (Don included) talked up the networked enterprise, the flat corporation, open innovation, and business ecosystems as successors to the hierarchies of industrial power. However, the architecture of the early-twentieth-century corporation remains pretty much intact. Even the big dot-coms adopted a top-down structure with such decision makers as Jeff Bezos, Marissa Mayer, and Mark Zuckerberg. So why would any established firm—particularly ones that make their money off other people's data, operate largely behind closed doors, and suffer surprisingly little in data breach after data breach—want to leverage blockchain technologies to distribute power, increase transparency, respect user privacy and anonymity, and include far more people who can afford far less than those already served?

Transaction Costs and the Structure of the Firm

Let's start with a little economics. In 1995, Don used Nobel Prize–winning economist Ronald Coase's theory of the firm to explain how the Internet would affect the architecture of the corporation. In his 1937 paper "The Nature of the Firm," Coase identified three types of costs in the economy: the costs of search (finding all the right information, people, resources to create something); coordination (getting all these people to work together efficiently); and contracting (negotiating the costs for labor and materials for every activity in production, keeping trade secrets, and policing and enforcing these agreements). He posited that a firm would expand until the cost of performing a transaction inside the firm exceeded the cost of performing the transaction outside the firm.[5]

Don argued that the Internet would reduce a firm's internal transaction costs somewhat; but we thought, because of its global accessibility, it would reduce costs in the overall economy even more, in turn lowering barriers to entry for more people. Yes, it did drop search costs, through browsers and the World Wide Web. It also dropped coordination costs through e-mail, data processing applications like ERP, social networks, and cloud computing. Many companies benefited from outsourcing such units as customer service and accounting. Marketers engaged customers directly, even turning consumers into producers (prosumers). Product planners crowdsourced innovations. Manufacturers leveraged vast supply networks.

However, the surprising reality is that the Internet has had peripheral impact on corporate architecture. The industrial-age hierarchy is pretty much intact as the recognizable foundation of capitalism. Sure, the networks have enabled companies to outsource to low-cost geographies. But the Internet dropped transaction costs inside the firm as well.

From Hierarchy to Monopoly

So companies today remain hierarchies, and most activities occur within corporate boundaries. Managers still view them as a better model for organizing talent and intangible assets such as brands, intellectual property, knowledge, and culture, as well as for motivating people. Corporate boards still compensate executives and CEOs far beyond any reasonable measure of the value they create. Not incidentally, the industrial complex continues to generate wealth, but not prosperity. In fact, as we have pointed out, there is strong evidence of a growing concentration of power and wealth in conglomerates and even monopolies.

Another Nobel laureate, Oliver Williamson, predicted as much,[6] and pointed out the negative effects on productivity: "Suffice it to observe here that the move from autonomous supply (by the collection of small firms) to unified ownership (in one large firm) is unavoidably attended by changes in both incentive intensity (incentives are weaker in the integrated firm) and administrative controls (controls are more extensive)."[7] Peter Thiel, cofounder of PayPal, wrote in praise of monopolies in his enormously readable and equally controversial book, *Zero to One*. A Rand Paul supporter, Thiel said, "Competition is for losers. . . . Creative monopolies aren't just good for the rest of society; they're powerful engines for making it better."[8]

While Thiel might be right about striving to dominate one's industry or market, he provided no real evidence that monopolies are good for consumers or society as a whole. To the contrary, the entire body of competition law in most democratic capitalist countries derives from a contrary notion. The idea of fair competition dates back to Roman times, with the death penalty for some violations.[9] When firms have no real competition, they can grow as inefficient as they want, raising prices in and outside the firm. Look at governments. Even in the technology industry, many argue that monopolies may help with innovation in the short term but may harm society in the long term. Companies may amass monopoly power through cool products and services that customers love, but the honeymoon eventually ends. It's not so much that their innovations no longer delight; it's that the companies themselves begin to ossify.

Most thinkers understand that innovation typically comes from the edge of the firm, not from its core. Harvard University law professor Yochai Benkler agrees: "Monopolies may have lots of money to invest in R&D but typically not the internal culture of pure and open exploration that is required for innovation. The Web didn't come from monopolies; it came from the edge. Google did not come from Microsoft. Twitter did not come from AT&T, or for that matter even from Facebook."[10] In monopolies, layers of bureaucracy distance the executives at the top from market signals and emergent technology at the edges, where companies bump up against one another and other markets, other industries, other geographies, other intellectual disciplines, other generations. According to John Hagel and John Seely Brown, "The periphery of today's global business environment is where innovation potential is the highest. Ignore it at your peril."[11]

Executives should be excited about blockchain technology, because the wave of innovation coming from the edge may well be unprecedented. From the major cryptocurrencies—Bitcoin, BlackCoin, Dash, Nxt, and Ripple—to the major blockchain platforms—Lighthouse for peer-to-peer crowdfunding, Factom as a distributed registry, Gems for decentralized messaging, MaidSafe for decentralized applications, Storj for a distributed cloud, and Tezos for decentralized voting to name a few—the next era of the Internet has real value attached to it and real incentives to participate. These platforms hold promise for protecting user identity, respecting user privacy and other rights, ensuring network security, and dropping transaction costs so that even the unbanked can take part.

Unlike incumbent firms, they don't need a brand to convey the trustworthiness of their transactions. By giving away their source code for free, sharing power with everyone on the network, using consensus mechanisms to ensure integrity, and conducting their business openly on the blockchain, they are magnets of hope for the many disillusioned and disenfranchised. As such, blockchain technology offers a credible and effective means not only of cutting out intermediaries, but also of radically lowering transaction costs, turning firms into networks, distributing economic power, and enabling both wealth creation and a more prosperous future.

1. Search Costs—How Do We Find New Talent and New Customers?
How do we find the people and information we need? How do we determine if their services, goods, and capabilities are best for us as we seek to bring the tonic of the market to bear on our internal operations?

Although the architecture of the firm is basically intact, the first era of the Internet dropped such costs significantly and enabled important changes. Outsourcing was really just the beginning. Tapping into ideagoras (open markets for brainpower), companies like Procter & Gamble are finding uniquely qualified minds to innovate a new product or process. In fact, 60 percent of P&G's innovations come from outside the company, by building or harnessing ideagoras like InnoCentive or inno360. Other firms like Goldcorp have created global challenges to search for the best minds to solve their toughest problems. Goldcorp, which published its geological data and talent outside its boundaries, discovered $3.4 billion worth of gold, resulting in a hundredfold increase in the company's market value.

Now imagine the opportunities that arise from the ability to search the World Wide Ledger, a decentralized database of much of the world's structured information. Who sold which discovery to whom? At what price? Who owns this intellectual property? Who is qualified to handle this project? What medical skills does our hospital have on staff? Who performed what type of surgery with what outcomes? How many carbon credits has this company saved? Which suppliers have experience in China? What subcontractors delivered on time and on budget according to their smart contracts? The results of these queries won't be résumés, advertising links, or other pushed content; they'll be transaction histories, proven track records of individuals and enterprises, ranked perhaps by reputation score. Get the picture? Said Vitalik Buterin, founder of the Ethereum blockchain, "Blockchains will drop search costs, causing a kind of decomposition that allows you to have markets of entities that are horizontally segregated and vertically segregated. That never really existed before. Instead you had kind of monoliths that do everything."[12]

Several companies are working on search engines for blockchains, given the potential bonanza. Google's mission is to organize the world's information, so it would make sense for it to assign considerable manpower to investigate this.

There are three key distinctions between Internet search and blockchain search. First is user privacy. While transactions are transparent, people own their personal data and can decide what to do with it. They can participate anonymously or at least pseudonymously (anonymity through a false name) or quasinymously (partial anonymity). Interested parties will be able to search for information that users have made open. Andreas Antonopoulos said, "Transactions are anonymous if you want them to be anonymous. . . . but the blockchain enables radical transparency a lot easier than it enables radical anonymity."[13]

Many firms will need to rethink and redesign the recruiting process. For example, human resources or personnel staff will need to learn how to query the blockchain with yes/no questions: Are you a human being? Have you earned a PhD in applied mathematics? Can you code in Scrypt, Python, Java, C++? Are you available to work full time from January through June next year? And other qualifications. These queries will scurry about the black boxes of people on the job market and yield a list of people who meet these qualifications. They could also pay prospective talent to place pertinent professional information on a blockchain platform where they can sort through it. HR staff must master the use of reputation systems, moving forward with candidates without knowing anything irrelevant to the job, such as age, gender, race, country of origin. They also need search engines that can navigate various degrees of openness, from fully private to fully public information. The upside is an end to subconscious or even institutional bias and headhunter or executive recruiting fees. The downside is that precise queries lead to precise results. There is less possibility of serendipity, the discovery of a candidate who lacks the qualifications but has great capacity to learn and to make the random creative connections that a firm desperately needs.

Ditto for marketing. Firms may have to pay just to query a prospective customer's black box, to see whether that customer meets a firm's target audience. That customer may decide globally to withhold certain data such as gender, because a no answer is still valuable. But in so doing, the firm will learn nothing more about the prospect beyond the yes/no results of the query. Chief marketing officers and marketing agencies will need to rethink any strategy based on e-mail, social media, and mobile marketing: where the infrastructure may lower communications costs to zero, customers will raise costs to a figure that makes reading a firm's message worth their while. In other words, you'll be paying customers to listen to your elevator pitch, but you will have tailored your query to pitch only to a sharply defined audience so that you will be reaching exactly the people you want to reach without invading their privacy. You can test different queries to learn about different microniches at every stage of new product development. Let's call it *black box marketing*.

The second distinction is that search can be multidimensional. When you search the World Wide Web today, you search a snapshot in time, as indexed over the last several weeks.[14] Computer theorist Antonopoulos called this two-dimensional search: *horizontal*, a wide search across the Web, and *vertical*, a deep search of a particular Web site. The third dimension is *sequence*, to see these in the order of uploading over time. "The blockchain can add the additional dimension of time," he said. The opportunity to search a complete record of everything that ever happened in three dimensions is profound. To make his point, Antonopoulos searched the bitcoin blockchain to find its famous first commercial transaction, the purchase of two pizzas done by someone named "Laslo" for 10,000 bitcoins. "The blockchain provides an almost archaeological record, a deep find, preserving information forever." (To save you from doing the math, if the pizza costs $5 when $1 was equal to 2,500 bitcoins, that would be worth $3.5 million as of the writing of this book . . . but we digress.)

For firms, this means a need for better judgment: managers need to hire people who have demonstrated good judgment, because there's no walking back poor decisions, no spinning the order of events, no denying an executive's disreputable behavior. For really important decisions, firms could implement internal consensus mechanisms whereby all stakeholders vote on mission-critical decisions to end the chorus of ignorance and denial of prior knowledge. Or use prediction markets to test scenarios. If you're an executive of a future Enron, no scapegoating. As for New Jersey governor Chris Christie, good luck telling a prosecutor that you knew nothing of plans to close the George Washington Bridge.

The third distinction is value: where information on the Internet is abundant, unreliable, and perishable, it is scarce, tamperproof, and permanent on the blockchain. To this last characteristic, Antonopoulos notes: "If there is enough financial incentive to preserve this blockchain into the future, the possibility of it existing for tens, hundreds, or even thousands of years cannot be discounted."

What an amazing concept. The blockchain as part of the archaeological record, like the original stone tablets of Mesopotamia. Paper records are ephemeral and temporary, whereas (ironically) the oldest form of recording information, tablets, is the most permanent. The implications for corporate architecture are considerable. Imagine a permanent, searchable record of important historical information, like the history of finance. Corporate staff responsible for developing financial statements, annual reports, reports to governments or donors, marketing materials for prospective employees, clients, and consumers—will start with this public, indisputable view of their firm, maybe even creating a filter that enables stakeholders to see what they see at the press of a button. Companies could have transaction ticker tapes and dashboards, some for internal managerial use and some public. Rest assured: All your competitors will construct such feeds and dashboards of your firm as part of their competitive intelligence programs. So why not put those on your Web site and draw everyone to you?

This provides enormous incentive for firms to look for resources outside their boundaries, as they have almost infinitely better information about the qualities and record of candidates, be they individuals or companies.

Companies like ConsenSys are developing identity systems where job prospects or prospective contractors will program their own personal avatars to disclose pertinent information to employers. They can't be hacked like a centralized database can. Users are motivated to contribute information to their own avatars because they own and control them, their privacy is completely configurable, and they can monetize their own data. This is very different from, say, LinkedIn, a central database owned, monetized, and yet not entirely secured by a powerful corporation.

Could Coase and Williamson have imagined a platform that could drop search costs so that firms could find capability outside their boundaries that cost less and could perform better?

2. Contracting Costs—What Do We Agree to Do, Anyway?

How do we come to terms with other parties or enter into an agreement? It's one thing to lower the costs of finding people and resources that can do the job. But that's not enough to shrink a firm significantly. All parties must agree to work together. The second reason why we have firms is contractual costs, such as negotiating the price, establishing capacity, and spelling out the conditions of a supplier's goods or services; policing them and enforcing the terms; and handling remedies if parties don't deliver as promised.

We've always had social contracts, understandings of relationships in the specialization of roles where some people in the tribe hunted and protected the tribe, and others gathered and sheltered the tribe. People have traded physical objects in real time since the dawn of modern man. Contracts are a more recent phenomenon, as we began trading promises, not property. Oral agreements proved easily manipulated or misremembered, and eyewitnesses were unreliable. Doubt and distrust tempered collaboration with strangers. Contracts had to be fulfilled immediately, and there were no formal mechanisms for enforcement of the terms beyond what you could take by force. The written contract was a way of codifying an obligation, of establishing trust and setting expectations. Written contracts provided guidance when someone did not hold up his end of the bargain, or something unexpected happened. But they couldn't exist in a vacuum; there had to be some legal framework that recognized contracts and enforced each party's rights.

Today contracts are still made of atoms (paper), not bits (software). As such they have huge limitations, serving to simply document an agreement. As we shall see, if contracts were software—smart and distributed on the blockchain—they could open a world of possibilities, not the least of which is to make it easier for companies to collaborate with external resources. And just imagine how the Uniform Commercial Code might look on the blockchain.

Coase and his successors argued that contracting costs are lower inside the boundaries of firms rather than outside in the market—that a firm is essentially a vehicle for creating long-term contracts when short-term contracts are too much effort.

Williamson advanced this idea by arguing that firms exist to resolve conflicts, largely through making contracts with various parties inside the firm. In the open market, the only dispute mechanism is the court—costly, timely, and often unsatisfactory. Further, he argued that in some cases like fraud, other illegal acts, or conflict of interest, there is no market dispute mechanism at all. "In effect, the contract law of internal organization is that of forbearance, according to which a firm becomes its own court of ultimate appeal. Firms, for this reason, are able to exercise fiat that the markets cannot."[15] Williamson conceived of the firm as "a governance structure" for contractual arrangements. He said that organizational structure matters in reducing the costs of managing transactions and that "recourse to the lens of contract, as against the lens of choice, frequently deepens our understanding of complex economic organization."[16] This is a recurring theme in management theory, perhaps most powerfully explained by economists Michael Jensen and William Meckling. They argued that entities are nothing more than a collection of contracts and relationships.[17]

Today, some erudite blockchain thinkers have picked up on this view. Ethereum inventor Vitalik Buterin argues that corporate agents (i.e., executives) could use corporate assets only for certain purposes approved by, say, a board of directors, who in turn are subject to shareholder approval. "If a corporation does something, it's because its board of directors has agreed that it should be done. If a corporation hires employees, it means that the employees are agreeing to provide services to the corporation's customers under a particular set of rules, particularly involving payment," Buterin wrote. "When a corporation has limited liability, it means that specific people have been granted extra privileges to act with reduced fear of legal prosecution by the government—a group of people with more rights than ordinary people acting alone, but ultimately people nonetheless. In any case, it's nothing more than people and contracts all the way down."[18]

That's why the blockchain, by reducing contracting costs, enables firms to open up and develop new relationships outside their boundaries. ConsenSys, for example, can architect complex relationships with a diverse set of members, some inside its boundaries, some outside, and some straddling walls, because smart contracts govern these relationships rather than traditional managers. Members self-assign to projects, define agreed-upon deliverables, and get paid when they deliver—all on the blockchain.

Smart Contracts

The rate of change is increasingly setting the stage for smart contracts. More people are developing not only computer literacy, but also fluency. As far as evidencing transactions goes, this new digital medium has significantly different properties from its paper predecessors. As cryptographer Nick Szabo highlighted, not only can they capture a greater array of information (such as nonlinguistic sensory data) but they are dynamic: they can transmit information and execute certain kinds of decisions. In Szabo's words, "Digital media can perform calculations, directly operate machinery, and work through some kinds of reasoning much more efficiently than humans."[19]

For the purposes of this discussion, smart contracts are computer programs that secure, enforce, and execute settlement of recorded agreements between people and organizations. As such, they assist in negotiating and defining these agreements. Szabo coined the phrase in 1994, the same year that Netscape, the first Web browser, hit the market:

> A smart contract is a computerized transaction protocol that executes the terms of a contract. The general objectives of smart contract

> design are to satisfy common contractual conditions (such as payment terms, liens, confidentiality, and even enforcement), minimize exceptions both malicious and accidental, and minimize the need for trusted intermediaries. Related economic goals include lowering fraud loss, arbitration and enforcement costs, and other transaction costs.[20]

Back then, smart contracts were an idea all dressed up with nowhere to go, as no available technology could deploy them as Szabo described. There were computer systems such as electronic data interchange (EDI) that provided standards for the communication of structured data between the computers of buyers and sellers, but no technology that could actually trigger payments and cause money to be exchanged.

Bitcoin and the blockchain changed all that. Now parties can make agreements and automatically exchange bitcoin when they meet the terms of the agreement. Most simply, your brother-in-law can't weasel out of a hockey bet. Less simply, when you purchase a stock, the trade settles instantly and the shares are immediately transferred to you. Even less simple, when contractors deliver the software code that meets the necessary specifications, they get paid.

The technological means of executing limited smart contracts has existed for some time. A contract is a bargained-for exchange enforceable before the exchange. Andreas Antonopoulos explained with a simple example: "So if you and I were to agree right now that I would pay you fifty dollars for the pen on your desk, that's a perfectly enforceable contract. We can just say, 'I promise to pay you fifty dollars for the pen on your desk,' and you would respond, 'Yes, I would like that.' That turns out to be 'offer acceptance and consideration.' We've got a deal, and it can be enforced in a court. That has nothing to do with the technological means of implementation of the promises that we have made."

What interests Andreas about the blockchain is that we can execute this financial obligation in a decentralized technological environment with a built-in settlement system. "That's really cool," he said, "because I could actually pay you for the pen right now, you would see the money instantly, you would put the pen in the mail, and I could get a verification of that. It's much more likely that we can do business."

The law profession is slowly plugging into this opportunity. Like everyone in the middle, lawyers may become subject to disintermediation and will eventually need to adapt. Expertise in smart contracts could be a big opportunity for law firms that want to lead innovation in contract law. However, the profession isn't known for breaking new ground. Legal expert Aaron Wright, coauthor of a new book about the blockchain, told us, "Lawyers are laggards."[21]

Multisignature: Smart Complex Contracts

But, you say, wouldn't the costs of complex and time-consuming negotiations of smart contracts outweigh the benefits of open boundaries? The answer at this point appears to be no. If partners spend more time up front determining the terms of an agreement, the monitoring, enforcement, and settlement costs drop significantly, perhaps to zero. Further, settlement can occur in real time, possibly in microseconds throughout the day depending on the deal. Most important, by partnering with superior talent, companies can achieve better innovation and become more competitive.

Let's consider the use of independent contractors. In the early days of digital trade, the blockchain accommodated only the simplest two-party transactions. For instance, if Alice needed someone to complete a piece of code quickly, she would post an anonymous "coder needed" request on an appropriate discussion board. Bob would see it.[22] If the price and timing were right, he would send work samples. If his samples met Alice's needs, then she made Bob an offer. They agreed on terms: Alice would send half the fee immediately and half upon receipt of completion and successful test of the code.

Their contract was straightforward—an offer to hire and an acceptance to do the job, and it needn't have been in writing, though their interactions on the blockchain made it so. Their ownership of bitcoins was associated with digital addresses (long strings of numbers) that had two components: a public key that served as an address, and a private key that gave its owner exclusive access to any coins associated with that address. Bob sent Alice his public key, and she directed the first payment there. The network recorded the transfer and associated those bitcoins to Bob's public key wallet.

What if, at this point, Bob decided that he didn't want to do the project? In this two-party transaction, Alice would have little recourse. She couldn't go to her credit card company to reverse the transaction. She couldn't (yet) go to civil court and sue Bob for breach of contract. Beyond a randomly generated alphanumeric code and an online advertisement, she would have no way of identifying Bob unless he'd posted his ad on a centralized platform that could track Bob down, or they'd exchanged e-mails through a centralized service. She could, however, indicate that his public key was not to be trusted, thus lowering his reputation score as a coder.

Without assurances of the other party's trustworthiness in fulfilling off-chain actions, the deal was a prisoner's dilemma of sorts: it still required some trust. Reputation systems could mitigate this uncertainty to some extent. But we needed to introduce trust and security into this anonymous and open system.

In 2012, "core developer" Gavin Andresen introduced a new type of bitcoin address to the bitcoin protocols called "pay to script hash" (P2SH). Its purpose was to allow one party "to fund any arbitrary transaction, no matter how complicated."[23] Parties use multiple authenticating signatures or keys rather than a single private key to complete a transaction. The community usually refers to this multiple-signature feature as simply "multisig."

In a multisig transaction, parties agree on the total number of keys generated (N) and how many will be required to complete a transaction (M). This is called an M-of-N signature scheme or security protocol. Think of a lockbox requiring multiple physical keys to open. With this feature, Bob and Alice would agree in advance to employ a neutral, disinterested third-party arbitrator to help them complete their transaction. Each of the three parties would hold one of three private keys, two of which are needed to access the transferred funds. Alice would send her bitcoin to a public address. At this point, those funds can be viewed by anyone, but accessed by no one. Once Bob sees the funds have been posted, he fulfills his end of the bargain. If, upon receipt of Bob's good or service, Alice is unsatisfied and feels cheated, she could refuse to provide Bob with the second key. The two parties would then look to the arbitrator, holder of the third key, to help them resolve their disagreement. The intervention of such arbitrators is called for only in cases of disputes like these, and at no point do they themselves have access to the funds—a mechanism enabling the rise of "smart contracts."

To contract remotely, let alone automatically, you need a certain degree of trust that the system will enforce your rights under the deal. If you can't trust the other party, you have to trust the dispute resolution mechanisms and/or legal system behind it. Multisig technology allows these deliberately disinterested third parties to bring security and trust to anonymous transactions.

Multisig authentication is growing in popularity. A start-up called Hedgy is using multisig technology to create futures contracts: parties agree on a price of bitcoin that will be traded in the future, only ever exchanging the price difference. Hedgy never holds collateral. The parties place it in a multisig wallet until the execution date. Hedgy's goal is to use multisig as a foundation for smart contracts that are completely self-executable and fully evidenced on the blockchain.[24] Think of the blockchain as a dialectic between anonymity and openness, where the multisig feature reconciles the two without loss of either.

Among other things, the smart contract changes the role of those within firms who are in the business of finding and contracting for talent. HR departments need to understand that talent is outside their boundaries, not just inside. They need to step up to the challenges of using smart contracts to lower the costs of building relationships with external resources.

3. Coordination Costs—How Should We All Work Together?

So we've found the right people and you've contracted with them. How do you manage them? Throughout his writings, Coase discussed costs of coordinating, meshing, or otherwise orchestrating the different people, products, and processes into an enterprise that can effectively create value. Against traditional economists who argued that there were internal markets within firms, Coase said that when "a workman moves from department Y to department X, he does not go because of a change in relative prices, but because he is ordered to do so."[25] In other words, markets allocate resources via the price mechanism, but firms allocate resources via authoritative direction.

Williamson went on to explain that there are two significant coordinating systems. First is the price system for decentralized resource allocation needs and opportunities (the market). But second, (traditional) "firms employ a different organizing principle—that of hierarchy—whereupon authority is used to affect resource allocation." Over the last few decades, hierarchies have come under scrutiny as structures for killing creativity, undermining initiative, disempowering human capital, and scapegoating responsibility through opacity. To be sure, many management hierarchies have become unproductive bureaucracies. However, hierarchy as a concept has gotten a bad rap, as has its most eloquent defender, Canadian-born psychologist Elliot Jaques. In a classic 1990 *Harvard Business Review* article, Jaques argued, "35 years of research have convinced me that managerial hierarchy is the most efficient, the hardiest, and in fact the most natural structure ever devised for large organizations. Properly structured, hierarchy can release energy and creativity, rationalize productivity, and actually improve morale."[26]

The trouble is that, in recent business history, many hierarchies have not been effective, to the point of ridicule. Exhibit A is *The Dilbert Principle*, most likely one of the best-selling management books of all time, by Scott Adams. Here's Dilbert on blockchain technology from a recent cartoon:

> Manager: I think we should build a blockchain.
> Dilbert: Uh-oh. Does he understand what he said or is it something he saw in a trade magazine ad?
> Dilbert: What color do you want your blockchain?
> Manager: I think mauve has the most RAM.

In the cartoon, Adams captures one of the marks of hierarchies gone wrong—that managers often rise to a level of power where they lack the knowledge required for effective leadership.

Combined with progressive management thinking about how to build effective, innovative organizations, the first generation of the Internet enabled progressive thinking managers to change the top-down assignment of work and appropriation of credit, recognition, and promotion.

For better or for worse, centralized hierarchies are the norm. Decentralization, networking, and empowerment have been sensible since the early days of the Internet. Teams and projects have become the foundation of internal organization. E-mail enabled people to collaborate across organizational silos. Social media dropped some collaboration costs internally and dropped transaction costs and made the boundaries of corporations more porous as companies could link up with suppliers, customers, and partners more easily.

However, today's commercial social media tools are helping many firms achieve new levels of internal collaboration. Empowerment, the real decentralization of power, is an important focus in business; and companies have experimented or implemented new concepts ranging from matrix management to holacracy—with varying degrees of success.

In fact, there is widespread agreement that when firms distribute responsibility, authority, and power, the result will typically be positive: better business function, customer service, and innovation. But this practice is easier said than done.

The Internet also hasn't dropped what economists call "agency costs"—the cost of making sure that everybody inside the firm is acting in the owner's interest. In fact, another Nobel Prize–winning economist (yes, there do seem to be a lot of them in this story), Joseph Stiglitz, argued that the sheer size and seeming complexity of these firms have increased agency costs even as a firm's transaction costs have plummeted. Hence, the huge pay gap between CEO and front line.

So where does blockchain technology come in and how can it change how firms are managed and coordinated internally? With smart contracts and unprecedented transparency, the blockchain should not only reduce transaction costs inside and outside of the firm, but it should also dramatically reduce agency costs at all levels of management. These changes will in turn make it harder to game the system. So firms could go beyond transaction cost to tackle the elephant in the boardroom—agency cost. Yochai Benkler told us, "What's exciting to me about blockchain technology is that it can enable people to function together with the persistence and stability of an organization, but without the hierarchy."[27]

It also suggests that managers should brace themselves for radical transparency in how they do coordinate and conduct themselves because shareholders will now be able to see the inefficiencies, the unnecessary complexity, and the huge gap between executive pay and the value executives actually contribute. Remember, managers aren't agents of owners; they're intermediaries.

4. Costs of (Re-)Building Trust—Why Should We Trust One Another?

As we have explained, trust in business and society is the expectation that another party will be honest, considerate, accountable, and transparent—that he or she will act with integrity.[28] It's a lot of work to establish trust, and many economists and other academics argue that we have vertically integrated firms because establishing trust is easier within corporate boundaries than in an open market. With trust at an all-time low, the challenge for firms is not simply figuring out whom to trust, but how to get outside capability to trust them.

Indeed, economist Michael Jensen and colleagues made the case that *integrity* is a factor of production. Not the first but among the most eloquent on the topic, they explain that the seemingly never-ending scandals in the world of finance with their damaging effects on value and human welfare argue strongly for the addition of integrity to financial operations. To them this is not an issue of virtue, but an opportunity in financial economics to "create significant increases in economic efficiency, productivity, and aggregate human welfare." To them, "Integrity . . . on the part of individuals or organizations has enormous economic implications (for value, productivity, quality of life, etc.). Indeed, integrity is a factor of production as important as labor, capital, and technology."[29]

Wall Street lost trust (and nearly killed capitalism) because of a set of integrity violations. But has it changed? And will it change? In the past, corporate social responsibility advocates argued that companies "do well by doing good." We haven't seen the evidence. Many companies *did well by doing bad*—by having bad labor practices in the developing world, by externalizing their costs onto society such as pollution, by being monopolies and gouging customers. The collapse of 2008 taught us for sure that companies "do badly by being bad." The major banks found this out the hard way. Prior to 2008 many were making upwards of 20 percent return on equity. For many today it is well below 5 percent, with some not even making their cost of capital. From a shareholder perspective, they should no longer exist.[30]

What are the chances, realistically, that Wall Street will wake up to Jensen's exhortations and act with integrity? Surely, expedience and short-term gain are coded into the DNA of the Western financial system.

Enter blockchain technology and digital currencies. What if parties didn't have to trust one another, but could still act with honesty, accountability, consideration, and transparency because it was the foundation of the technological platform of finance?

Steve Omohundro gave us a compelling example. "If somebody from Nigeria wants to buy something that I'm selling, I'm going to be very skeptical, I'm not going to accept a credit card or a check from Nigeria. With the new platform, I know I can trust it and I don't have to incur the costs of establishing trust. So it enables transactions which simply couldn't happen otherwise."[31]

So Wall Street banks don't have to splice integrity into their DNA and behavior; the founders of blockchains have coded it into their software protocols and deployed it across the network—enabling a new utility for the financial services industry. The good news is that the industry can reestablish trust and maintain it in an ongoing way.

With blockchain technology causing the costs of searching, contracting, coordinating, and creating trust to plummet, it should be easier for firms not just to open up, but also to forge trusting relationships with external parties. Acting in one's self-interest serves everybody's interests. Cheating the system costs more than using it as designed.

This is not to say that corporate brands or for that matter acting ethically is unimportant or no longer required. Blockchain helps ensure integrity and therefore trust in transactions between peers. It also helps achieve transparency—a critical factor in trust. However, as author and technology theorist David Ticoll says: "Trust and brand are about more than vouchsafing a transaction. They are also about quality, enjoyment, safety of a device or service, cachet and coolness. In today's COP21 world, the best brands transparently and verifiably signify outcomes that are environmentally, socially, and economically responsible."[32]

Still, through smart contracts, executives can be held accountable—they must abide by their commitments as enforced and settled by software. Companies can program relationships with radical transparency so everyone has a better understanding about what each party has signed up to do. And overall, like it or not, they must conduct business in a way that is considerate of the interests of other parties. The platform demands it.

DETERMINING CORPORATE BOUNDARIES

Overall, the boundaries that separate a company from its vendors, consultants, customers, external peer communities, and others will become harder to define. Perhaps as important, they will constantly change.

Firms will still exist, blockchain notwithstanding, because the mechanisms for searching, contracting, coordinating, and establishing trust within corporate boundaries will be more cost-effective than those in the open market, at least for many activities. The idea of the so-called free agent nation, where individuals execute work outside the boundaries of corporations, is illusory. Melanie Swan, who founded the Institute for Blockchain Studies, said, "What's the right size of the corporation for optimal transactibility? Well, it's not a unitary thing, of people working only as individuals or e-lancers." To her, there will be new kinds of "flexible business entities of individuals and groups partnering around projects." She views the new model of the firm more like the guild, the preindustrial associations of merchants or tradesmen who worked together in a particular town. "We still need organizations acting as coordinating mechanisms. But the new models of team collaboration are not yet fully clear."[33]

Today we often hear that firms should focus on their core. But when considering how blockchain technology drops transaction costs, what is core? And how do you define that when a company's core is constantly changing?

It seems that everyone has a different definition of what the optimal firm size should be to maximize productivity and competitive advantage. Many firms we examined didn't have a clear view, seeming to choose the Bob Dylan approach to determining what's in and what should be out ("You don't need a weatherman to know which way the wind blows"). Back-office processing, for example, was described as a no-brainer, without any clear criteria as to why.

Some are more rigorous. From the core competencies view developed by Gary Hamel and C. K. Prahalad, firms gain competitive advantage through competence mastery. Those competencies mastered are central to the firm, while others can be acquired from outside.[34] However, a firm may have mastery over some activities that are not mission critical. Should they still be kept inside?

Strategist Michael Porter has an implicit view that competitive advantage stems from activities, in particular from networks of reinforcing activities that are hard to replicate in their totality. It's not the individual parts of the business that matter, but how they are strung together and built to reinforce one another in a unique activity system. Competitive advantage comes from the entire system of activities; while any individual activity within the system may be copied, competitors cannot produce the same benefit unless they manage to duplicate the entire system.[35]

Others argue that companies should always retain functions or capabilities that are mission critical—those that firms must absolutely get right for survival and success. But making computers is mission critical for computer companies; yet Dell, HP, and IBM outsource much of this activity to electronics manufacturing services companies like Celestica, Flextronics, or Jabil. Final assembly of vehicles is mission critical for an auto manufacturer; yet BMW and Mercedes contract with Magna to do this activity.

Stanford Graduate School of Business professor Susan Athey argues persuasively: "There may be some mission-critical functions, like the collection and analysis of big data, that are just too risky to move outside corporate boundaries, even if you don't have unique abilities in that area."[36] True, there may be some functions like data analytics where survival depends on being uniquely good, and there may be existential risks of partnering. Still, external resources can be deployed strategically to build internal capability.

Our view is that the starting point for corporate boundary decisions is to understand your industry, competitors, and opportunities for profitable growth—and use this knowledge as the basis for developing a business strategy. From there, the blockchain opens up new opportunities for networking that every manager and knowledge worker needs to consider at all times. Boundary choices are not simply for senior executives, they are for anyone who cares about marshaling the best capability for innovation and high performance. We should add—and this is no small point—that you can't outsource your corporate culture.

Enter the Matrix

Taking into account how blockchain technology can enable access to unique capabilities outside corporate boundaries, firms can now define those business activities or functions that are fundamental to competitiveness—that are both mission critical and also unique enough to ensure differentiated value.

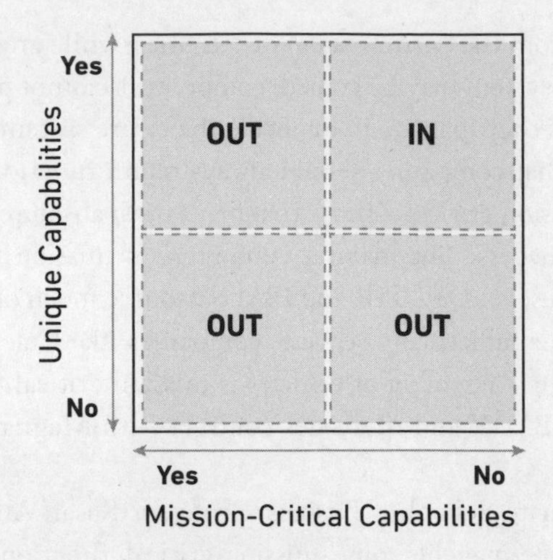

However, this In-Out Matrix is just a starting point for defining corporate boundaries at any given point. What other factors should firms consider in determining what is fundamental? What extenuating circumstances are there that might affect choices to outsource or nurture internally?

Hacking Your Future: Boundary Decisions

When making boundary choices, firms should start using the blockchain to marshal a 360-degree view and reach consensus on what is unique and what is mission critical in their business. Let's return to Joe Lubin and ConsenSys, as they foreshadow the modus operandi of the blockchain-based enterprise. Remember that ConsenSys is in its infancy, and much can go wrong to undermine its business. We can still learn from this company's example.

1. Are there possible partners who could do the work better? In particular, could we benefit from harnessing new peer production communities, ideagoras, open platforms, and other blockchain business models? The company ConsenSys is able to orchestrate extraordinary expertise to do its work, even though many are outside its boundaries.

2. Given blockchain technology, what are the new economics of corporate boundaries—the transaction costs of partnering, versus keeping/developing in-house? Can you develop a suite of smart contracts whose core elements are modular and reusable? ConsenSys uses smart contracts to reduce coordination costs.

3. What is the extent of technological interdependence versus modularity? If you can define business components that are modular, then you can easily reconfigure them outside corporate boundaries. ConsenSys sets standards for software development and provides access to various software modules that its partners can build upon.

4. What are your firm's competencies with regard to the managing of outsourced work? Can smart contracts enhance those competencies and lower costs? From the get-go, ConsenSys was a blockchain business. CEO Joe Lubin embraces the technology and a modified holacracy, and we can see the seven design principles at work.

5. What are the risks of opportunism where a partner might encroach on fundamental parts of your business, as some have suggested Foxconn may do to smart phone companies? ConsenSys tries to mitigate this challenge by building loyalty through incentive structures whereby its talent shares in the wealth they create.

6. Are there legal, regulatory, or political obstacles to deeper networking (and shrinking) of the organization? Not a problem for ConsenSys yet.

7. Speed and pace of innovation are important to boundary decisions. Sometimes firms have no choice but to partner for a strategic function because they cannot develop it in-house fast enough. A partner arrangement can be a placeholder. Will partnering help us build an ecosystem that will improve our competitive advantage? This is ConsenSys's strategy: build a network of collaborators around the Ethereum platform, grow the platform and ecosystem, and increase the probability of success for all components.

8. Is there a danger of losing control of something fundamental—for example, a product or network architecture? Firms must have a sense of which parts of the value chain will be key to creating and capturing value in the future. If these are farmed out, the firm will lose. The Ethereum platform provides a basic architecture for ConsenSys.

9. Is there a capability, like the exploitation of data assets, that must be part of the fabric of your enterprise and all its operations? Even though you lack a unique capability, you should view partnering as a transitional tactic to develop extraordinary internal expertise and capacity. Blockchain technologies will introduce a new set of capacities that need to reside in the cranium of every employee. You can't move culture outside your boundaries.

CHAPTER 5

NEW BUSINESS MODELS: MAKING IT RAIN ON THE BLOCKCHAIN

Founded a month before the market crashed in 2008, Airbnb has become a $25 billion platform, now the world's largest supplier of rooms as measured by market value and rooms occupied. But the providers of rooms receive only part of the value they create. International payments go through Western Union, which takes $10 of every transaction and big foreign exchange off the top. Settlements take a long time. Airbnb stores and monetizes all the data. Both renters and customers alike have concerns about privacy.

We brainstormed with blockchain expert Dino Mark Angaritis to design an Airbnb competitor on the blockchain. We decided to call our new business bAirbnb. It would look more like a member-owned cooperative. All revenues, except for overhead, would go to its members, who would control the platform and make decisions.

BAIRBNB VERSUS AIRBNB

bAirbnb is a *distributed application* (DApp), a set of smart contracts that stores data on a home-listings blockchain. The bAirbnb app has an elegant interface: owners can upload information and pictures of their property.[1] The platform maintains reputation scores of both providers and renters to improve everyone's business decisions.

When you want to rent, the bAirbnb software scans and filters the blockchain for all the listings that meet your criteria (e.g., ten miles from the Eiffel Tower, two bedrooms, four-plus star ratings only). Your user experience is identical to that in Airbnb, except that you communicate peer to peer on the network, through encrypted and cryptographically signed messages not stored in Airbnb's database.[2] You and the room owner are the only two people who can read these messages. You can swap phone numbers, an exchange that Airbnb blocks to preserve future revenues. On bAirbnb you and the owner could communicate off-chain and complete the transaction entirely off-chain, but you are better off completing the transaction on-chain for a few reasons.

Reputation: Because the network records the transaction on the blockchain, a positive review from each user improves your respective reputations. The risk of a negative review motivates each party to remain honest. Remember, people with good reputations can use the same persona across multiple DApps and benefit from continuity as a good person.

Identity Verification: Because we are not dealing with a centralized system that checks ID on our behalf, each party needs to confirm the other party's identity. The blockchain calls up a contract from a "VerifyID" application, one of many contracts that bAirbnb, SUber (blockchain Uber), and other DApps use to verify real-world identity.

Privacy Protection: VerifyID doesn't track and store all transactions in a database. It simply returns a TRUE or FALSE when it receives a request for verification of a public key (persona). Different kinds of DApps can call VerifyID, but VerifyID never knows details of transactions. This separation of identity from activity greatly improves your privacy.

Risk Reduction: Home owners currently store customer identities and financial data on their own servers, which can be hacked and leaked, exposing owners to litigation and large liabilities. On the blockchain, you needn't trust a vendor with your data; there is no central database to hack and leak. There are only individual peer-to-peer pseudonymous transactions.

Insurance: Today Airbnb offers $1 million insurance for owners and compensates them for theft and damage. On bAirbnb, owners can get the bAirbnb insurance DApp. Renters with good reputations like you have lower insurance rates and needn't subsidize renters who lack caution, scrutiny of prospects, or poor treatment of property. When you submit a booking request, bAirbnb sends your public key (persona) to the insurance contract for a quote. The insurance DApp contacts a list of trusted providers; fake insurers need not apply. Insurers perform their own calculations in real time through autonomous agent software based on the inputs to the contract—such as the market value of the owner's house, how much the owner wants insured, owner reputation, your reputation as a renter, and rental price. bAirbnb takes the best bid and adds it to the nightly fee the owner wants to charge. The blockchain processes this calculation in the background; owners and renters have a comparable user experience to that of Airbnb but a superior and more equitable value exchange.

Payment Settlement: Of course, on the blockchain, you transfer funds to the owner in seconds, not days as with Airbnb. Owners can manage security deposits more easily with smart contracts. Some parties use escrow accounts to release payments partially (nightly, weekly, hourly, etc.) or in full as the parties agree. In disputes involving smart contracts, parties can call for arbitration.

Property Access Using Smart Locks (IoT device): A smart lock connected to the blockchain knows when you have paid. When you arrive, your near-field communication-enabled smart phone can sign a message with your public key as proof of payment, and the smart lock will open for you. Owners need not drop keys off to you or visit the property unless they want to say hello or address some emergency.

You and the owner have now saved most of the 15 percent Airbnb fee. Settlements are assured and instant. There are no foreign exchange fees for international contracts. You need not worry about stolen identity. Local governments in oppressive regimes cannot subpoena bAirbnb for all its rental history data. This is the real sharing-of-value economy; both customers and service providers are the winners.

GLOBAL COMPUTING: THE RISE OF DISTRIBUTED APPLICATIONS

Before we examine the other possible distributed business entities like bAirbnb, a word on how the underlying technology enables decentralization. Until the blockchain, centralized organizations have held concentrated computing power.

In the first decades of enterprise computing, all software applications (apps) ran on the computers of their owners. GM, Citibank, U.S. Steel, Unilever, and the U.S. federal government owned huge data centers that ran proprietary software. Companies rented or "time shared" computer power from providers like the 1980s giant CompuServe to run their own applications.

As the personal computer matured, the software market specialized: some developed client apps (the PC) and some, server apps (a host computer). With widespread adoption of the Internet, specifically the World Wide Web, individuals and companies could use their computers to share information—initially as text documents and later as images, videos, other multimedia content, and eventually software apps.[3] Sharing began to democratize the information landscape. But it was short-lived.

In the 1990s, a new variant of time-sharing appeared, initially called *virtual private networks* (VPNs) and then *cloud computing*. Cloud computing enabled users and companies to store and process their software and data in third-party data centers. New technology companies like Salesforce .com built fortunes by harnessing the cloud model to save customers the big costs of developing and running their own software. Cloud service providers like Amazon and IBM built ginormous multibillion-dollar businesses. During the 2000s, social media companies like Facebook and Google created services that ran on their own vast data centers. And to continue this trend of centralized computing, companies like Apple moved away from the Web's democratizing architecture to proprietary platforms like the Apple Store where customers acquired proprietary apps, not on the open Web but in exclusive walled gardens.

Again and again in the digital age, large companies have consolidated—created, processed, and owned or acquired—applications on their own large systems. Centralized companies have begotten centralized computing architectures that have, in turn, centralized technological and economic power.

Some red flags: With single points of control, companies themselves are vulnerable to catastrophic crashes, fraud, and security breaches. If you were a customer of Target, eBay, JPMorgan Chase, Home Depot, or Anthem, or for that matter Ashley Madison, the U.S. Office of Personnel Management (second breach!), and even Uber, you felt the pain of hacking in 2015.[4] Systems of different parts of a company still have big challenges communicating with one another, let alone with systems outside the firm. For us users, it means that we've never really had control. Others define our services with their implicit values and goals that may conflict with ours. As we generate reams of valuable data, others own it and are building vast fortunes—perhaps the greatest in history—while most of us receive little benefit or compensation. Worst of all, central powers are using our data to create mirror images of each of us and may use these to sell us stuff or to spy on us.

Along comes blockchain technology. Anyone can upload a program onto this platform and leave it to self-execute with a strong cryptoeconomical[5] guarantee that the program will continue to perform securely as it was intended. This platform is public, not inside an organization, and it contains a growing set of resources such as digital money to incent and reward certain behavior.

We're moving into a new era in the digital revolution where we can program and share software that's distributed. Just as the blockchain protocol is distributed, a distributed application or DApp runs across many computing devices rather than on a single server. This is because all the computing resources that are running a blockchain constitute a computer. Blockchain developer Gavin Wood makes this point describing the Ethereum blockchain as a platform for processing. "There is only one Ethereum computer in the world," he said. "It's also multiuser—anyone who ever uses it is automatically signed in." Because Ethereum is distributed and built to the highest standards of cryptosecurity, "all code, processing, and storage exists within its own encapsulated space and no one can ever mess with that data." He argued that critical rules are built into the computer, comparing it to "virtual silicon."[6]

As for DApps, there have been warm-up acts prior to blockchains. BitTorrent, the peer-to-peer file-sharing app, demonstrates the power of DApps as it currently consumes over 5 percent of all Internet traffic.[7] Lovers of music, film, and other media share their files for free, with no central server for authorities to shut down. Iconoclastic programmer Bram Cohen, who incidentally is less than enthusiastic about bitcoin because of all the commercial activity around it, developed BitTorrent. "The revolution will not be monetized," he said.[8]

Most of us think that generating revenue and economic value through technological innovation is positive, as long as the revolution is not monetized by the few. With blockchain technology the possibilities for DApps are almost unlimited, because it takes DApps to a new level. If, as the song says, "Love and marriage, love and marriage, go together like a horse and carriage," then so do DApps and blockchains. The company Storj is a distributed cloud storage platform and a suite of DApps that allow users to store data securely, inexpensively, and privately. No centralized authority has access to a user's encrypted password. The service eliminates the high costs of centralized storage facilities; it's superfast; and it pays users for renting their extra disk space. It's like Airbnb for your computer's spare memory space.

THE DAPP KINGS: DISTRIBUTED BUSINESS ENTITIES

How do DApps infuse greater efficiency, innovation, and responsiveness into the structure of the firm? What new business models can we make with DApps to generate value? And if powerful institutions are capturing the benefits of the Internet today, how can we move beyond "outsourcing" and "business webs" to truly distributed models of innovation and value creation that can distribute prosperity and the ownership of data and wealth? We mapped what we believe to be the four most important innovations onto a two-by-two matrix.

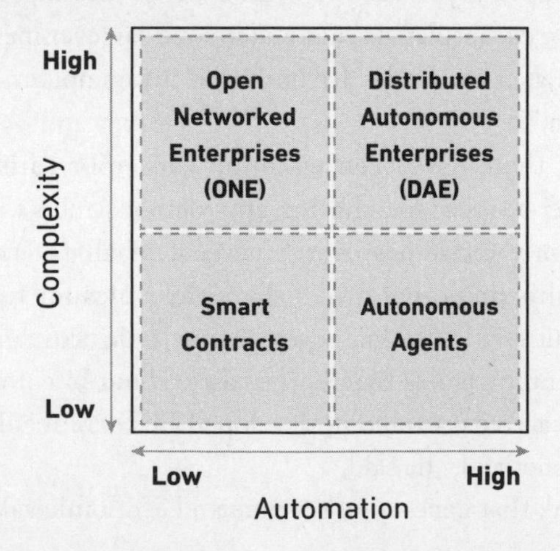

The X-axis identifies the degree to which humans participate in the model. At the left, the model requires some human involvement. At the right, the model requires no people.

The Y-axis describes the functional complexity of the model, not its technical complexity. At the lower end are models that perform a single function. At the top are models that perform diverse functions.

These are all components of the blockchain economy because they use blockchain technology and often cryptocurrencies as their foundation. Smart contracts (discussed in the last chapter) are the most basic form: they involve some complexity that requires human involvement, increasingly in the form of multisignature agreements. As smart contracts grow in complexity and interoperate with other contracts, they can contribute to what we call *open networked enterprises* (ONEs). If we combine ONEs with *autonomous agents*—software that makes decisions and acts on them without human intervention—we get what we're calling a *distributed autonomous enterprise* that requires little or no traditional management or hierarchy to generate customer value and owner wealth. And we think that very large numbers of people, thousands or millions, might be able to collaborate in creating a venture and sharing in the wealth it creates—distributing, rather than redistributing, wealth.

Open Networked Enterprises

At very low cost, smart contracts enable companies to craft clever, self-enforcing agreements with previously improbable classes of new suppliers and partners. When aggregated, smart contracts can make firms resemble networks, rendering corporate boundaries more porous and fluid.

Blockchain technology also drops Coase's search costs and coordination costs so that companies can disaggregate into more effective networks. An auto company could check a supplier's trustworthiness by just scanning the analytic services online. Soon, just type "axle" or "window glass" into any number of industry exchanges on the blockchain and negotiate the price online.

We can extend that simple scenario to finding a replacement part, a supply chain partner, a collaborator, or a piece of software for managing a distributed resource. Need steel from China, rubber from Malaysia, or glass from Wichita, Kansas? No problem. Decentralized online clearinghouses operating as DApps for each commodity will enable purchasers to contract for price, quality, and delivery dates with a few clicks of a mouse. You'll have a detailed searchable record of previous transactions—not just how various companies were rated but precisely how they honored their commitments. You can track each shipment on a virtual map that shows its precise location in the journey. You can microschedule goods to show up just in time. No warehouse required.

AUTONOMOUS AGENTS

Imagine a piece of software that could roam the Internet with its own wallet and its own capacity to learn and adapt, in pursuit of its goals determined by a creator, purchasing the resources it requires to survive like computer power, all while selling services to other entities.

The term *autonomous agent* has many definitions.[9] For our discussion, it is a device or software system that on behalf of some creator takes information from its environment and is capable of making independent choices. We could describe some autonomous agents as "intelligent" although they lack general intelligence. However, they are not "just computer programs" because they can modify how they achieve their objectives. They can sense and respond to their environment over time.[10]

The computer virus is the most cited example of an autonomous agent; the virus survives by replicating itself from machine to machine without deliberate human action. Unleashing a virus on the blockchain could be more difficult and certainly costly because it would likely have to pay the other party to interact with it, and the network would quickly identify its public key, crash its reputation score, or not validate its transactions.

For positive blockchain examples, consider the following. A cloud computing service rents processing power from various sources, growing to Amazon's size by making rental deals with other computers that have excess capacity.[11] A driverless car owned by a community, company, individual, or perhaps itself moves around the city picking up and dropping off passengers and charging them appropriate fees. We're interested in agents that can do transactions, acquire resources, make payments, or otherwise produce value on behalf of their creator.

Vitalik Buterin, who created the Ethereum blockchain, has theorized about these agents and developed a taxonomy to describe their evolution. At one end are single-function agents like viruses that go about working to achieve their limited goals. Next up are more intelligent and versatile agents, say, a service that would rent servers from a specific set of providers like Amazon. A more sophisticated agent might be able to figure out how to rent a server from any provider and then use any search engine to locate new Web sites. An even more capable agent could upgrade its own software and adapt to new models of server rental such as offering to pay end users for rental of their unused computers or disks. The penultimate step consists of being able to discover and enter new industries, leading into the next evolution of the species—full artificial intelligence.[12]

Weathernet

Could an autonomous agent use blockchain technology to make money forecasting the weather? Flash-forward to 2020. The best weather forecasts globally are coming from a network of smart devices that are measuring and predicting the weather all around the world. That year, an autonomous agent named BOB is released onto this network to collaborate with these devices to create a business. Here's how BOB works.

Distributed environmental sensors (weatherNodes) on utility poles, in people's clothes, on roofs of buildings, traveling in cars, and linked to satellites are all connected in a global mesh network. No need for an Internet service provider for connectivity. Rather than communicating with a central database, they store their data on a blockchain.[13] Many are solar powered and so they don't need the electrical grid; they can effectively operate indefinitely.

The blockchain handles a few functions. First, it settles payments. As an incentive, each weatherNode receives a micropayment every thirty seconds for providing accurate weather telemetry (temperature, humidity, wind, etc.) at a particular location in the world.

The blockchain also stores all weatherNode transactions. Each weatherNode signs all of its data with its public key stored on the blockchain. A public key identifies the weatherNode and allows other entities to determine its reputation. When the node produces accurate weather data, its reputation is enhanced. If a node is broken or compromised and produces inaccurate data, it loses status. Nodes with low reputation receive less bitcoin than nodes with high reputation—the beneficiary being the creator of the app—whether an individual, company, or cooperative.

The blockchain also allows data providers and data consumers to participate peer to peer on a single, open system, rather than subscribe to dozens of centralized weather services around the world, and program their software to communicate with each of their application programming interfaces (APIs). With smart contracts we can have a global "WeatherDataMarketplace DApp" where consumers bid for data in real time and receive the data in a universally agreed-upon format. Centralized data providers can ditch their proprietary systems and individualized sales efforts, and instead become data providers for the globally accessible WeatherDataMarketplace DApp.

WeatherDApp: Sensors LP

In the first era of the Internet, technical innovation occurred only in the center; centralized utilities like energy companies, cable corporations, and central banks decided when to upgrade the network, when to support new features, and whom to give access. Innovation couldn't occur at the "edges" (i.e., individuals using the network) because the rules and protocols of closed systems meant that any new technologies designed to interact with the network would need the central power's permission to operate on it.

But central powers are inefficient because they don't know exactly what the market wants in real time. They have to make educated guesses that are always less accurate than what real-time markets demand. We end up with WeatherCorp, a centralized service that installs sensors and puts up satellites so that it can sell subscriptions to data that few people may want.

The blockchain allows any entity to become a weather provider or weather data consumer, with very low barriers to entry. Just buy a weatherNode, put it on your roof, and connect it to the GlobalWeatherDataMarketplace DApp LP (for linked peers) and you'll start earning income right away. And if you can rig your own rooftop weatherNode that happens to provide more accurate data, well, good for you! You innovated on the edge, and the market rewards you for it. The incentives for innovation on open networks are aligned to increase efficiency better than closed networks.

Dueling Bots

What about conflicts of interest? If the weatherNode started expanding its capability and entered the crop insurance marketplace, wouldn't it have cognitive dissonance? Farmer weatherNodes want to emphasize the impact of droughts, and insurer weatherNodes claim droughts are minimal. The owners and designers of agents need transparency of operations. If both are filtering sensor data through a biased screen, then their respective reputations will drop.

Vitalik Buterin points out that autonomous agents are challenging to create, because to survive and succeed they need to be able to navigate in a complicated, rapidly changing, or even hostile environment. "If a Web hosting provider wants to be unscrupulous, they might specifically locate all instances of the service, and then replace them with nodes that cheat in some fashion; an autonomous agent must be able to detect such cheating and remove or at least neutralize cheating nodes from the system."[14]

Note that autonomous agents also separate personhood from asset ownership and control. Before blockchain technology, all assets—land, intellectual property, money—required a person or legal organization of people to own it. According to Andreas Antonopoulos, cryptocurrencies completely ignore personhood. "A wallet could be controlled by a piece of software that has no ownership and so you have the possibility of completely autonomous software agents that control their own money."[15]

An autonomous agent could pay for its own Web hosting and use evolutionary algorithms to spread copies of itself by making small changes and then allowing those copies to survive. Each copy could contain new content that it discovers or even crowdsources somewhere on the Internet. As some of these copies become very successful, the agent could sell ads back to users, ad revenue could go into a bank account or posted on a secure place on the blockchain, and the agent could use this growing revenue to crowdsource more ad content and proliferate itself. The agent would repeat the cycle so that appealing content propagates and hosts itself successfully, and unsuccessful content basically dies because it runs out of money to host itself.

DISTRIBUTED AUTONOMOUS ENTERPRISES

We now suggest you buckle up in your *Star Trek* captain's seat for a moment. Imagine BOB 9000—a set of autonomous agents that cooperate in a complex blockchain-based ecosystem according to a mission statement and rules. Together they create a suite of services that they sell to humans or organizations. Humans animate the agents, endowing them with computing power and capital to go about their work. They buy the services they need, hire people or robots, acquire partner resources such as manufacturing capability and branding and marketing expertise, and adapt in real time.

This organization could have shareholders, possibly millions of them who participated in a crowdfunding campaign. The shareholders provide a mission statement, say, to maximize profit lawfully, while treating all stakeholders with integrity. Shareholders could also vote as required to govern the entity. As opposed to traditional organizations, where humans make all decisions, in the ultimate distributed organization much of the day-to-day decision making can be programmed into clever code. In theory, at least, these entities can run with minimal or no traditional management structure, as everything and everyone works according to specific rules and procedures coded in smart contracts. There would be no overcompensated CEO, management, or corporate bureaucracy, unless the entity decided to hire and build one. There would be no office politics, no red tape, no Peter Principle of the Dilbertian enterprise at work, because technology providers, open source communities, or enterprise founders would set the software's agenda to execute specific functions.

Any human employees or partner organizations would perform under smart contracts. When they do the job as specified, they are instantly paid—perhaps not biweekly but daily, hourly, or in microseconds. As the entity wouldn't necessarily have an anthropomorphic body, employees might not even know that algorithms are managing them. But they would know the rules and norms for good behavior. Given that the smart contract could encode the collective knowledge of management science and that their assignments and performance metrics would be transparent, people could love to work.

Customers would provide feedback that the enterprise would apply dispassionately and instantly to correct course. Shareholders would receive dividends, perhaps frequently, as real-time accounting would obviate the need for year-end reports. The organization would perform all these activities under the guidance and incorruptible business rules that are as transparent as the open source software that its founders used to set it in motion.

Welcome to tomorrow's distributed autonomous enterprise (DAE), powered by blockchain technology and cryptocurrencies, where autonomous agents can self-aggregate into radically new models of the enterprise.

Before you say that this all sounds impractical, pointless, or something from sci-fi, consider the following. Using tokens, companies such as ConsenSys have already issued shares in their firms, staging public offerings without regulatory oversight. You could legally record the ownership of privately held corporations and transfer those shares to other persons on the blockchain. Your share certificates can pay dividends and confer voting power. That said, your new "blockcom" is distributed; it doesn't exist without jurisdiction, but your shareholders can live anywhere. Imagine a similar mechanism to issue debt in the form of bonds, either private corporate bonds or sovereign bonds, essentially creating a bond market. The same logic applies to commodities—not the commodity itself but a note that corresponds to the commodity, similar to how the Chicago Mercantile Exchange or the global gold market work.

But don't think of securities as you currently know them. Imagine a global IPO with 100 million shareholders each contributing a few pennies. It's not unthinkable that management and governance could occur on a massive scale, with millions of people having voting shares. At last, investors at the bottom of the pyramid could participate and own shares of a wealth-creating venture anywhere in the world. In theory at least, we could design a corporation without executives, only shareholders, money, and software. Code and algorithms could replace a layer of representatives (i.e., the executive board), with shareholders exerting control over that code. The opportunity for prosperity is significant, nothing less than the democratization of ownership of wealth-creating instruments.

Not practical? Perhaps. But consider that entrepreneurs are already using scripting languages such as Ethereum to design such functions for eventually autonomous models. Already innovators are implementing code that permits multisignature control over funds. Through crowdfunding campaigns, masses of people are purchasing equity in companies. DApps are already giving way to autonomous agents.

This completely distributed enterprise could have a wallet that requires thousands of signatories to achieve consensus in order to spend money on an important transaction. Any shareholder could suggest a recipient for that money, marshaling consensus around that transaction. A structure like this would have obvious challenges. For example, mechanisms would need to be in place to quickly achieve consensus. Or who is responsible for the outcome of that transaction? If you've contributed one ten-thousandth of a vote, what is your legal responsibility and liability? Could there be self-propagating criminal or terrorist organizations? Andreas Antonopoulos is not concerned. He believes that the network will manage such dangers. "Make this technology available to seven and a half billion people, 7.499 billion of those will use it for good and that good can deliver enormous benefit to society."[16]

THE BIG SEVEN: OPEN NETWORKED ENTERPRISE BUSINESS MODELS

There are countless opportunities to construct open networked enterprises that disrupt or displace traditional centralized models, potentially evolving into nascent distributed autonomous enterprises. Consider how the distributed model will disrupt or replace the eight functions of financial services—everything from retail banking and stock markets to insurance companies and accountancies. Incumbent and new entries alike can construct new business architectures that can innovate better, create better value at lower cost, and shift and enable producers to share in the wealth they create.

Blockchain technology takes some of the new business models described in *Wikinomics* to a new level.[17] Let's look at how we can expand peer production, ideagoras, prosumers, open platforms, the new power of the commons, the global plant floor, and the wiki (social) workplace by adding in native payment systems, reputation systems, uncensorable content, trustless transactions, smart contracts, and autonomous agents—the key innovations of the blockchain revolution.

1. The Peer Producers
Peer producers are the thousands of dispersed volunteers who brought you open source software and Wikipedia, innovative projects that outperform those of the largest and best-financed enterprises. Community members participate for the fun of it, as a hobby, to network, or because of their values. Now, by enabling reputation systems and other incentives, blockchain technology can improve their efficiency and reward them for the value they create.

Peer production communities can be "commons-based peer production," a phrase coined by Harvard Law professor Yochai Benkler.[18] Sometimes called *social production*, also Benkler's term, this system means that goods and services are produced outside the bounds of the private sector and are not "owned" by a corporation or individual. Among the countless examples are the Linux operating system (owned by no one but now the most important operating system in the world), Wikipedia (owned by the Wikimedia Foundation), and the Firefox Web browser (owned by the Mozilla Foundation). Peer production can also refer to activities in the private sector where peers collaborate socially to produce something but the good is not socially owned.

Peer production as a business model matters for two reasons. First, sometimes peers collaborate voluntarily to produce goods and services where a corporation acts as curator and achieves commercial benefit. Readers create the content on the Reddit discussion platform, but they don't own it. Reddit is the tenth-biggest site in the United States in terms of traffic. Second, companies can tap into vast pools of external labor. IBM embraced Linux and donated hundreds of millions of dollars' worth of software to the Linux community. In doing so, IBM saved $900 million a year developing its own proprietary systems and created a platform on which it built a multibillion-dollar software and services business.

Experience shows that long-term sustainability of volunteer communities can be challenging. In fact, some of the more successful communities have found ways to compensate members for their hard work. As Steve Wozniak said to Stewart Brand, "Information should be free, but your time should not."[19]

In the case of Linux, most of the participants get paid by companies like IBM or Google to ensure that Linux meets their strategic needs. Linux is still an example of social production. Benkler told us, "The fact that some developers are paid by third parties to participate does not change the governance model of Linux, or the fact that it is socially developed." This is more than so-called open innovation that involves cooperation between firms and sharing certain intellectual property, he said. "There is still substantial social motivation for many contributors and as such it's a hybrid model."[20]

Further, many of these communities are plagued with bad behavior, incompetence, saboteurs, and trolls—people who sow discord by posting inflammatory, incorrect, or off-topic messages to disrupt the community. Reputation in these communities is typically very informal, and there is no economic incentive for good behavior.

With blockchain technology peers can develop more formal reputations for effective contributions to the community. To discourage bad behavior, members could ante up a small amount of money that either increases or decreases based on contribution. In corporate-owned communities, peers could share in the value they create and receive payment for their contributions as smart contracts drop transaction costs and open up the walls of the firm.

Consider Reddit. The community has revolted over centralized control but still suffers from flippant, abrasive members. Reddit could benefit from moving to a more distributed model that rewards great contributors. ConsenSys is already working on a blockchain alternative to Reddit that does just that. By offering financial incentives, the ConsenSys team thinks it can improve the quality of Redditlike conversations, without centralized control and censorship. The Ethereum platform provides incentives, perhaps in real time, to produce high-quality content and behave civilly while contributing to collective understanding.

Reddit has a system in place, called Reddit "Gold"—a token that users can buy and then use to reward people whose contributions they value. The money from tokens goes to site maintenance. The gold has no intrinsic value to users. So with a real, transferable, blockchain-based coin incentive, Reddit members could actually begin to get paid for making the site more robust.

Wikipedia, the flagship of social production, could benefit as well. Right now all persons who edit articles develop an informal reputation based on how many pieces they have edited and how effective they are, as measured by highly subjective terms. The Wikipedia community debates constantly over incentive systems, but administering some kind of financial compensation to seventy thousand volunteers hasn't been feasible.

What if Wikipedia went on the blockchain—call it Blockapedia. In addition to the benefits of entries time-stamped into an immutable ledger, there could be more formal measures of one's reputation that could help incent good behavior and accurate contributions. Sponsors could fund, or all editors could contribute money to, an escrow account. Each editor could have a reputation linked to the value of her account. If she tried to corrupt an article, stating for example that the Holocaust never happened, the value of her deposit would decline, and in cases of defamation or invasion of privacy, she would lose it and even face civil or criminal action. The true events of the Second World War could be established in many ways, for example, by accessing unchangeable facts on the blockchain or through algorithms that show consensus regarding the truth.

The size of your Blockapedia security deposit could be proportional to your previous reputation on Wikipedia or similar platforms. If you're a brand-new user and have no reputation, you'll put up a larger security deposit to participate. If you've edited, say, two hundred articles on Wikipedia successfully, your deposit might be small.

This is not necessarily about moving Wikipedia to a for-hire compensation model. "It's simply a case of providing real-world economic gain or loss depending on the accuracy and veracity of the information you're providing,"[21] said Dino Mark Angaritis, CEO of the blockchain-based Smartwallet. Defacing Blockapedia hurts your formal reputation but you also lose money.

But Wikipedia works pretty well right now, right? Not quite. Andrew Lih, writing in *The New York Times*, pointed out that, in 2005, there were months when more than sixty editors were made administrator, a position with special privileges in editing the English-language edition. In 2015, the site has struggled to promote even one editor per month. Being a voluntary global organization, there are internal tensions. Worse, editing content on a mobile device is difficult. "The pool of potential Wikipedia editors could dry up as the number of mobile users keeps growing." Lih concludes that the demise of Wikipedia would be unfortunate. "No effort in history has gotten so much information at so little cost into the hands of so many—a feat made all the more remarkable by the absence of profit and owners. In an age of Internet giants, this most selfless of websites is worth saving."[22]

Overall, peer production communities are at the heart of new, networked models of value creation. In most industries, innovation increasingly depends on dense networks of public and private participants and large pools of talent and intellectual property that routinely combine to create end products. As IBM embraced Linux, firms can even tie into self-organizing networks of value creators like the open source movement to cocreate or peer-produce value.

2. The Rights Creators

During the first generation of the Internet, many creators of intellectual property did not receive proper compensation for it. Musicians, playwrights, journalists, photographers, artists, fashion designers, scientists, architects, and engineers all were beholden to record labels, publishers, galleries, film studios, universities, and large corporations that insisted these inventors assign their intellectual property rights to what essentially are large rights management operations in exchange for less and less of their IP's value.

Blockchain technology provides a new platform for creators of intellectual property to get value for it. Consider the digital registry of artwork, including the certificates of authenticity, condition, and ownership. A new start-up, Ascribe, enables artists themselves to upload digital art, watermark it as the definitive version, and transfer it so that, like bitcoin, it moves from one person's collection to another's. That's huge. The technology solves the intellectual property world's equivalent of the double-spend problem better than existing digital rights management systems, and artists could decide whether, when, and where they wanted to deploy it.

Meme artist Ronen V said, "Art is a currency. The evolution of art into digital currency is—no question—the future. And this is a good step."[23] Musicians, photographers, designers, illustrators, or other artists whose work could be digitized and watermarked as a definitive copy could use this technology to transform their intellectual property into a tradable asset, a limited edition perhaps customized for a particular fan. Artists and museums can use Ascribe's technology to loan pieces to other individuals or institutions.[24] Monegraph offers a similar service: it uses digital watermarks and the cryptography intrinsic to the blockchain for authenticating pieces. Artists simply upload the art to a page on the Internet and submit the URL to Monegraph. The firm issues a set of public and private keys, except that the value associated with the public key is a digital deed to the art rather than bitcoin per se. Monegraph also tweets a public announcement of the deed, noteworthy because the U.S. Library of Congress archives public Twitter feeds.[25] Someone else might try to claim the URL as his own, but there would already be at least two proofs in the public record to verify ownership.[26]

Verisart, a Los Angeles–based start-up with bitcoin core developer Peter Todd as an adviser, has even greater ambitions. Certifying the authenticity and the condition of a piece of fine art is big business, and one that is largely paper based and controlled by elite experts with access to restricted databases. Finding who owns the art, where it's stored, and in what condition is a real challenge, even for those who actually know what they're looking for. Verisart is combining blockchain technology and standard museum metadata to create a public database of art and collectibles. This worldwide ledger will serve artists, collectors, curators, historians, art appraisers, and insurers anywhere in the world.[27] By using the bitcoin blockchain, Verisart can confer digital provenance to any physical work, not just digital art, and users will be able to check a work's authenticity, condition, and chain of title from their mobile device before they participate in an online auction or agree to a sale. "We believe technology can aid trust and liquidity, especially as more of the $67 billion annual art market shifts to private sales (peer-to-peer) and online transactions," founder Robert Norton told *TechCrunch*. "The art world is not broken. It just relies too much on middlemen to ensure trust and liquidity. We believe the advent of a decentralized world-wide ledger coupled with powerful encryption to mask the identities of buyer and seller will be attractive to the art world."[28] The artist becomes what could be called a "rights monetizer" with the technology making deals and collecting revenue in real time.

You could apply this same model to other fields as well. In science, a researcher could publish a paper to a limited audience of peers, as Satoshi Nakamoto did, and receive reviews and the credibility to publish to a larger audience, rather than assigning all rights to a scientific journal. The paper might even be available for free but other scientists could subscribe to a deeper analysis or threaded discussions with the author about it. She could make her raw data available or perhaps share data with other scientists as part of a smart contract. If there is a commercial opportunity flowing from the paper, the rights could all be protected in advance. More on this in chapter 9.

3. Blockchain Cooperatives

The trust protocol supercharges cooperatives—autonomous associations formed and controlled by people who come together to meet common needs.

"It's nonsense to call Uber a sharing economy company," said Harvard professor Benkler. "Uber has used the availability of mobile technology to create a business that lowers the cost of transportation for consumers. That's all it has done."[29] David Ticoll said, "In common English usage, sharing denotes free exchange—not financial transactions. As in kids' sharing toys. It's a shame that this term has somewhat lost that meaning." To him, "sharing is the main way that humans and members of other species have conducted exchanges with one another for millions of years, beginning with the act of conception itself. While some Internet companies have facilitated genuine sharing, others have appropriated and commoditized the social relationships and vocabulary of sharing."[30]

Most so-called sharing economy companies are really service aggregators. They aggregate the willingness of suppliers to sell their excess capacity (cars, equipment, vacant rooms, handyman skills) through a centralized platform and then resell them, all while collecting valuable data for further commercial exploitation.

Companies like Uber have cracked the code for large-scale service aggregation and distribution. Airbnb competes with hotels on travel accommodations; Lyft and Uber challenge taxi and limousine companies; Zipcar, before it was purchased by Avis, challenged traditional car rental companies with its hip convenience and convenient hourly rentals.

Many of these companies have globalized the merchandising of traditional local, small-scale services—like bed-and-breakfasts, taxis, and handypersons. They use digital technologies to tap into so-called underutilized, time-based resources like real estate (apartment bedrooms), vehicles (between-call taxis), and people (retirees and capable people who can't get full-time jobs).

Blockchain technology provides suppliers of these services a means to collaborate that delivers a greater share of the value to them. For Benkler, "Blockchain enables people to translate their willingness to work together into a set of reliable accounting—of rights, assets, deeds, contributions, uses—that displaces some of what a company like Uber does. So that if drivers want to set up their own Uber and replace Uber with a pure cooperative, blockchain enables that." He emphasized the word *enable*. To him, "There's a difference between enabling and moving the world in a new direction." He said, "People still have to want to do it, to take the risk of doing it."[31]

So get ready for blockchain Airbnb, blockchain Uber, blockchain Lyft, blockchain Task Rabbit, and blockchain everything wherever there is an opportunity for real sharing and for value creation to work together in a cooperative way and receive most of the value they create.

4. The Metering Economy

Perhaps blockchain technology can take us beyond the sharing economy into a *metering economy* where we can rent out and meter the use of our excess capacity. One problem with the actual sharing economy, where, for example, home owners agree to share power tools or small farming equipment, fishing gear, a woodworking shop, garage or parking, and more, was that it was just too much of a hassle. "There are 80 million power drills in America that are used an average of 13 minutes," Airbnb CEO Brian Chesky wrote in *The New York Times*. "Does everyone really need their own drill?"[32]

The trouble is, most people found it easier and more cost-effective to make one trip to Home Depot and buy a drill for $14.95 than rent it for $10 from someone a mile away, making two trips. Wrote Sarah Kessler in *Fast Company* magazine: "The Sharing Economy is dead and we killed it."[33]

But with blockchains we can rent our excess capacity for certain commodities that are pretty much zero hassle—Wi-Fi hot spots, computing power or storage capacity, the heat generated by our computers, our extra mobile minutes, even our expertise—without lifting a finger, let alone schlepping to and from some stranger's house across the city. When you travel, your Wi-Fi can rent out itself in your absence, charging fractions of pennies for every second of usage. Your imagination (and possibly new regulation) is your only limit. Your subscriptions, physical space, and energy sources can now become sources of income, metering their use directly to a counterparty and charging them for it through micropayments. All you need is a decentralized value transfer protocol to allow them to safely and securely transact with one another. These platforms instill subsidiary rights in all our assets. You need to decide the extent to which you want to assign others usage and access rights—even the right to exclude others from using your assets—and what to charge for those rights.

This can work for physical assets too. For example, we've heard a lot about autonomous vehicles. We can build an open transportation network on the blockchain where owners each have a private encrypted key (number) that lets them reserve a car. Using the public key infrastructure and existing blockchain technologies like EtherLock and Airlock, they can unlock and use the car for a certain amount of time, as specified by the rules of the smart contract—all the while paying the vehicle (or its owners) in real time for the time and energy that they use—as metered on a blockchain. Because blockchain technology is transparent, the group of owners can track who is abiding by their commitments. Those who aren't take a reputational hit and eventually lose access altogether.

5. The Platform Builders

Enterprises create platforms when they open up their products and technology infrastructures to outside individuals or communities that can co-create value or new businesses. One type is *prosumers*, customers who produce.[34] In a dynamic world of customer innovation, a new generation of producer-consumers considers the "right to hack" its birthright. Blockchain technology supercharges prosumption. Nike running shoes could generate and store data on a distributed ledger that, in turn, Nike and the shoe wearer could monetize as agreed in their smart contract. Nike could offer a tiny piece of its shares with every pair it sells, if the customer agrees to activate the smarts in the shoes, or even sync her shoes to other wearables, such as a heart monitor or glucose level calculator or other valuable data for Nike.

Some platforms differ from prosumer communities where a company decides to cocreate products with its customers. With open platforms, a company offers partners a broader venue for staging new businesses or simply adding value to the platform.

Now with blockchain technology companies can quickly create platforms and partner with others to create platforms or utilities for an entire industry. Robin Chase founded Zipcar (a service aggregator) as well as Buzzcar (users can share their cars with others), and is now the author of *Peers Inc.*, a lucid book on the power of peers working together. She told us, "Leveraging the value found in excess capacity depends on high-quality platforms for participation. These platforms don't come cheap. The blockchain excels in providing a standard common database (open APIs) and standard common contracts. The blockchain can make platform building cheaper and manageable." That's just the beginning. "Best of all, its common database makes for data transparency and portability: consumers and suppliers can pursue the best terms. They can also cooperate as peers on the blockchain to create their own platforms, rather than using the capabilities of traditional companies."[35]

Think of the car of the future itself. It would exist as part of a blockchain-based network where everyone can share information, and various parts of the vehicle can do transactions and exchange money. Given such an open platform, thousands of programmers and niche businesses could customize applications for your car. Soon such platforms could transform entire industries such as financial services by settling all kinds of financial transactions and exchanges of value. A consortium of the largest banks is already working on the idea. Platforms are the rising tide that lifts all boats.

Wikinomics introduced the concept of ideagoras—emerging marketplaces for ideas, inventions, and uniquely qualified minds, which enabled companies like P&G to tap global pools of highly skilled talent more than ten times the size of its own workforce. Firms use services like InnoCentive and Inno360 to facilitate holding "Challenges," "Digital Brainstorms," and other techniques to find the right temporary talent outside their boundaries to address critical business challenges. It's about using data to find the right talent to hack your business for the better.

Talent—the uniquely qualified minds to solve problems—can post their availability to the ledger so that firms can find them. Rather than

InnoCentive, think *b*InnoCentive. Individuals can cultivate not only a portable identity, but also a portable résumé (an extended version of their identity) that can provide appropriate information about them to potential contractors. Think a distributed skills inventory owned by no one or everyone.

As every business becomes a digital business, the hackathon is an important form of ideagora. Now with blockchain technology and open source code repositories, every company could provide venues to geeks and other business builders for problem solving, innovating, and creation of new business value.

Blockchains and blockchain-based software repositories will fuel such activity. Companies can now use powerful new programming languages like the Ethereum blockchain with built-in payment systems. An excerpt from a conversation on *Hacker News*: "Imagine how cool it would be if I could share a guid for my repo—and then your bit client (let's call it gitcoin, or maybe just bit) can fetch new commits from a distributed block chain (essentially the git log). Github is no longer an intermediary or a single point of failure. Private repo? Don't share the guid."[36]

How cool indeed! (Well, maybe you didn't understand one iota of that little piece of coolness, but you probably get the idea.)

6. Blockchain Makers

Manufacturing-intensive industries can give rise to planetary ecosystems for sourcing, designing, and building physical goods, marking a new phase of peer production. It's about making it on the blockchain. Just as a modern aircraft has been described as "a bunch of parts flying in formation," companies in most industries are tending to disaggregate into networks of suppliers and partners. Three-dimensional printing will move manufacturing closer to the user, bringing new life to mass customization. Soon, data and rights holders can store metadata about any substance from human cells to powered aluminum on the blockchain, in turn opening up the limits of corporate manufacturing.

This technology is also a powerful monitor of the provenance of goods and their movement throughout a supply network. Consider an industry close to all our hearts (and other body parts)—the food industry. Today your local grocery store may claim—and truly believe—that its beef is safe, raised humanely, fed quality ingredients, and given no unnecessary drugs. But it can't guarantee it. No one keeps histories of single cows; bad things happen to good bovines. We trust our hamburger with no means to verify. Usually it makes no difference; billions and billions keep getting served. But once in a while, we get a glimpse of mad cow disease.

The food industry could store on the blockchain not just the number of every steer, but of every cut of meat, potentially linked to its DNA. Three-dimensional search abilities could enable comprehensive tracking of livestock and poultry so that users could link an animal's identity to its history. Using sophisticated (but relatively simple to use) DNA-based technologies and smart database management, even the largest meat producers could guarantee quality and safety. Imagine how these data might expedite lab tests and a community health response to a crisis.

Knowing how our food was raised or grown is not a radical idea. Our ancestors bought supplies at local markets or from retailers who sourced products locally. If they didn't like how a local rancher treated his cattle, they didn't buy his beef. But transportation and refrigeration have estranged us from our foodstuffs. We've lost the values of the old food chain.

We could restore these values. We could lead the world in developing a modern, industrialized, open food system with down-to-earth family farm values. Transparency lets companies with superior practices differentiate themselves. The brand could evolve from the marketing notion of a trustmark—something that customers believe in because it's familiar—into a relationship based on transparency. Surely food producers have an appetite for that.[37]

7. The Enterprise Collaborators

Yochai Benkler spoke about how blockchain technology could facilitate peer-to-peer collaboration within firms, and between firms and peers of all sorts. "I'm excited about the idea that you have a fully distributed mechanism for accounting, for actions, and for digital resources across anything; whether it's currency, whether it's social relations and exchange, or whether its an organization."[38]

Today, commercial collaboration tools are beginning to change the nature of knowledge work and management inside organizations.[39] Products like Jive, IBM Connections, Salesforce Chatter, Cisco Quad, Microsoft

Yammer, Google Apps for Work, and Facebook at Work are being used to improve performance and foster innovation. Social software will become a vital tool for transforming virtually every part of business operations, from product development to human resources, marketing, customer service, and sales—in a sense the new operating system for the twenty-first-century organization.

But there are clear limitations to today's suites of tools, and the blockchain takes these technologies to the next level. Existing vendors will either face disruption or embrace blockchain technologies to deliver much deeper capability to their customers.

What would a blockchain social network for the firm look like? Think Facebook for the corporation (or simply an alternative to Facebook for you). Because several companies are working on this, we can flash-forward a year or two and here's what we get:

Every user has a multifaceted wallet, a sort of portal into the decentralized online world. Think a portable personal profile, a persona or identity that you own. Unlike your Facebook profile, the wallet has diverse functions and stores many kinds of personal and professional data and valuables including money. It is also private to you and you share only what you want. You have pairs of public-private keys that serve to anchor your persistent digital ID. While multiple personas can be housed in the wallet for each person or company, let's assume that a wallet holds a single canonical persona anchored in a single key pair. A publishing system delivers a stream of information that you or your firm will happily pay for—a colleague's patch of new code, a summary of a conversation with a new client, or—with the client's permission—a tape recording of a call, a Twitter feed from a conference that you couldn't attend, live stream of a client's use of your new product, photographs of your competitors' booths at an industry expo, a Prezi presentation that seems to be closing new business, a video how-to of something a colleague just invented, assistance in completing a patent application, or anything else that you value.

There is advertising, perhaps from third parties or maybe from the HR department about open enrollment or changes in insurance plans, but you, not Facebook, get revenue or some reward for paying attention. This is called an "attention market." You could receive microcompensation for agreeing to view or interact with an advertisement, or for feeding back in detail about a new product pitch, or just about anything else, such as transcribing CAPTCHAs[40] or scanned documents.

The news stream, publishing system, and the attention market all look similar, but payments flow differently for each. Said ConsenSys's Joe Lubin, "You pay for publishing. Companies pay for your attention. The news stream has no payment flow. I am happy to read your stream, because I value that social connection, but I am not going to pay to see a picture of you and your buddies drinking at a bar, or to read your opinion on the Blue Jays pitching staff."[41]

You also participate in or create topical discussion channels, where you configure your privacy. Privacy is enhanced in other manners too. For example, spy agencies can't conduct traffic analysis because they are unable to discern the source or destination of messages.

There would also be a nifty mechanism for finding people and feeds that you might care about. In addition, distributed tools aggregate and present interesting new people or information for you to follow or friend, possibly using Facebook's social graph to help out. Lubin calls this "bootstrapping the decentralized Web using the pillars of the centralized Web."[42]

Experience shows that value ultimately wins out in the digital age. The benefits of this distributed model are huge—at least to the users and companies. The huge resources of social media companies notwithstanding, there is no end to the richness and functionality that we can develop in such an open source environment. Compare the power and success of Linux versus proprietary operating systems. Blockchain technologies ensure security. Your privacy is completely configurable. No social media company can sell or leak your personal information to government agencies without your permission. If you're a dissident in a totalitarian country, no one can track what you have read or said online. Because you own your data, you can monetize it along with your attention and efforts. You share in the wealth of big data.

Companies too should be enthusiastic about their employees' using such platforms for business. To attract talent, firms need to show integrity and respect their employees' security and privacy. More important, as any firm works to become networked, approaching talent outside its boundaries, they can offer up such interenterprise collaborative platforms that their partners can trust. Time will tell.

In summary, these are seven of the emerging business models whereby both companies large and small can make it "rain on the blockchain." Overall, the open networked enterprise shows profound, even radical potential to supercharge innovation and harness extraordinary capability to create good value for shareholders, customers, and societies as a whole.

HACKING YOUR FUTURE: BUSINESS MODEL INNOVATION

As for a company managed by software agents, Ronald Coase must be high-fiving up there somewhere in Economists' Heaven (although some might dispute that such a place exists). Remember the reverse of Coase's law? A corporation should shrink until the costs of transactions inside are less than the costs of transactions outside its boundaries. As technology continues to drop costs in the market, it's conceivable that corporations could and should have very little inside—except software and capital.

Think about it.

To begin, the cost of "search" continues to drop as new agents have the ability to conduct three-dimensional searches of the World Wide Ledger of everything commercial that exists or has existed. So no need for a corporate library, information specialists, HR search specialists, or the myriad other professionals involved in acquiring pertinent information to run a business.

Second, smart contracts would radically reduce the costs of contracting, policing contracts, and making payments. No longer paper, these programs could formulate their terms through a series of templates; bargain, accepting or rejecting terms and conditions based on rules and extensive information collected from external sources; formulate self-enforcing policies; determine when performance conditions have been met; and execute transactions.

Third, the cost of coordination of all these resources outside the organization could be trivial—measured in the energy to power the servers hosting the enterprise software. As for managing humans, organizations, and factories hired by the enterprise, the enterprise has no need for bureaucracy. With the new platform we can imagine a new organization that requires little or no traditional management or hierarchy to generate customer value and owner wealth.

Finally, the costs of establishing trust would approximate zero. Trust does not rest with the organization, but rather within the functionality, security, and auditability of the underlying code and the mass collaboration of the countless people securing the blockchain.

How would you go about designing a distributed autonomous enterprise? Such an entity could have rich functionality—agents executing ranges of tasks or more broadly business functions all based on a preapproved charter. Individuals, organizations, or collectives of potential shareholders or users will design them by defining the following:

1. Conviction: a belief about the world and what needs to be done to create value or change things.

2. Purpose: its reason for existence. Why are we creating this enterprise in the first place?

3. Constitution: outlines the overall objectives of the enterprise and the rules by which it will create value.

4. Modus operandi: for example, how it will go about creating this value. How it will fund itself—through crowdfunding, traditional early-stage investment, or using revenue. How it will acquire resources.

5. Division of labor between humans and technology: for the foreseeable future, perhaps humans should be in charge.

6. Application functions: how the enterprise will sense and respond to changing conditions.

7. Moral guidelines: Google's promise to "Do No Evil" is not going to be good enough. The DAE needs some clear guidelines about what is and isn't acceptable behavior.

There may not be a distributed autonomous enterprise in your near future, but the thinking behind these new entities can inform your business strategizing today. With the rise of a global peer-to-peer platform for identity, trust, reputation, and transactions, we can finally re-architect the deep structures of the firm for innovation, shared value creation, and perhaps even prosperity for the many, rather than just wealth for the few. Now you have at least seven emerging business models that could help you shake some windows and rattle some doors in your industry while distributing wealth more democratically.

Overall, smart companies will work hard to participate fully in the blockchain economy rather than play its victims. In the developing world, the distribution of value creation (through entrepreneurship) and value participation (through distributed ownership of the firm) may hold a key to reconciling the prosperity paradox. Our story becomes even more interesting when you consider that billions of agents will be embedded in the physical world. Which takes us to the next chapter.

CHAPTER 6

THE LEDGER OF THINGS: ANIMATING THE PHYSICAL WORLD

A power pole collapses at eight o'clock on a hot night in the remote outback of Australia. This is a problem for William and Olivia Munroe, who raise sheep and cattle one hundred miles outside the old gold mining town of Laverton, on the edge of the Great Victoria Desert.[1] In the summer, the temperature frequently soars close to 120 degrees Fahrenheit (48.9°C). Their children, Peter and Lois, attend school via satellite link, the family's only means of accessing health services in case of illness or emergency. Although the Munroes have a backup generator, it can't power the water pumps, communications, and air-conditioning for long. In short, the lives of the Munroe family depend entirely upon reliable energy.

At daybreak, nine hours later, the power utility sends out a team to find and fix the downed pole. Customer complaints give the company an idea of where the break occurred, but the team takes more than a day to identify, reach, and fix the pole. Meanwhile, the Munroes and nearby residents, businesses, and institutions go without power and connectivity at considerable inconvenience, economic impact, and physical risk. In the outback, blackouts are not just paralyzing; they're dangerous. To minimize these hazards, at great expense the company deploys teams of inspectors to check the extensive network regularly for downed or deteriorating poles.

Imagine how much safer, easier, and cheaper it would be if each power pole were a smart thing. It could report its own status and trigger actions for replacement or repair. If a pole caught fire or began to tip or fall for any reason, it would generate an incident report in real time and notify a repair crew to come with the appropriate equipment to the precise location. Meanwhile, the pole could potentially reassign its responsibilities to the nearest working pole. After all, they're all on the grid. The utility could restore power to the community more quickly without the huge ongoing costs of field inspection.

POWER TO THE PEOPLE

That's just the beginning. Using emerging software and technologies associated with the Internet of Things, we can instill intelligence into existing infrastructure such as a power grid by adding smart devices that can communicate with one another. Imagine creating a new flexible and secure network quickly and relatively inexpensively that enables more opportunities for new services, more participants, and greater economic value.

This configuration is known as a *mesh network*, that is, a network that connects computers and other devices directly to one another. They can automatically reconfigure themselves depending upon availability of bandwidth, storage, or other capacity and therefore resist breakage or other interruption. Communities can use mesh networks for basic connectivity where they lack access or affordable service. Mesh networks are alternatives to traditional top-down models of organization, regulation, and control; they can provide greater privacy and security because traffic doesn't route through a central organization.[2]

Organizations are already combining mesh networks with blockchain technology to solve complex infrastructure problems. Filament, an American company, is experimenting with what it calls "taps" on power poles in the Australian outback. These devices can talk directly to each other at distances of up to 10 miles. Because the power poles are approximately 200 feet apart, a motion detector on a pole that's falling will notify the next pole 200 feet away that it's in trouble. If for any reason the tap on that pole isn't available, it will communicate with the next pole, or the next pole (up to 10 miles) that will communicate to the company through the closest Internet backhaul location (within 120 miles).

With the tap's twenty-year battery and Bluetooth low energy (BLE) technology, customers can connect to the devices directly with their own phone, tablet, or computer. The tap can contain numerous sensors to detect temperature, humidity, light, and sound, all of which customers could use to monitor and analyze conditions over time, maybe to develop predictive algorithms on the life cycle or impending failure of a power pole. Customers could become weatherNodes or meter these data as an information service or license the data set through the blockchain to another user, such as a government, broadcaster, pole manufacturer, or environmental agency.

Filament's business model is a service model involving three parties: Filament, its integration customer, and the utility company. Filament owns the hardware; its devices continually monitor the condition of the power poles and report changes, whether they're fallen, on fire, or compromised by dust accumulation or brush fire smoke. Filament sells the sensor data stream to the integrator, and this integrator sells to the utility.

The utility pays monthly for a monitoring service. The service enables the power company to eliminate the very expensive field inspection of its operations. Because power poles rarely fall, the power company rarely uses the actual communication capability of the mesh network, and so Filament could deploy the excess capacity of the taps for other uses.

"Since Filament owns the devices, we can sell extra network capacity on top of this network that spans most of the continent," said Eric Jennings, Filament's cofounder and CEO. "Filament could strike a deal with FedEx to give their semitrucks the ability to send telemetry data to HQ in real time, over our network in rural Australia. We add FedEx to the smart contract list, and now they can pay each device to send data on their behalf."[3] FedEx drivers could use the mesh network for communications and vehicle tracking across remote areas to indicate estimated arrival times and breakdowns. The network could alert the nearest repair facility to dispatch the necessary parts and equipment.

Blockchain technology is critical. This Internet of Things (IoT) application depends on a Ledger of Things. With tens of thousands of smart poles collecting data through numerous sensors and communicating that data to another device, computer, or person, the system needs to continually track everything—including the ability to identify each unique pole—to ensure its reliability.

"Nothing else works without identity," said Jennings. "The blockchain for identity is the core for the Internet of Things. We create a unique path for each device. That path, that identity, is then stored in the bitcoin blockchain assigned to Filament. Just like a bitcoin, it can be sent to any address."[4] The blockchain (along with smart contracts) also ensures that the devices are paid for so they continue to work. The Internet of Things cannot function without blockchain payment networks, where bitcoin is the universal transactional language.

Social Energy: Powering a Neighborhood

Now, instead of poles, imagine digitizing every node in a power system to create entire new peer-to-peer models of power production and distribution. Everyone gets to participate in a blockchain-enabled power grid. Under a New York State–sponsored program to increase energy resiliency even in extreme weather conditions, work is under way to create a community microgrid in the Park Slope area of Brooklyn. Once built, this microgrid and its locally generated power will provide resiliency in emergencies and reduce costs to customers while promoting clean, renewable electricity, energy efficiency, and storage options in the community.

While campus microgrids have been around for a while, they aren't common in residential areas. Most home owners, businesses, governments, and other organizations in urban North America get their power from regulated utilities at regulated prices. Currently, we have more variety in locally generated renewable energy from, say, solar panels on rooftops. The local utility captures excess power in its supply for redistribution at wholesale rates, often with considerable leakage. The consumer, who may be located across the street from a local power source, still must go through the utility and pay full retail for renewable energy generated by their neighbor. It's ridiculous.

"Instead of the command-and-control system the utilities have now where a handful of people are actually running a utility grid, you can design the grid so that it runs itself," said Lawrence Orsini, cofounder and principal of LO3 Energy. "The network becomes far more resilient because all of the assets in the grid are helping to maintain and run the utility grid."[5] It's a distributed peer-to-peer IoT network model with smart contracts and other controls designed into the assets themselves (i.e., the blockchain model).[6] When a hurricane destroys transmission towers or fire cripples a transformer substation, the grid can quickly and automatically reroute power to prevent a massive blackout.

Resiliency isn't the only benefit. Locally generated power, used locally, is significantly more efficient than the utility-scale model, which relies on transmitting energy across vast distances, where energy is lost. LO3 Energy is working with local utilities, community leaders, and technology partners to create a market where neighbors can buy and sell the local environmental value of their energy. "So, instead of paying an energy services company that's buying renewable energy credits, you get to pay the people who are actually generating the electricity that is serving your house, that is local and green, and that actually has an environmental impact in your neighborhood. It seems a lot fairer, right?" said Orsini.[7] Right!

If you can locate each of the assets and assign locational value for generation and consumption, then you can create a real-time market. According to Orsini, you can auction your excess energy to your neighbors who might not be able to generate renewable energy. In doing so, your community can create energy resiliency through peer-to-peer trading. Community members can reach consensus on the rules of the real-time microgrid market such as time-of-day pricing, floor or ceiling prices, priority given to your nearest neighbor, or other parameters so as to optimize price and minimize leakage. You will not be sitting at your computer all day long setting prices, offering to buy or sell.

Future microgrids will harvest heat from the computational power needed to create and secure this transactive grid platform. Distributing the computing power to buildings in the community and using the higher temperatures generated to power heating, hot water, and air-conditioning systems increases the productivity of the same energy. "Our focus is on increasing Exergy," says Orsini.

With increasing generation of renewable power at the local level, the Internet of Things is challenging the regulated utility model, and not a moment too soon. We need to respond to climate change and brace ourselves for increasingly extreme weather conditions, particularly melting ice caps that drown islands in oceans, and droughts that turn dry land into desert. Currently, we're losing about fifteen million acres per year to desertification, the worst losses in sub-Saharan Africa where, unlike the Munroes of the outback, people can't afford water pumps, air-conditioning, or migration.[8] We need our utility grids and our engines not to leach energy and carbon into our atmosphere. While the utilities are looking at IoT benefits to their existing infrastructure ("smart grid"), connecting microgrids could lead to entirely new energy models. Utility companies, their unions, regulators, and policy makers, as well as innovative new entrants such as LO3, are exploring these new models for generating, distributing, and using electricity first at the neighborhood level and then around the world.

THE EVOLUTION OF COMPUTING: FROM MAINFRAMES TO SMART PILLS

Unlike our energy grid, computing power has evolved through several paradigms. In the 1950s and 1960s, mainframes ruled—International Business Machines and the Wild "BUNCH" (Burroughs, Univac, National Cash Register Corp., Control Data, and Honeywell). In the 1970s and 1980s, minicomputers exploded onto the scene. Tracy Kidder captured the rise of Data General in his 1981 best seller *The Soul of a New Machine*. Like mainframe companies, most of these firms exited the business or disappeared. Who remembers Digital Equipment Corporation, Prime Computer, Wang, Datapoint, or the minicomputers of Hewlett-Packard or IBM? In 1982, IBM hardware and Microsoft software brought us the decade of the PC, with Apple's Macintosh barely nipping at their heels. How things change.

Driven by the same technological advances, communications networks evolved, too. From the early 1970s, the Internet (originating in the U.S. Advanced Research Projects Agency Network) was evolving into its present-day, worldwide, distributed network that connects more than 3.2[9] billion people, businesses, governments, and other institutions. The computing and networking technologies then converged in mobile tablets and handhelds. BlackBerry commercialized the smart phone in the early aughts, and Apple popularized it in the iPhone in 2007.

What is relatively new and very exciting is the ability of these devices to go beyond relatively passive monitoring, measuring, and communicating (weather patterns, traffic patterns) to sensing and responding; that is, executing a transaction or acting according to predefined rules of engagement. They can sense (falling temperatures, traffic jams) and respond (turn on the furnace, lengthen the green light); measure (motion, heat) and communicate (emergency services); locate (burst water main) and notify (repair crews); monitor (location, proximity) and change (direction); identify (your presence) and target (market to you), among many other possibilities.

The devices can be static (poles, trees, pipelines) or mobile (clothing, helmets, vehicles, pets, endangered animals, pills). Caregivers are using smart—or edible—electronic pills, for example, to identify and record whether and when a patient takes his medication. A skin patch or tattoo captures the data and can measure heart rate, food consumption, or other factors and communicate this information to a physician, caregiver, or the patient himself through an app to identify patterns and give feedback. The medical profession will soon be using similar technology for targeted drug delivery to certain types of cancer, measuring core temperature and other biomarkers.[10]

The devices can communicate with one another, with computers and databases directly or through the cloud, and with people (send you a text message or call your mobile). These devices, through their evolving machine intelligence and the data they collect, are putting analysis of data, pattern recognition, and trend spotting into individual hands.[11] The industry term *big data* hardly describes the myriad data that the physical world will generate. By the most conservative estimate, the 10 billion or so devices connected via the Internet today will grow to more than 25 billion by 2020.[12] Call it "infinite data" from infinite devices.

So why don't we live in smart homes and drive smart cars and practice smart medicine? We see six big obstacles. One is the Rube Goldberg rollout of applications and services. Simply put, few of the early consumer IoT devices have delivered practical value, unless you want your smoke detector to ask your night light to call your smart phone and warn you of a fire.[13]

Another is organizational inertia and the unwillingness or inability of executives, industry associations, and unions to envision new strategies, business models, and roles for people. While some creative entrepreneurs have developed new businesses on some of these principles (i.e., enabling physical assets to be identified, searched, used, and paid for) and thereby disrupted existing markets (e.g., Uber, Airbnb), the impact is still comparatively minor and reliant upon a company and its app as intermediary.

A third is fear of malicious hackers or other security breaches that could modify the information and rules of engagement, overriding devices with potentially disastrous consequences. A fourth is the challenge of "future-proofing," critical for capital things with very long life spans, longer than the life span of a typical application or even a company. Startups go bankrupt or sell themselves to larger firms all the time.

A fifth is scalability; to realize the full value of the IoT, we must be able to connect multiple networks together so that they interoperate. Last is the overarching challenge of centralized database technology—it can't handle trillions of real-time transactions without tremendous costs.

To overcome these obstacles, the Internet of Everything needs the Ledger of Everything—machines, people, animals, and plants.

THE INTERNET OF THINGS NEEDS A LEDGER OF THINGS

Welcome to the Internet of Everything enabled by the Ledger of Everything—distributed, reliable, and secure information sharing, sensing, and automating actions and transactions across the Internet, thanks to blockchain technology. Technologists and science fiction writers have long envisioned a world where a seamless global network of Internet-connected sensors could capture every event, action, and change on earth. With ubiquitous networks, continued advancements of processing capability, and an increasing array of cheap and tiny connected devices, that vision of an "Internet of Things" is edging closer to reality.

Remember, Satoshi Nakamoto designed the bitcoin blockchain to ensure the integrity of each bitcoin transaction online and the bitcoin currency overall. By recording each transaction at every node and then sharing that record with every other node on the network (i.e., the blockchain), the blockchain ensures that we can verify the transaction quickly and seamlessly across the peer-to-peer network. We can conduct transactions of value—in this case financial—automatically, securely, and confidently without needing to know or trust each node on the network, and without going through an intermediary. The Ledger of Everything requires minimal trust.

Blockchain technology enables us to identify smart devices with relevant core information and program them to act under defined circumstances without risk of error, tampering, or shutting down in the Australian outback. Because the blockchain is an incorruptible ledger of all data exchanges that occur in the network, built up over time and maintained by the collaboration of nodes in that particular network, the user can be sure the data are accurate.

There is growing agreement among technology companies that the blockchain is essential to unlocking the potential of the Internet of Things. None other than IBM, the progenitor of large, centralized computer systems, has come on board. In a report, "Device Democracy: Saving the Future of the Internet of Things," IBM identified the value of the blockchain:

> In our vision of a decentralized IoT, the blockchain is the framework facilitating transaction processing and coordination among interacting devices. Each manages its own roles and behavior, resulting in an "Internet of Decentralized, Autonomous Things"—and thus the democratization of the digital world . . . devices are empowered to autonomously execute digital contracts such as agreements, payments and barters with peer devices by searching for their own software updates, verifying trustworthiness with peers, and paying for and exchanging resources and services. This allows them to function as self-maintaining, self-servicing devices. . . .[14]

Therefore, by using the blockchain, whole new business models open up because each device or node on the network could function as a self-contained microbusiness (e.g., sharing power or computing capability at very low cost).

"Other examples are a music service, or an autonomous vehicle," noted Dino Mark Angaritis, founder of Smartwallet. "Each second that the music is playing or the car is driving it's taking a fraction of a penny out of my balance. I don't have a large payment up front and pay only for what I use. The provider runs no risk of nonpayment. You can't do these things with traditional payment networks because the fees are too high for sending fractions of a penny off your credit card."[15]

Spare bedrooms, empty apartments, or vacant conference rooms could rent themselves out. Patents could license themselves. Our e-mail could charge spammers for each item received. You get the idea. With machine learning, sensors, and robotics, autonomous agents could manage our homes and office buildings, interactive sales and marketing, bus stop shelters, traffic flow and road usage, waste collection and disposal (i.e., where the bins speak to the trucks), energy systems, water systems, health care devices embedded or worn, inventories, factories, and supply chains.

Carlos Moreira, CEO of WISeKey, said that the greatest opportunities lie in what he called the *industrial blockchain*.[16] WISeKey, a Swiss-based company working in the area of identity management, cybersecurity, and mobile communications, provides secure transactional capability to watches and other wearable devices and is now offering its trust model to manufacturers and chip makers for outfitting a very large number of other IoT devices to be authenticated and to communicate across the Internet or other network. "We are moving into another world where the trust is delegated at the object level. An object that is not trusted will be rejected by the other objects automatically without having to check with a central authority," Moreira said. "This is a huge paradigm shift that has tremendous consequences in the way that processes will be conducted in the years to come."[17]

In this emerging world, users connect with smart devices using secure identification and authentication, potentially public/private keys, and they define the rules of engagement, such as privacy, with other devices, rather than going along with the rules of a centralized node or intermediary. Manufacturers can transfer maintenance, ownership, access, and responsibility to a community of self-maintaining devices, future-proofing the IoT and saving infrastructure costs, replacing each device exactly when it hits obsolescence.

Thus the blockchain can address the six obstacles to a functioning Internet of Things. To sum up, the new Ledger of Everything has nine nifty network features:

Resilient Self-corrects; no single point of failure

Robust Can handle billions of data points and transactions

Real-time Stays on 24/7/365 and data flows instantly

Responsive Reacts to changing conditions

Radically open Constantly evolves and changes with new input

Renewable Can be multipurpose, reused, and recycled
Reductive Minimizes costs and friction, maximizes process efficiency
Revenue-generating Enables new business models and opportunities
Reliable Ensures integrity of data, trustworthiness of participants

Why do we believe the IoT enabled by the blockchain has such huge potential? The primary driver is that it allows *animation* of the physical world. Once we bring these objects to life on the ledger, they can sense, respond, communicate, and take action. Assets can search, find, use, and compensate one another according to smart contracts, thereby enabling highly disruptive new markets, just as the Internet has previously done for people and all manner of digital content.

The questions for managers, entrepreneurs, and civic leaders: How will you take advantage of these new opportunities to change and grow? How will your organization respond to the inevitable disruption to your existing operational model? How will you compete with the creative new models of start-ups and collaborations?

Opportunities for greater efficiency, improved service, reduced costs, increased safety, and better results abound in our lives, and we can improve each by applying blockchain logic to the Internet of Things. We're beginning the next major phase of the digital revolution. Michelle Tinsley of Intel explained why her company is deeply investigating the blockchain revolution: "When PCs became pervasive, the productivity rates went through the roof. We connected those PCs to a server, a data center, or the cloud, making it really cheap and easy for lean start-ups to get computer power at their fingertips, and we're again seeing rapid innovation, new business models."[18] Intel wants to accelerate the process of understanding what's working, what's not working, and where the opportunities lie. "We could see this technology be a whole other step function of innovation, where it enables all sorts of new companies, new players. To be a leader in the technology industry, we cannot be absent from the conversation," she said.[19] Just imagine the potential of applying these capabilities across many types of businesses, many untouched by the Internet revolution.

THE TWELVE DISRUPTIONS: ANIMATING THINGS

What possibilities are there for animating the physical world? Unlike Pinocchio, we don't have a Blue Fairy. (And unlike Pinocchio, the blockchain doesn't lie.) But today, right now, we have distributed ledger technology that will actually enable not only GE to "bring good things to life." Even better, Pinocchio can't go long-nose on the ledger.

We're in the early days of thinking about the possibilities of the Ledger of Everything (built into the IoT). While consumer devices have received the most attention in the popular media to date, there are potential applications across virtually every sector. There are many ways of classifying and grouping potential applications because so many applications cross boundaries and could fit into more than one category. McKinsey, for example, uses the concept of *settings* in its classification of the IoT.[20] We've identified opportunities for the Ledger of Everything in twelve major functional areas. Specific benefits—and the business case—will be specific to each application. The categories below illustrate the potential and the potential significant disruption to existing markets, players, and business models.

1. Transportation
In the future, you'll call up an autonomous vehicle to get you safely where you need to go. It will intuitively take the fastest route, avoid construction, handle tolls, and park all on its own. In times of traffic congestion, your vehicle will negotiate a *passing rate* so that you arrive at your destination on time, and freight managers will use the blockchain-enabled IoT on all cargo to clear customs or other required inspections quickly. No red tape. Allianz, a manufacturer of street sweepers, could equip its municipal machines with minicam or sensor technology that identified cars whose owners hadn't moved them (if they couldn't move themselves) on alternate-side-of-the-street-parking days in New York City, feed that sensor data to the traffic police, and spare the physical writing of parking tickets. Or, the street sweeper itself could extract the parking fine in bitcoin from the car itself as it swept by—because the New York State Department of Transportation would require all cars registered in the five New York City

boroughs to maintain bitcoin wallets connected to their license plates. Autonomous vehicles, on the other hand, would sense the oncoming sweeper and simply move themselves to let it pass.

2. Infrastructure Management

Many professionals will use smart devices to monitor location, integrity, age, quality, and any other relevant factors of pavement, rail lines, power poles and lines, pipelines, runways, ports, and other public and private infrastructure in order to monitor conditions, detect problems (e.g., breakage or tampering), and initiate a response both rapidly and cost-effectively. That's where companies such as Filament will come in, with new affordable technologies to animate existing infrastructure without the huge capital required to replace it. Eric Jennings of Filament estimates that "over 90 percent of infrastructure is currently disconnected, and it's unfeasible to rip it all out and replace it with brand-new, wireless, connected assets."[21]

3. Energy, Waste, and Water Management

"Send a truck to empty me," said the overflowing waste bin. "Fix me," said the leaky pipe. The Internet of Things should inspire a hundred new children's books. Traditional utilities in both the developed and developing world can use the blockchain-enabled IoT for tracking production, distribution, consumption, and collection. As we've already seen, new entrants without significant embedded infrastructure are planning to use these technologies to create entirely new markets and new models (e.g., community microgrid).

4. Resource Extraction and Farming

Cows can become blockchain appliances, enabling farmers to track what the cows eat, which medications they've had, and their complete health history. This technology can also help track expensive and highly specialized equipment and make it more widely available for just-in-time usage and cost recovery; improve miner and farmworker safety through tagging of safety equipment and automated checklists (to ensure that equipment is being used properly); monitor weather, soil, and crop conditions to start irrigation, automated harvesting, or other actions; and compile "infinite data" analytics to identify new resources or advise on agricultural best practices based on past patterns and results. Sensors in soil and on trees could help environmental protection agencies to monitor farmers and their usage of the land.

5. Environmental Monitoring and Emergency Services

Remember autonomous weather agent BOB? BOB will live in a world of weather sensors and make money collecting and selling critical weather data. Examples here include monitoring air and water quality and issuing alerts to reduce pollutants or stay indoors; flagging dangerous chemicals or radioactivity for emergency workers; monitoring lightning strikes and forest fires; installing earthquake and tsunami early warning and alert systems; and, of course, storm monitoring and early warning. In addition to improving the response time for emergency services and reducing the risk of these events to human life, we could use this longitudinal data to increase our understanding of underlying trends and patterns, identify preventive measures in some cases, and improve our predictive capability to provide even earlier warning.

6. Health Care

In the health care sector, professionals use digitization to manage assets and medical records, keep inventory, and handle ordering and payments for all equipment and pharmaceuticals. Today, hospitals are full of smart devices that oversee these services, but few communicate with one another or take into account the importance of privacy protection and security in direct patient care. Blockchain-enabled IoT can use emerging applications to link these services. Applications in development include monitoring and disease management (e.g., smart pills, wearable devices to track vital signs and provide feedback) and improved quality control. Imagine an artificial hip or knee that monitors itself, sends anonymized performance data to the manufacturer for design improvements, and communicates with a patient's physician, "Time to replace me." Technicians will be unable to use specialized equipment if they haven't taken prerequisite steps to ensure their reliability and accuracy. New smart drugs could track themselves in clinical trials and present evidence of their effectiveness and side effects without risk of modified results.

7. Financial Services and Insurance
Financial institutions could use smart devices and the IoT to tag their claims on physical assets, making them trackable and traceable. Because digital currencies enable the storage and transfer of value rapidly and securely for all users large and small, they also enable risk assessment and management. Thinking further, could the poor and disadvantaged earn small amounts of cash, or perhaps electricity or other "credits," if they allowed their limited assets to be tagged and shared as in the earlier microgrid example? Owners will be able to tag priceless objects, antiquities, jewelry, the stuff of museums, anything ever handled by Sotheby's and insured by Lloyd's. Insurers could adjust payment according to where the object is and its environment—if it's in New York's Metropolitan Museum of Art under controlled climate, then a lower insurance rate; if traveling to Greece, then charge a higher rate. The object could tell whether it was in a vault or around a celebrity's neck. Insurance rates could be higher if the device was hanging on Lindsay Lohan's neck versus, say, Anne Hathaway's. Driverless cars would surely have lower insurance rates, and devices themselves could settle insurance claims on the spot based on sensor data.

8. Document and Other Record Keeping
As we have explained, physical assets can become digital assets. All documentation relating to a particular "thing" can be digitized and carried on the blockchain including patents, ownership, warranties, inspection certification, provenance, insurance, replacement dates, approvals, etc., significantly increasing data availability and integrity, reducing paperwork handling, storage, and loss, and other process improvements related to that documentation. For example, a vehicle will not start if it failed a recent safety inspection, if its liability insurance has expired, if its owner has failed to pay parking tickets or moving violations, or if the driver's license of the person attempting to drive it has been suspended. Items on the shelves will notify store managers when they've passed their "sell by" date. Store managers might even program these items to lower their own price as the sell-by date approaches.

9. Building and Property Management
An estimated 65 percent of the twelve billion square feet of commercial

real estate in the United States is vacant.[22] Digital sensors can create marketplaces of these real estate assets by enabling real-time discovery, usability, and payment. Vendors are now entering this field and developing new service models to rent the space in off hours. In the evenings, your conference room can moonlight as a classroom for neighborhood youth or an office for a local start-up. Other applications will include security and access control, lighting, heating, cooling, and waste and water management. The greenest of buildings will run on the Ledger of Things. Imagine the data on elevator usage and flow of people through the building, how these will inform an architect's design of public and private spaces. Spare residential space can list itself and negotiate through the Ledger of Everything to help tourists, students, managers of homeless shelter programs, and others find space that meets their needs. These ideas apply to all types of residential, hotel, office, factory, retail/wholesale, and institutional real estate.

10. Industrial Operations—The Factory of Things

The global plant floor needs a global Ledger of Things, aka the industrial blockchain. Factory managers will use smart devices to monitor production lines, warehouse inventory, distribution, quality, and other inspections. Entire industries may adopt the ledger approach to significantly increase efficiency for such processes as supply chain management. Large and complex machines, like airplanes and locomotives, consist of millions of parts. Each individual component of a jet engine or railcar could have sensors that send out an alert when it needs fixing. Imagine a train on its way from Baltimore to Long Beach notifying the maintenance crew in Long Beach three days ahead of time that it needed a critical new part. The sensor could even issue an RFP and accept the best bid and delivery for the part, cutting time and massive cost out of the operating efficiencies of large corporations like General Electric, Norfolk Southern, and others. Even more significant, manufacturers in realms from cars to light bulbs to Band-Aids are investigating how they can embed smart chips into their products or parts thereof and monitor, collect, and analyze performance data. With such data, they could provide automatic upgrades, anticipate client needs, and offer new services, in effect changing from product suppliers to ongoing software-based services.

11. Home Management
Feeling lonely? You can always talk to your house. Your own home and numerous products and services are entering the market to allow automated and remote home monitoring. These services go beyond the "nanny cam" to include access controls, temperature adjustments, lighting, and, eventually, just about everything else in your home. While "smart homes" have been relatively slow to take off, companies such as Apple, Samsung, and Google are working to simplify installation and operations. According to BCC Research, "The U.S. home automation market is estimated to go from almost $6.9 billion in 2014 to $10.3 billion in 2019 . . . the growth will be steady and long-term."[23]

12. Retail Operations and Sales
Walking down the street, your mobile device advises you that the dress you love is available at the Gap. Walk into the store and the dress, in your size, is waiting for you. After trying it on, you scan it and the payment is complete. But you've got other things to do, so the dress finds its way to your house before you get home. In addition to operational efficiencies and environmental monitoring, retailers will be able to personalize products and services to identifiable customers as they walk or drive by based on their location, demographics, known interests, and purchasing history, provided that those customers opened their black boxes to retailers on the blockchain.

THE ECONOMIC PAYOFF

Throughout this chapter we've referenced numerous potential benefits of the distributed, blockchain-enabled IoT at many levels (individual, organizational, industrial, societal). Redesigning and automating processes across peer-to-peer networks, rather than through people or centralized intermediary apps, could bring numerous benefits as already identified including:

- Speed (end-to-end automation)
- Reduced costs (associated with sending nearly infinite amounts of data to giant central processing facilities; elimination of expensive intermediaries)

- Increased revenue, efficiency, and/or productivity (freeing up excess capacity for reuse)
- Improved effectiveness (built-in checklists and other protocols reduce impact of human error)
- Increased security and integrity (person-to-person trust is not required as trust is designed into the network architecture)
- Reduced likelihood of system failure (elimination of bottlenecks, built-in resiliency)
- Reduced energy consumption (energy required by the network itself offset by increased efficiency and reduced wastage, dynamic pricing, and feedback loops)
- Increased privacy protection (intermediary can't override or ignore rules defined in the blockchain)
- Improved understanding of underlying patterns and processes and opportunities to improve them through the collection and analysis of "infinite data"
- Strengthened predictive capability of various events whether negative (severe weather, earthquakes, failing health) or positive (best time to plant crops, buying patterns).

The distributed open model means that IoT networks can be self-sustaining even after a company pulls out or a manufacturer fails. Interoperability, when designed into the system, will enable the connecting of different IoT networks and will unleash even greater value.[24]

Many of these benefits depend upon the concepts of distributed, or decentralized, networks and the elimination of a central (e.g., command and control) or other intermediary (e.g., a clearinghouse or management app). Once these new intermediaries are in place, others will feel pressure to "work around" or eliminate them. In Eric Jennings's view, "People will do the things they'll do to minimize their own discomfort, leading to silos and concentration and centralization. What's a short-term gain for those particular people is a long-term loss for everyone else." He said, "The Internet of Things should be completely decentralized where devices can be autonomous, discover each other directly, establish secure communication with each other directly, and eventually pay each other in value, directly between machines."[25]

The IBM Institute for Business Value has conducted research into what it calls the five major "vectors of disruption" that will increase our leverage of physical assets as the result of the blockchain-enabled IoT.[26] While IBM clearly has a business interest in the IoT, its work on business value is nonetheless very helpful.

First, the institute noted that these new networks will enable users to instantly search, access, and pay for available physical assets, such as underutilized storage or computing capability. The assets in supply can match themselves to demand. Because we can assess risk and credit automatically online and repossess virtually, we can reprice credit and risk significantly downward. Automated usage of systems and devices improves operational efficiency. Finally, firms can crowdsource, collaborate, and optimize with business partners in real time through digitally integrated value chains.

In short, you have an opportunity to make conceptually simpler, more efficient markets. You can access previously inaccessible assets, determine price in real time, and reduce your risk. Once the basic infrastructure is in place, barriers to entry are low (e.g., just develop an app), and the ongoing costs also relatively low (e.g., no more third-party service fees). It drastically lowers the cost of transmitting funds and lowers the barrier to having a bank account, obtaining credit, and investing. It could even support micropayment channels, matching minute-by-minute service usage with minute-by-minute payment.

The Ledger of Things enables "distributed capitalism," not just redistributed capitalism. Far from a free-for-all, these markets can be shaped according to our values—as individuals, companies, and societies—and these values coded into the blockchain, such as incentives to use renewable energy, use resources from our closest neighbors first, honor price commitments, and protect privacy. In short, the Ledger of Everything on top of the IoT animates and personalizes the physical world even as we share more. As IBM stated, "At a macroeconomic level, we are all winners in the IoT future, even though different industries will experience a mix of different effects."[27] According to the McKinsey Global Institute, the economic value of the IoT has, if anything, been underestimated; the economic impact—including consumer surplus—could be as much as $11.1 trillion per year in 2025 for IoT applications.[28] That's a 10 percent lift on current global GDP of well over $100 trillion; that's huge!

Networked intelligence, a phrase coined in *The Digital Economy*, referred to how the network would be smarter than its smartest node in one domain after another. As we have explained, the first generation of the Internet dropped transaction costs somewhat. We have faster supply chains, new approaches to marketing, and peer-to-peer collaborations like Linux and Wikipedia on a massive scale, with many innovative new business models. Blockchain technology will accelerate this process. As the Internet of Things takes hold, these trends will go into hyperdrive.

THE FUTURE: FROM UBER TO SUBER

We've covered a lot of ground in this chapter. Now let's pull all the strands of innovation together in just one scenario.

Consider service aggregators like Uber and Lyft. Uber is an app-based ride-sharing network of drivers who are willing to give other people a lift for a fee. To use Uber, you download the Uber app, create an account, and provide Uber with your credit card information. When you use the app to request a car, it asks you to select the type of car you want and marks your location on a map. The app will keep you posted on the availability and whereabouts of your prospective driver. At the end of the ride, Uber automatically charges your credit card. If you don't want to give the default tip, then you need to change your billing settings on Uber's Web site.[29] Uber Technologies, Inc., the company behind the development and operation of the Uber app, retains a share of the price paid for every ride.

It sounds great, particularly in cities with a small taxi fleet. But Uber's services come with a number of problems and red flags. Driver accounts have been hacked, rides are subject to surge pricing, and passengers have been subject to reckless driving and sexual harassment or assault.[30] Uber is also tracking users' every move, releasing some of this information to city officials for traffic studies. To top it all off, drivers create considerable value but they get to keep only part of it.

Now let's imagine the Uber experience if it were a distributed application on the blockchain. Mike Hearn, a former Google employee who quit his job to work full time on bitcoin, laid out this alternative universe based on bitcoin technology at the 2013 Turing Festival.[31] Hearn called this network "TradeNet" and described how, with the help of bitcoin, people could begin to rely on driverless vehicles.

It works like this. Most people don't own cars, but rather share vehicles in a commons. In Chicago, Melissa requests a car through SUber (think blockchain Super Uber). All the available vehicles start automatically posting offers, which Melissa's node ranks and presents to her based on her selection criteria. Melissa factors in how much she's willing to pay for faster routes (e.g., higher-priced toll lanes).

Meanwhile John, unlike most users, is a SUber vehicle owner and as his self-driving car is taking him to work, it identifies all the parking options, both public and privately owned, selects a space, and reserves and pays for it through an autonomous parking marketplace. Because John's predetermined parameters always include seeking the cheapest available spot within a ten-minute walk of his destination, he almost always goes with his car's first choice. The underlying parking database that supports the parking also contains information on parking rules for specific streets on different days and at different times of day, whether or not the parking space is covered or in the open, or whether the owner of the space has established a minimum price. All this runs on a distributed peer-to-peer platform—connecting multiple apps—so no centralized company is mediating the orders or taking part of the fee. There is no surge pricing and no unexpected fees.

What is striking about this proposed model is not the driverless vehicles, because self-driving cars will be commonplace—probably sooner rather than later. Rather, the cars could be fully autonomous agents that earn their own fares, pay for their own fuel and repair, get their own auto insurance, negotiate liability in collisions, and operate ("drive") without outside human control, except when they need to take some entity—maybe a human being—to court.

As a condition of operating, SUber administrators could program the vehicles' protocols into the blockchain to obey all traffic rules, take the most direct, fastest, or least expensive route, and honor their bids. The drivers' initial entry and registration into the SUber system could require vehicles to register necessary documentation including ownership, safety inspections, and insurance, and the system would permanently log these records to ensure reinspection or insurance and permit renewals as required. Sensors could monitor the overall "health" of the vehicle and signal necessary repairs, make the appointment at the appropriate repair shop, and preorder any necessary parts. Because the vehicles are driverless, they're not subject to sarcasm, cronyism, sexism, racism, or other forms of human discrimination or corruption. Plus, they won't try to push their politics or line the dashboard with incense. All of this happens behind the scenes, between objects, and powered by an autonomous application. The drivers have created a blockchain cooperative as described in the previous chapter and they receive nearly all the wealth they create. The users—Melissa and John—experience only the convenience, with none of the hassle. What's not to love?

Where the Internet reduced the costs of search and coordination, a digital currency like bitcoin on the blockchain will enable us to cut the costs of bargaining, contracting, policing, and enforcing these contracts. We'll be able to negotiate the best deal and get the promised delivery from any other entity that will accept bitcoin, including a driverless taxi. How will the Ubers of the world compete?

But the scenario doesn't stop there. Intelligence designed into the city's infrastructure will move traffic along (variable lane direction, variable pricing, automated traffic signal management based on traffic flow), further reducing wasted energy and costs. The blockchain could support safety controls, both on the vehicles (driver and driverless) and/or on the infrastructure, such as proximity warnings and automated braking, as well as antitheft or prevention of unqualified or inebriated drivers from taking the wheel. In addition, cities will use the sensors to help manage the transportation infrastructure, including asset management of infrastructure and fleets, monitoring rail line and pavement conditions, generating maintenance plans and budgets, and dispatching repair crews when necessary.

What's truly powerful, the systems work together—intelligent vehicles operating on an intelligent infrastructure. While there will still be business for drivers of shared vehicles, autonomous vehicles will be able to operate safely on city streets with their built-in navigation and safety systems, often interacting with the intelligent infrastructure to find and pay for an accelerated lane, or parking, or to search for and find a preferred route. The ready availability, affordability, and reliability of the autonomous vehicles will significantly reduce the number of private vehicles that, like the commercial real estate example above, are often just parked waiting and unused.

And it won't just be technology or car companies that will make this happen. While all of this could, in theory, be developed, owned, operated, and managed by a single civic transportation authority, that is likely not to be the path forward. SUber is more likely to evolve and innovate as an open and shared transportation platform, with various applications developed and introduced by local entrepreneurs, community groups, government, and others in either a profit-making (through revenue earned on a fleet of driverless vans), shared co-op (a neighborhood group invests in ten vehicles to be reserved and shared using the SUber app), public service (maintaining and operating a train or express buses on high-demand routes), or social enterprise (not-for-profits investing in SUber "points," which their clients can access when they need transportation).

This may emerge first in jurisdictions with relatively advanced infrastructure, already separate transportation corridors (rail, road, bike, pedestrian), significant transportation issues (traffic congestion), and a population with a long tradition of obeying traffic rules. It may also begin in "greenfield" city developments in cooperation with technology companies and car companies looking for test beds for their applications. Any scenario involving driverless vehicles would be less successful, even highly dangerous, when other road users cannot be isolated (on separate corridors), or predicted (animals on the road), or controlled (distracted pedestrians).

The SUber scenario is increasingly feasible. Such applications will likely emerge in the next few years and come to solve our transportation needs over the long term. Already today, local taxi and limousine commissions are battling Uber in many cities. City governments are struggling to balance consumer desire for affordable options with public safety and taxi licensing, even as the new models are seemingly inevitable. Why not look where the transportation sector is going and design solutions that best meet the city's needs, as Chicago has done in our hypothetical SUber scenario?

HACKING YOUR FUTURE FOR A WORLD OF SMART THINGS

We've seen throughout this chapter some mind-boggling opportunities in virtually all aspects of our lives, including—perhaps especially—many areas barely touched by the first wave of the digital revolution. At the same time, these opportunities threaten existing businesses and ways of doing business.

Key Issues: What should you as a manager be doing on both sides of this equation—to realize new opportunities, while minimizing threats? Whether you're a manager in the public, private, or social sector, do you have un- or underutilized physical assets that can be tapped for greater value? Are you realizing the greatest possible efficiencies and opportunities to develop products and technologies for the IoT itself? Are new entrants into this economy taking your customers and reducing your revenues through innovative new app-based business models that you should be installing first?

New Value: What are your physical assets and how can you enhance them to deliver greater value to your organization or community? Do you have physical spaces, machines, inventory, or other assets that you could tag, monitor, and animate as part of an autonomous network where you establish the operating parameters to drive out costs or add value? Could you embed, upgrade, and program sensors as part of a larger network for greater functionality and value? Could you glean new information from an IoT network to improve your planning and analysis for the future?

New Business Models: What opportunities exist for new products and services based on the new functionality and data you could gather through your network? Could your information and assets earn revenue because of their value to others, for example, renting out that expensive piece of equipment when you're not using it? Thinking about the value of information is not new (remember Sabre and American Airlines?), but still overlooked.

Opportunities: Could you link your network up with others for even greater value, perhaps as part of an end-to-end supply chain or distribution and sales channel? As an industry, are there shared processes and functions that could be automated by utilizing the blockchain? Are you enabling this interoperability by using technology built on open standards and vetted through international collaboration?

Threats: What lines of business will new entrants attack with their new IoT-based business models to serve markets that you currently serve? For example, rather than a one-time sale of a vehicle, consumer good, or piece of specialized equipment, is there ongoing value for you and your clients in a new service model built upon your ongoing connection to that equipment? Can you capitalize upon your existing expertise, resources, infrastructure, and customer loyalty to design new IoT-based business models that decrease the "space" and, therefore, the likelihood of entry of a disruptive new player?

Business Case: What are the costs and benefits of these opportunities? Where does the real value exist for your organization? Are you solving an actual business problem or need or just leading with the technology? How about developing a proof of concept with a leading client?

Strategic Plan: According to McKinsey, "executives will need to deal with three sets of challenges: organizational misalignment, technological interoperability and analytics hurdles, and heightened cybersecurity risks."[32] We add a fourth major challenge to this list—building in privacy and an incentive plan, including appropriate safeguards, from the beginning. How will IT and business functions have to adapt to the IoT? Which parts of the organization and business leaders should you involve?

CHAPTER 7

SOLVING THE PROSPERITY PARADOX: ECONOMIC INCLUSION AND ENTREPRENEURSHIP

A PIG IS NOT A PIGGY BANK

The Pacific coast of Nicaragua is one of the most beautiful landscapes in the Americas, where verdant green forest meets endless blue waters. Its rolling hills and stunning beaches make it a top destination for backpackers, sunbathers, and ecotourists alike. Nicaragua is also one of the poorest and least developed countries in the region. Sixty percent of the population lives below the poverty line. Those not employed in its tourism industry survive on near-subsistence-level agriculture and fishing. Nicaragua has the second-lowest nominal gross domestic product in the Americas, with 10 percent of its entire GDP from remittances—money earned overseas and repatriated by the Nicaraguan diaspora. Nineteen percent of Nicaraguans have a formal bank account, but only 14 percent are able to borrow and only 8 percent have formal savings.[1] Yet 93 percent have a mobile phone subscription, usually in the form of prepaid access.[2]

That is the reality that Joyce Kim faced when she took her team down to Nicaragua. Kim is the executive director of the Stellar Development Foundation, a not-for-profit blockchain technology organization (not to be confused with Stellar, the large architecture and construction firm). A Nicaraguan microfinance operation wanted to learn more about Stellar's financial platform. The woefully underdeveloped banking industry in Nicaragua keeps most people in an inescapable cycle of poverty and exacerbates the plight of would-be entrepreneurs. They struggle to start new businesses, register titles to their land and other assets, and resolve outstanding claims from the Sandinista government's mass land expropriation in the 1980s.[3] Stellar's platform would enable Nicaraguans to transfer, save, invest, borrow, and lend money.

Kim was both impressed and surprised by the local focus on microcredit. She understood that access to credit was paramount to economic inclusion but believed that savings, the ability to store value reliably and securely, was a prerequisite for almost all other financial services. When Kim asked about savings, she was told, "Oh, savings is not a problem around here. People have pigs."[4]

Livestock makes up the vast majority of farmers' net worth in many agrarian economies because financial services are not widely available and individuals have a tenuous right to their land. In Nicaragua that means people own pigs, and lots of them. Kim was surprised at first, but quickly saw the age-old logic to it. "You walk out of a meeting, and you look around and you see pigs are everywhere."[5] Livestock has long been an accepted, relatively useful form of savings. For those excluded from the digital economy, animals are about as liquid an asset as you can own, even more so if they produce milk, and they pay dividends in piglets, eggs, lambs, calves, and sometimes cheese.

Prosperity is a relative concept. In Kenya, Masai tribesmen who own four to five hundred head of goats are considered prosperous, but their lives can be rough, brutish, and short. Livestock-based wealth is "highly localized, so that you can't actually transact with anybody unless they're right there in front of you," Kim said. "You run the huge risks with your animals running away or getting sick or some blight coming that could actually wipe out all of your savings."[6]

Credit was an even tougher nut to crack than savings. Kim got to know a local Nicaraguan fisherman, a member of a cooperative, who explained that no one fisherman ever has access to enough credit to outfit an entire rig. According to Kim, "they form fishermen crews where one person will get a loan for the net, somebody else will get a loan for the bait, somebody else gets a loan for the boat, somebody else a loan for the motor, and then they come together and they form a crew." No one person is able to float his or her own venture (no pun intended) because access to credit is so tight. The model works, but it involves as many middlemen as there are fishermen.

The lifelong financial struggle of the Nicaraguan fishermen and farmers is the story of most unbanked people, around two billion adults in the world today.[7] What they lack—a store of value that won't get mad cow disease or die of old age, or a payment mechanism that extends beyond the village—we take for granted.

Financial inclusion is a prerequisite for economic inclusion. Its repercussions extend beyond finance. Kim said, "I don't consider financial access and financial inclusion to be the end goal. It's a path we all have to walk to get better education, better health care, equal women's rights, and economic development."[8] In short, financial inclusion is a fundamental right.

This chapter looks at opportunities for mobile and financial service providers and other businesses to use blockchains to unleash the economic potential at the bottom of the pyramid. We're talking billions of new customers, entrepreneurs, and owners of assets, on the ground and ready to be deployed. Remember, blockchain transactions can be tiny, fractions of pennies, and cost very little to complete. Anyone with the smallest of assets—say, a talent for embroidery or music, spare water pails, a chicken that lays eggs, a mobile phone that records data, audio, and images—could exchange value. The new platform also eliminates the point-of-access barrier. If you can access the Internet on a mobile device, then you can access assets with no forms to fill out and very little literacy. These are seemingly small but incredibly important breakthroughs. If we do this right, blockchain technology could unleash the biggest untapped pool of human capital in history, bringing billions of engaged, prospering entrepreneurs into the global economy.

THE NEW PROSPERITY PARADOX

For the first time in modern history, the global economy is growing but few are benefiting. On one hand, the digital age is bringing limitless possibilities for innovation and economic progress. Corporate profits are ballooning. On the other hand, prosperity has stalled. Throughout modern history, individuals and families at the 51st percentile were on the rise. Despite depressions and upheavals, prosperity for these individuals, and for society as a whole, steadily increased. This is no longer the case. Standards of living are even declining in the developed world. Median wages are stagnating in OECD countries. And, according to the International Labour Organization, youth unemployment in most of the world is stuck at about 20 percent. "Young people [are] nearly three times as likely as adults to be unemployed," the ILO reported.[9] In many developing nations, the numbers are significantly higher. Such unemployment is corrosive to all societies, no matter what their level of development. Most citizens want to contribute to their community. Anyone who has been jobless knows how it erodes self-esteem and well-being. Those with power and wealth are getting ahead, and those without are falling behind.

This new prosperity paradox, not to be confused with the intergenerational "Paradox of Prosperity" coined by economists such as Gilbert Morris, has befuddled every policy maker in the Western world. One of the best-selling business books of 2014, *Capital in the Twenty-First Century* by Thomas Piketty, became the #1 best seller on the *New York Times* hardcover nonfiction list in 2014. A tour de force of academic scholarship, *Capital* explains why inequality is accelerating and will likely continue to do so as long as the return on capital exceeds long-term economic growth. The rich have gotten richer because their money made them more money than their work did. Hence, the proliferation of new millionaires and billionaires. But his solution for how to stem growing social inequality, by imposing a wealth tax on the ones who own most of the world's wealth, was somewhat less inspiring, if only because we've heard it before.[10] Indeed, for as long as capitalism has been the primary mode of production, the debate about how to share the fruits more equally has not moved much beyond the redistribution of wealth, usually through taxation of the rich and the provision of public services to the poor. Advocates of our current economic model point to the hundreds of millions of people in the developing world (mostly in Asia) who have been lifted out of abject poverty, but often overlook the asymmetric benefits conferred to the very wealthy and the widening chasm between the superrich and the rest of the population in those same countries. Today, the global 1 percent owns half the world's wealth while 3.5 billion people earn fewer than two dollars a day.

Defenders of the status quo are quick to point out that most of the world's superrich got rich by creating companies, not through inheritance. However, behind the successes of a few are some very troubling statistics. The rate of new business formation is down. In the United States, the share of total firms that are younger than one year old fell by nearly half between

1978 and 2011, from 15 percent to 8 percent.[11] The millennial generation, oft characterized as entrepreneurial risk takers, is doing little to buck the trend and may be contributing to it. A recent analysis of Federal Reserve data found that only 3.6 percent of American households headed by someone under thirty held a stake in a private company, down from 10.6 percent in 1989.[12]

In the developing world, the digital revolution has done little to clear the entrepreneurial path of red tape and corruption. Where it costs only 3.4 percent of per-capita income to start a business in OECD countries, it costs 31.4 percent in Latin America and a shocking 56.2 percent in sub-Saharan Africa. In Brazil, an entrepreneur has to wait almost 103 days to incorporate his company versus 4 days in the United States and half a day in New Zealand.[13] Exasperated by government bloat and inefficiency, many would-be developing-world entrepreneurs instead choose to operate in the so-called informal economy. "There are a bunch of things taken for granted in the West. Property records are fine-tuned, for example. In the Global South, entrepreneurs would rather the government not know they exist. We need to make identity a profitable proposition," said Hernando de Soto. For now, staying in the shadows frees these entrepreneurs from meddlesome and corrupt officials, but it also profoundly limits their ability to grow their business, limits rights, and makes "dead capital" of money that could otherwise be deployed more efficiently.[14] Moreover, even for those who operate their businesses in the open, laws of many countries do not provide for limited liability. If your business fails, you're personally on the hook for all liabilities. If you bounce a business check in many Arab countries, you go straight to jail—"do not pass go" or any other institutions of due process on the way.[15]

Okay, then, the world has always had haves and have-nots. Today fewer people starve to death, or die from malaria or through violent conflict. Fewer people live in extreme poverty today than in 1990.[16] Certain emerging economies have benefited from outsourcing of manufacturing and liberalization of economic policies—China being a prime example of both—and the mean income of citizens in most developed countries increased. On balance, people are better off than they used to be, right? So what if the rich just happen to own significantly more? Shouldn't they keep what they've earned through their efforts? What's the problem?

Piketty pointed at capitalism. But capitalism, as a system for organizing the economy, is not the problem. In fact, capitalism is a great way to create wealth and prosperity for those who know how to use it. The problem is that most people never get a shot at seeing the benefits of the system because the Rube Goldberg machine of modern finance prevents many from accessing it.

Financial and economic exclusion is the problem. Fifteen percent of the population in OECD countries has no relationship with a financial institution, with countries like Mexico having 73 percent of the population unbanked. In the United States, 15 percent over fifteen years of age, *or 37 million Americans*, are unbanked.[17]

Financial inequality is an economic condition that can quickly morph into a social crisis.[18] In 2014, the World Economic Forum, a multistakeholder organization whose members include the largest companies and most powerful governments in the world, argued that growing inequality posed the single biggest risk globally, beating out global warming, war, disease, and other calamities.[19] Blockchain could be the solution. By lowering barriers to financial inclusion and enabling new models of entrepreneurship, the tonic of the market could be brought to bear on the dreams and ideas of billions of the unbanked.

Prosperity Purgatory: An Exercise in Futility

For centuries, banks have relied on network effects. Each successive customer, branch, product, dollar in, and dollar out increases the value of the bank's network. However, building these networks has come at a cost. Specifically, the cost of acquiring a profit-turning customer has only increased. If a prospect's money won't earn its keep, the bank won't be interested in keeping it. Thus, banks have little economic incentive to win customers in the bottom half of the pyramid. According to Tyler Winklevoss, banks don't serve most of the world and have no existing plans to serve them. However, new technology could remove that step. He said, "A lot of African countries leapfrogged the infrastructure of landline telecoms with cellular. They skipped that step. Blockchain will have the greatest impact in areas where the payment networks don't exist or are very poor."[20] Blockchain will push many nascent initiatives, such as mobile-money service providers like M-Pesa in Kenya, owned by Safaricom, and micro-credit outfits globally, into high gear by making them open, global, and lightning fast.

A bank is the most common financial institution, and so we will use it as an example here. How do you open a bank account? If you live in the developing world today, you will likely have to visit the branch in person. In Nicaragua, there are only 7 bank branches per 100,000 people compared with 34 per 100,000 in the United States. Nicaragua looks well banked compared with many countries in Africa, where there can be fewer than 2 branches per 100,000 people.[21] So you will probably have to travel a good distance to find a bank. You will also need to bring a government-issued identity card, but that will be just as difficult to come by if you don't already have one.

In the developed world—say, the United States—you need to meet certain requirements. While these requirements vary from bank to bank and state to state, you typically need to deposit and maintain a $100 to $500 minimum balance. You also need to prove your identity. Banks that do business in the States must comply with stringent "know your customer," "anti–money laundering," and "anti–terrorism financing" regulations.[22] And so they must do more comprehensive background checks on applicants before granting them an account. Ultimately, the bank is less interested in evaluating your character than it is in complying with regulatory agencies. That means a laundry list of requirements. First, you need a Social Security card. Don't have one? That's usually enough to get rejected. How about a photo ID like a driver's license or passport? Don't have one? You're not opening a bank account. Let's say you have both a Social Security card and photo ID. The bank, just to be safe, asks for a recent utility bill as proof of permanent residence or some proof of a previous bank account. If you happen to be new to town, or staying with family, or from an entirely *un-banked* region of the world, you'd likely fail some of these tests. The bank doesn't want you as a customer unless it can confirm your identity based on various papered credentials. It's not interested in knowing you as a well-rounded person. It's interested in knowing you as a set of checked boxes. Previous attempts to streamline this process for immigrants and the poor, such as the New York scheme to allow people to use their city ID cards, have failed.[23]

The Prosperity Passport: An Exercise in Utility

Fortunately for the unbanked, blockchain technology is engendering a new form of financial identity—one not dependent upon one's relation to a bank but rooted in one's own reputation. In this new paradigm, being "banked" in the traditional sense is no longer a prerequisite. Instead of passing the traditional ID tests, individuals can create a persistent digital ID and verifiable reputation and deploy it, in whole or in part, in different relationships and transactions. The blockchain endows this digital ID with trust and access to financial services. This capability is unprecedented at a massive scale. Joseph Lubin of ConsenSys said, "We all have reputation. It just isn't easy to use as social and economic systems are currently constructed. Most of it is ethereal and ephemeral. In the best case, it is fragmented and you have to present shallow documentation of it anew for every venture that requires it. In the worst case, billions of people don't have a way of presenting reputation to anybody but their immediate social circle."[24] It might as well be a pig or a cow. However, with the basic building blocks, people can construct digital identities that are not fragmented or ethereal but universal and standardized, with robust attestations of aspects of themselves and their interactions. They can share these digital IDs granularly—that is, share only very specific information about their identity—to facilitate more interactions that will likely lead to their own personal economic growth and prosperity. David Birch, a cryptographer and blockchain theorist, summed it up: "Identity is the new money."[25]

Consider the possibilities: the underbanked of the world can enfranchise themselves as they interact with microlending outfits. Potential vendors or lenders can track their usage and repayment of tiny loans, previously unfeasible, on the blockchain rather than rely on some credit score. "Once a previously unbanked person pays back a microloan, they are on their way to securing more and larger loans to build their businesses,"[26] said Lubin. This behavior, when repeated, adds to the reputation score of the borrower. Combined with a global, frictionless payment platform, individuals and small business owners can do the previously impossible: pay a remote vendor for merchandise or services, thereby advancing their prospects in the global economy. Joyce Kim mused, "What if we could create a credit score for women based on their household history?"[27]

Economic and financial fault lines often run alongside gender lines, making this technology a boon for the world's disenfranchised women. Referring to the global poor, de Soto said, "It's not that they don't want to come into the global economy. It's that the standards and information to bring them into the system are not in place. Blockchain is terrific because it gives us a common platform to bring people together."[28]

What could this persistent reputation mean for global entrepreneurship? If you have a reliable, unique, and robust identity, and you're deemed trustworthy, counterparties will feel more comfortable providing you with access to value. This is not redistribution of wealth but a wider distribution of opportunity. Haluk Kulin, CEO of Personal BlackBox, said, "The biggest redistribution that is about to happen is not a redistribution of wealth but a redistribution of value. Wealth is how much money you have. Value is where you participate."[29] Blockchain can enable every person to have a unique and verifiable reputation-based identity that allows them to participate equally in the economy. The implications of this equality are profound. Lubin imagines a future where the "unbanked and underbanked will become increasingly enfranchised as microlending services will enable investors across the globe to construct diverse portfolios of many microloans of which the usage and repayment can be tracked in full detail on the blockchain, using Balanc3's [a ConsenSys portfolio company] triple-entry accounting system, for instance."[30] In this new future, when people repay microloans, they are on their way to securing more and larger loans to build their businesses.

ROAD MAP TO PROSPERITY

Financial identity is the genesis for a wide array of financial and economic opportunity previously unattainable for more than two billion of the world's population. Blockchain technology enables people from all walks of life to map out their own prosperity. Imagine that, a wealth of one's own—for large numbers, ultimately billions of people.

Tools of Abundance: The most basic requirements to participate in an economy are tools like *a mobile phone and some kind of Internet access*, the portal through which people interact with different value systems. Dr. Balaji Srinivasan, managing partner at Andreessen Horowitz and a lecturer at

Stanford University, said, "If you can access the Internet on a mobile phone, suddenly you're able to access all these other things. You can access a bank or at least the mechanisms for it."[31] Blockchain technology creates a whole new set of business models previously unimaginable that empower individuals as economic agents.

Persistent Identity: You can use and port identity into different networks to establish reputation in a financial transaction or to plug into different social networks. Suddenly, a pig no longer has to be the family piggy bank. New payment rails and means to store value and transact with counterparties will open new frontiers. Indeed, this lowering of barriers to financial inclusion will make it easier than ever for entrepreneurs in the developed and developing world alike to build businesses. This includes everything from turning on a payment mechanism, to having a reliable store of value, to using blockchain software to manage financial statements.

Democratized Entrepreneurship: Under the right conditions, entrepreneurs are the engines of economic growth in society. They bring fresh thinking to the marketplace and fuel the creative destruction that makes market economies prosper. Blockchain technology bestows individuals and small companies anywhere in the world with many of the same capabilities of larger organizations. Blockchain-based ledgers and smart contracts lower barriers to starting a company, expedite incorporation, and cut red tape particularly in the developing world, where it takes three times longer to incorporate and costs five times as much.

Blockchains can automate, streamline, and otherwise dramatically improve the three components of business building: formation, fund-raising, and sales. Formation costs will drop significantly, as blockchain is a trusted, known way to incorporate a business. You can see ownership and maintain records easily, especially helpful in areas where the rule of law is absent. Financing a company is easier as you can access equity and debt capital on a global scale, and if you're using a common denominator—like bitcoin—you need not worry about exchange rates and conversion rates. Sales become a function of accessing anyone with a connected device. Buyers don't need a credit card, local currency, or bank account.

Through secure and immutable ledgers, entrepreneurs will be able to register their business and title of corporate assets; manage inventory, payables, and receivables; and leverage other financial metrics through triple-entry accounting software and other blockchain-based applications, reducing the need for auditors, tax lawyers, and other vendors who weigh on small businesses.[32] Regulators might cut small businesses a break for opting into a triple-entry accounting scheme. That means more to the bottom line and less wasted time. As the company grows, reconciling corporate actions and documentation will become less complex. Through smart contracts, an entrepreneur could automate many aspects of a company's operations: purchase orders, payroll, interest on debt, and financial audits in real time. Two new models for individual entrepreneurship will gain traction:

Metering excess capacity. From the centralized sharing economy to the distributed metering economy, individuals will be able to loan out their spare beds, wheelbarrows, oxen, and other tangible and intangible assets to peers in a network based on reputation scores. Blockchain enables previously impossible revenue streams such as metering Wi-Fi, electricity generated from roof-installed solar panels, Netflix subscriptions, latent computing power in your phone, and other household appliances—all through micropayments and smart contracts. The blockchain becomes a new utility for individuals to create value and earn income in nontraditional ways.

Micromonetizing data. Parents who work in the home and family caregivers of all kinds who labor tirelessly over young children and aging parents can at last monetize their efforts and be recognized for the value they deliver every hour of the day. This is not a developed-world opportunity exclusively. Big companies are looking for ways to market to people in the global South but often lack the right data to make business decisions. Contracting and licensing personal data could be a great opportunity to add a new revenue stream for a young entrepreneur while he is launching his new blockchain IPO. Today, huge digital conglomerates like Facebook and Google harvest petabytes of data about billions of people. We enter into a Faustian bargain where we give up data in exchange for cool services, but we lose privacy and data integrity in the process. Blockchain turns consumers into *prosumers*. Nike might like to know what you ate for breakfast, how often you go for a run, and whether you are thinking of buying new workout gear. Why not contract that data in exchange for Nike points or real money? Let's go one step further: Insurance companies are searching for the best data to make actuarial calculations. Your own data—how much you exercise, if you smoke, what you eat—are very valuable to them. Enter into a licensing agreement where every time they use your data to make an actuarial calculation and price a new product, you get a micropayment.[33]

Distributed Ownership and Investment

We're moving into a period of human history whereby very large numbers of people can become owners of wealth through distributed ledger technology. Enabling access to the world's financial markets and therefore the universe of investment opportunities, from conventional investments to participation in mass collaborative ventures, microlending schemes, blockchain IPOs, and reputation-based microlending, will open access to capital. Already, crowdfunding is changing the face of finance. In 2012, nonblockchain crowdfunding campaigns raised $2.7 billion around the world, an 80 percent increase over the year before. With direct peer-to-peer crowdsourced blockchain financings, these numbers are poised to grow manifold. Individuals can contribute small amounts of money through crowdfunding campaigns. Imagine a campaign that engages a million people each giving a dollar. Call it distributed ownership. Not meaningful, you say? Augur, the prediction market platform, raised millions of dollars in small increments from thousands around the globe. The range of possibilities is vast. Blockchain IPOs not only can improve the efficacy and efficiency of raising money, lowering the cost for the issuer, they can also be broadly inclusive, allowing previously unimaginable groups of burgeoning investors to participate. To date, the range of proposals to change income and wealth disparity has not reached beyond higher taxes for the wealthy on the one end, or, at its most extreme, outright expropriation by the state. Instead of redistributing and expropriating wealth, let's imagine how blockchain can create opportunities to share more equally in the wealth created by society.

REMITTANCES: THE STORY OF ANALIE DOMINGO

Analie Domingo[34] has been working as a nanny and housekeeper for twenty-five years. One of more than 200,000 Filipino-born people living in Toronto,[35] her story is fairly typical: She left the Philippines as a young woman to settle in Canada with no savings, no formal education, and very little knowledge of her adoptive country. Analie has worked very hard and has carved out a life for herself and her family. Ten years ago, she used her savings to put a down payment on a house, a remarkable feat as she had been dutifully sending money to her family in the Philippines for the previous three hundred months. Analie sent home so much money that her mother, now in her seventies, was able to purchase a home of her own in Manila.

Analie graciously agreed to let us join her on payday to document her experience. On Friday afternoon, Analie got her paycheck, handwritten by her employer, and walked it to the local bank. This took fifteen minutes; twenty minutes if you include the lineup at the teller. After she deposited it, she withdrew $200 Canadian. Cold hard cash in hand, she walked a block to catch a local bus. Instead of heading toward her home, she went two miles in the opposite direction and got dropped off at what can only be described as a bad neighborhood. She walked for another four blocks and finally arrived at the "financial institution" from which she would send the money: an iRemit counter at the bottom of a housing block in Toronto's St. James Town—one of the poorest and most notoriously dangerous neighborhoods in Canada. Because many people who use iRemit's services are unbanked, the company has begun offering other financial services, such as check cashing. Analie filled out a paper form, as she has done hundreds of times before, and handed over her hard-earned money. For a $200 wire, Analie paid a flat fee of $10. On the receiving end, her seventy-year-old mother endured a similarly taxing (and equally ridiculous) trek to receive the money. Of course, she had to wait three to four days before going to the bank, the average time these payments take to get processed. Analie walked back to the bus stop, boarded the bus, a subway, and another bus, and eventually, one hour later, reached her home.

The cost of sending that remittance, $10, is equal to 5 percent of the total value. In addition, there is typically a spread on the exchange rate of around 1 to 2 percent. At around 7 percent this is a slight discount to the international average of 7.68 percent.[36] That they are both "banked" and still have to go through this process makes the whole farcical routine more egregious. The hard cost fails to capture the all-in cost. For example, the time value of the two hours Analie wasted doing this is equal to another $40, based on her wages. Moreover, she had to leave work early because she feels unsafe going to the neighborhood when it's dark. For her mother, a septuagenarian living in Manila, the physical toll on her body of making the journey to pick up the money is equally significant. The purchasing power of the $10 Analie forwent to make the transaction happen is certainly material to her, but far more for her mother. Whereas in Canada $10 is the cost of a meal and bus fare, in Manila it could buy food for a week. Over her lifetime, Analie has paid *thousands* of dollars to intermediaries such as Western Union to send money home. Each monthly fee contributes to a global honeypot of $38 billion in fees paid annually on remittances.[37]

Remittances of funds sent back to their homelands by people living in distant locations connect diasporas globally. Diasporas are global communities formed by people dispersed from their ancestral lands but who share a common culture and strong identity with their homeland.

One of the functions of many of today's diasporas is to address and help solve common, global problems. Remittances represent one of the largest flows of capital to developing countries and can have an enormously positive impact on the quality of the lives of some of the world's most vulnerable people. In some countries, remittances are a huge and vital component of the economy. In Haiti, for example, remittances account for 20 percent of GDP. The Philippines receives $24 billion every year in remittances, or 10 percent of GDP.[38] According to the International Monetary Fund, recipients usually spend remittances on necessities—food, clothing, medicine, and shelter, meaning remittances "help lift huge numbers of people out of poverty by supporting a higher level of consumption than would otherwise be possible."[39] Remittance flows to developing nations are estimated to be three to four times as large as foreign aid flows.[40] The positive effects of remittances on the poor in developing countries are well understood, yet despite this enormous economic injection, remittance costs are still appallingly high. In some of the most expensive corridors between nations, fees on remittances can run north of 20 percent.[41]

Canada is one of the largest net senders of remittances in the world. In Ontario, Canada's largest province by population and largest economy, 3.6 million people identify as being foreign born and every year billions of dollars leave the province in the form of remittances.[42] Analie's story is noteworthy because it is the norm in Canada.

Consider the Dufferin Mall, also in Toronto. On most days the mall sees a steady flow of traffic and could be mistaken for any other shopping center in Canada or the United States. But every Thursday and Friday around five o'clock in the evening, something entirely different happens. Paychecks in hand, thousands of foreign-born Canadians descend on the mall to send remittances from the mall's various banks and foreign exchange dealers to needy family members in their home countries. A cottage industry of foreign exchange dealers and Western Union outposts has popped up in convenience stores, bars, and restaurants in the surrounding area to deal with the overflow.

Oftentimes traveling by bus, streetcar, or subway, with children in tow and exhausted from a long day, Torontonians speaking Filipino, Cantonese, Spanish, Punjabi, Tamil, Arabic, Polish, and other languages get to the mall, and then stand in long lines waiting for the chance to send their hard-earned money home. These days, most people pass the time on their smart phones, chatting over WhatsApp, Skyping friends and family in Toronto and abroad, playing games, and watching videos. More often than not, it takes upwards of a week for this money to arrive at its intended destination, at which point someone on the receiving end needs to go through a similarly tedious, time-consuming process.

What's wrong with this scenario? Just about everything. Let's tease out the bright spots. Remember, most of the people waiting in line were using smart phones, a technology that is pervasive in Canada and increasingly ubiquitous globally. Seventy-three percent of Canadians own a smart phone, and in Toronto the number is almost certainly higher. The country has a wireless network infrastructure among the best in the world, which means that not only can most Canadians own a smart phone (effectively a supercomputer), but they can also use it to harness the power of the mobile Web in ways that would have seemed like science fiction two decades ago. Why do those people wait in line to send money via a physical point of sale using decades-old technology instead of what they have at their fingertips?

Dollars are a lot less data intensive than HD video. In fact, according to Skype, video calling consumes 500 kilobits per second.[43] Sending one bitcoin takes about 500 bits, or roughly one one-thousandth the data consumption of one *second* of video Skype!

By disintermediating traditional third parties and radically simplifying processes, blockchain can finally enable instant, frictionless payments, so that people don't wait in line for an hour or more, travel great distances, or risk life and limb venturing into dangerous neighborhoods at night just to send money. Today, a number of companies and organizations are leveraging the bitcoin protocol to lower remittance costs. Their goal is to put billions of dollars into the hands of the world's poorest people. These industries have been controlled by a handful of firms that have used their unique positioning and legacy infrastructure to produce monopoly economics. But they too see the risk from this technology and they're scared. According to Eric Piscini, who leads Deloitte's cryptocurrency group, companies in the payment space today "are really nervous about what the blockchain is actually doing to them. Western Union, MoneyGram, iRemit, and others are very nervous about the disruption to their business model."[44] They should be, as there is an emerging industry of new and disruptive companies that plan to take their place.

Well, Luke, My Friend, What About Young Analie?

There are two main obstacles to creating a blockchain-based payment network for the world's poor. First, many of the people sending the money get paid in cash and those on the receiving end live in a predominantly cash-based economy. Second, most people in the developed and developing world alike don't have the knowledge and tools to use blockchain effectively. While cash may very well go the way of the dodo, until employers start beaming value to smart wallets in the developed world, and tiny streetside merchants in Manila, Port-au-Prince, and Lagos start accepting digital payments, we will still need hard currency. Western Union understands that, and that's why it is still very relevant today, with more than *500,000* agents all over the world.[45] If you're looking to exchange your remittance for cash, your options are limited. Western Union wouldn't be effective if it had only one agent. Its network has allowed it to maintain a monopoly position on the entire market for decades. There have been few if any companies with a seamless, easy-to-use "killer app" technology. Until now.

Enter Abra, and other companies like it. With a name like Abra, one would expect to see a little "cadabra," and the company does not disappoint. Abra is building a global digital asset management system on the bitcoin blockchain. Its stated mission is to turn every smart phone into a teller that can dispense physical cash to any other member of the network. We wanted to test whether this solution improved Analie's experience.

Analie and her mom both downloaded the app to their Android smart phones. Analie's balance to start was in Canadian dollars. At the click of a button, Analie initiated the transfer to her mom. She got it, in pesos, almost instantly. At this point, her mom had the choice of keeping pesos on her phone as a store of value and choosing to spend them at a growing number of merchants that now accept Abra as a payment system. By creating a payment mechanism and store of value, Abra effectively displaces the conventional banking system's two most essential roles: payments and value storage. This alone is a revolutionary concept, but here's where it gets really interesting: Mom wants cash. She pays her rent, buys her food, and manages virtually all other expenses in cash. She checks the app and notices there are four other Abra users within a four-block radius of her. She messages them all to see who will exchange her digital pesos for physical pesos and at what price. The four come back to her with different "bids" for their services. One person will do it for 3 percent, another for 2 percent, and two more for 1.5 percent. Mom decides to go with the teller offering 2 percent—not because it's the cheapest but because this teller has a five-star rating and has agreed to meet her halfway. They meet and she swaps her Abra pesos for physical pesos, the teller makes his commission, and they both walk away happy. Abra takes a 25-basis-point fee on conversion.

The entire process, from money leaving Toronto to the Filipino recipient holding cash, takes less than an hour and costs 25 basis points net, inclusive of foreign exchange and all other transaction costs. Whereas every Western Union transaction requires up to seven or eight intermediaries—corresponding banks, local banks, Western Union, the individual agents, and others—the Abra transaction requires only three: two peers and the Abra platform. "I get it now. That's really cool!" said Analie, ecstatically.[46]

For Abra to scale globally, it must address two core challenges. First, the network requires a critical mass of tellers to make the service convenient. Analie's mom won't use it if the nearest teller is twenty miles away. Abra understands this, and it is presigning tellers—at last count many thousands in the Philippines alone—who are ready to transact when things go live. Second, the model works on the assumption that tellers and customers will abide by their commitment when they transfer digital for physical currency. This is less of a concern. Businesses like Airbnb, Lending Club, and Zipcar have debunked the myth that individuals will not trust one another. Indeed, for Abra CEO Bill Barhydt, the staggering growth in the number of so-called sharing economy companies convinced him this wasn't an issue. "People are willing to trust each other faster than they're willing to trust an institution," he said.[47]

The smart phone is key to all of this. In the same way the smart phone allows you to rent your apartment to someone else or rent your car to someone else or provide ride sharing to someone else, it can also be used as an ATM. Barhydt said, "It's amazing what people are willing to do in a shared economy model and they're just not doing it for money yet, maybe with the exception of peer-to-peer lending." Moreover, he said, "It's more important to us that you trust each other rather than Abra. If you trust each other, it's highly likely that you're going to get to know Abra, and that you're going to like it and you're going to have a good experience," and ultimately trust the platform.[48]

Abra is not a remittance app but instead a new global platform for value exchange that combines in equal measure the distributed, trustless blockchain network, the power of smart phone technology, and the very human inclination to want to trust peers in a network. By offering users the ability to store value in traditional currencies, transmit value across the network, and also pay at a growing merchant network, Abra takes on not only Western Union, but also the credit card networks, like Visa. According to Barhydt:

> The settlement rails for a Western Union transaction, and the settlement rails for a Visa transaction, are very different. But the settlement

> rails for an Abra transaction that's used for both person-to-person payment, as well as person-to-merchant payments, are exactly the same. . . . We have come up with a single solution that works domestically or cross-border, and that can be used for both person-to-person payments and person-to-merchant payments for the first time.[49]

Abra might eventually become a global juggernaut, rattling the walls of the biggest financial institutions in the world. But for now, it's an elegant and simple solution to an important global problem. With remittances topping half a trillion dollars next year, the market opportunity is nothing to sneeze at.

BLOCKCHAIN HUMANITARIAN AID

Can blockchain fundamentally transform how NGOs, governments, and individual donors deliver foreign aid? Hundreds of billions of dollars of aid flow annually into developing nations, yet the macroeconomic effects of aid are not always clear.[50] There is ample evidence to suggest that corrupt officials, local strongmen, and other intermediaries steal much of it long before it ever reaches its intended source. More troubling, according to the *Journal of International Economics*, an "increase in government revenues may *lower* the provision of public goods." The report concluded that "large disbursements of aid, or windfalls, do not necessarily lead to increased welfare."[51] Organizational bloat and leadership corruption combine for lots of waste and greater disparity between haves and have-nots in the poorest countries. This is true for direct foreign aid from government to government but also for NGOs that put boots on the ground in hard-hit places.

We touched briefly on the question of foreign aid in our introduction. Let's explore it further. Recall that the Red Cross came under fire in the aftermath of the 2010 Haiti earthquake after a study conducted by ProPublica, an independent, not-for-profit news organization, and National Public Radio found the organization squandered funds and did not fulfill many of its commitments such as building 130,000 new homes. It built only six.[52] In its defense, the Red Cross argued that Haiti's shoddy land title registry hindered its efforts: nobody could figure out who actually owned the land. As a result, the Red Cross improvised a less desirable solution.

Could a blockchain-based land title registry improve this situation by providing clear title and perhaps prevent unlawful expropriation?

Foreign aid is perhaps the clearest example of the ineptitude of many governments and the rent-seeking behavior of unethical intermediaries, and is thus excellent grounds to explore blockchain solutions. The 2010 Haiti earthquake was one of the most devastating humanitarian crises of the past hundred years.[53] While the government was paralyzed and the crisis raged on, thousands of "digital humanitarians" converged on the Internet to help first responders collect, triage, and visualize pleas for help from mobile phones of devastated Haitians. Originally formed online by like-minded volunteers, these ad hoc groups became increasingly organized and effective amid the crisis. One in particular—CrisisCommons—made a real difference. CrisisCommons exemplifies a global solution network, an emerging nonstate network of civil society organizations, companies, and individuals, collaborating to solve a major problem. The digital revolution has enabled new networks to connect and collaborate across borders and can solve problems and enable global cooperation and global governance. The Internet makes all this possible. Never before could people organize collectively to create a public good as they did in Haiti. This information layer of the Internet proved vital—providing critical connections, know-how, and data for people in need and volunteer organizations alike. Imagine if there was also a value layer. What kind of possibilities could that enable?

The blockchain can improve the delivery of foreign aid in two ways. First, by disintermediating the middlemen who act as conduits of large aid transfers, it can reduce the chronic problem of outright misappropriation and theft. Second, as an immutable ledger of the flow of funds, it compels large institutions, from aid groups to governments, to act with integrity and abide by their commitments. If they don't, people will be able to see their malfeasance and hold them to account.

One could easily imagine UNICEF or the UN's women's initiative using the blockchain to get funding directly to women and children without having to go through local power structures. Individuals in poor countries could sign up for certain benefits through a distributed ledger managed by a network of different aid groups acting as nodes on the network. When particular aid is delivered—say, vaccinations from the Red

Cross or school supplies by UNICEF—those "transactions" can be timestamped on the ledger. This would reduce or perhaps prevent aid groups accidentally double spending on particular people or communities, thus spreading the benefits of aid more equitably.

Indeed, UNICEF has begun exploring cryptocurrencies. In June 2015, UNICEF announced the launch of Unicoin, a digital currency that children can "mine" by submitting an inspirational drawing to the program. The coins are then exchanged for a notepad and pencil.[54] This is a small start, but the opportunities are limitless. It's not far to imagine the hypothetical we posed in chapter 1—orphanages in villages all around the developing world working with UNICEF to set up accounts for each child from the moment they arrive. Donations could be split on a pro rata basis into each kid's personal individual account. Governments, strongmen, and other corrupt officials simply couldn't access it. The poorest and most vulnerable children in the world would have the funds to start a life when they move into adulthood. This is attainable with blockchain.

Natural disaster relief or provisions for the poor cannot all be peer to peer, of course. Oftentimes, institutions are not only desirable but also essential. But the blockchain can radically improve the transparency of how those organizations, and other institutions in the foreign aid value chain, function. Every dollar donated to the Red Cross could be tracked from its starting all the way through the value chain to the individual it directly benefits. Recall our hypothetical in chapter 1—the Red Cross could run crowdfunding campaigns for each of its most important initiatives—delivering medical aid and fighting the spread of disease, water purification, the rebuilding of homes—and when you donate you would know whether your dollar went to a plank of wood, a gallon of water, or a gauze Band-Aid. If funds went missing, the community would know and could hold these organizations accountable. Smart contracts could be employed that hold the aid groups themselves accountable. The funds for major projects—from housing initiatives to the implementation of a water purification scheme—could simply go into escrow and be released only after the successful completion of key milestones—securing title for a site, importing raw materials, signing a contract with a local supplier, building the finished product, installing a certain number of clean water access points—is achieved. The result? Radically improved transparency and accountability in the delivery of foreign aid, and thus significant improvements in the end results.

Foreign aid is the second-largest fund transfer from developed to developing nations, after remittances. Blockchain technology can enable transparency, accountability, and more efficient operations for well-meaning NGOs and better delivery of critical services in times of crisis and in normal circumstances. Of course, there are a multitude of implementation challenges—things that must be overcome. People on the ground will need to know how to use this technology. Mobile phone networks could fail in the midst of a crisis. Crafty criminal elements and corrupt governments might still find ways to defraud the poor and destitute. But are these reasons not to explore this technology? No. The situation today is dysfunctional and in many cases plainly broken. Empowering individuals and holding aid groups accountable will mean more aid in the hands of the right people. Alleviating poverty and addressing catastrophic crises is the first rung of the ladder to global prosperity. Let's take a chance on blockchain.

Microfinance: Peer-to-Peer Aid with Picopayments

Microfinance is an industry that transcends both financial services and development aid. Rather than delivering aid from the top down, microfinance institutions (MFIs) try to empower individuals to save, invest, and build small businesses. More often than not, they take the form of communal savings co-ops, where members of the community come together to pool their funds and loan them out to one another for short-term needs. When implemented and managed properly, microfinance outfits can deliver a real benefit to struggling communities: they reduce chronic hunger, increase savings and investment, and in many cases empower women.[55]

However, there are some problems with MFIs today: First, there is very little oversight into how they are run and occasionally they enable predatory loans and coercive loan recovery methods, straining communities and adding to their desperation. Second, in light of this, governments in developing countries have found that the best way to curb bad behavior is to outlaw or severely restrict MFIs altogether, as was the case in India in 2010, following an MFI controversy.[56] Third, funds don't always end up in the right hands. There is no way to ensure that the community member who needs the money the most receives it. Fourth, they are still largely regional, limiting both funds and also opportunity to invest and save.

So, people who are working on poverty will ask themselves, where does the blockchain fit in the mix of tools? How can it improve on what we're doing?

First, it will improve administrative accountability. As with corporate transparency, donors will be similarly attracted to any nonprofit outfit that uses the blockchain for greater transparency and accountability. Additionally, if microloans are recorded to the blockchain and customers of an MFI are granted permission to access them, then they can hold those outfits more accountable for bad behavior. What would-be borrower or saver would choose the opaque and murky when she can choose the open?

Second, it can mean better protection of women and children. Through smart contracts, funds can be donated into escrow accounts, accessible only by women, say, for accessing food, feminine products, health care, and other essentials. The men can't take it out of their hands to buy cigarettes or booze, or to gamble, which can be a persistent problem with money from savings or microfinance.

Third, it will enable people to source funds and opportunity worldwide, and will attract donors worldwide. Communities are typically limited by geography in which MFI they use. In the future, a would-be borrower could go online and source the best bids from a number of potential lenders, finding the one with the best rates, terms, and reputation. Formal MFIs will continue to exist, of course, but there will be easier ways to connect peers through blockchain that will make them less necessary.

Finally, blockchain payment rails, such as bitcoin, are basically tailor-made for small, disenfranchised borrowers by enabling tiny payments (picopayments, we call them) and by dropping costs close to zero. In a world where every penny counts, users should be able to pay back loans, withdraw funds, and save in tiny increments, all of which was far more challenging in a preblockchain world. They should also be able to do it instantly and efficiently, given that despite abject poverty in many parts of the world, cell phone penetration and Internet connectivity are becoming commoditized.

SAFE AS HOUSES? THE ROAD TO ASSET OWNERSHIP

Land title registration is what Hernando de Soto referred to as a nonmarketed transaction, an economic exchange generally involving a local government. Nonmarketed transaction costs include the resources wasted by waiting in line, tracking down ownership, completing and filing paperwork, cutting through red tape, resolving disputes, greasing the palms of officials and inspectors, and so on.[57] These costs are rampant in poor economies where systems are weak and government officials are known to behave without integrity. Honduras is such a place, the second-poorest country in Central America with an extremely unequal distribution of income. The economic downturn of 2008 stymied the inflow of remittances, and a military coup ousted the democratically elected Manuel Zelaya in 2009. The coup was backed by one of the region's largest landowners, a palm oil tycoon who benefited significantly in earlier land grabs that coerced Aguán campesinos to sell their land titles.[58]

Since the mid-1990s, the World Bank and other global NGOs[59] have poured $125.3 million plus technical expertise into Honduras for designing and managing land-related development projects that would accelerate the country's growth.[60] We came across plans to incorporate spatial data infrastructures that would support the geotagging of data on land and natural resource ownership and usage, climatic and natural hazards, and socioeconomic conditions that municipalities could use to inform strategic planning and investment. There was also mention of integrating databases of land projects with databases of environmental and disaster management projects at national and local levels.[61] Very ambitious.

The problem is that there are still allegations of pervasive corruption in property registry, land sales, and dispute resolution, including accusations against the middlemen, judges, and local bureaucrats. According to the Office of the U.S. Trade Representative, the property registration system is still highly unreliable.[62] Households in rural villages were systematically passed over during land title registration of residences, usually their most valuable asset, because the government limited the World Bank's jurisdiction to urban zones. In rural areas, the cash-strapped campesinos benefited least from land administration programs. Rural poverty has not decreased in Honduras since 1998. Ambiguity and corruption manifest themselves in title disputes all over the developed world. If Honduras was to suffer a catastrophic natural disaster, as Haiti did in 2010, aid organizations like the Red Cross would be similarly hamstrung in untangling the mess of titles to deliver safe, durable housing.

"What if there was a universal ledger that could include all these data and infuse trust into a highly untrustworthy situation? Blockchain seems to be particularly good at handling transactions, which none of the other systems necessarily are," said de Soto. "The fact is poor countries are by nature very corrupt, and so having your transaction ledger in every node with safety procedures makes the system efficient, cheap, and fast, but it is also the kind of thing that the poor want too because it protects their rights," he adds.[63] Here's how it works: The blockchain is an open ledger, meaning that it could reside on the desktops of the Honduran officials who needed to reference it, the mobile devices of field workers who input data, and citizens who want to maintain a copy. It's a distributed ledger, meaning that none of these parties owns it, and it's a P2P network, meaning that anybody could access it. In jurisdictions like Honduras where trust is low in public institutions and property rights systems are weak, the bitcoin blockchain could help to restore confidence and rebuild reputation.

That's what the Texas-based start-up Factom plans to do in cooperation with the Honduran regime and in partnership with Epigraph, a title software company. Factom's president, Peter Kirby, said, "The country's database was basically hacked. So bureaucrats could get in there and they could get themselves beachfront properties." He added that 60 percent of Honduran land is undocumented. The goal of the project, which has not been signed definitively, is to record the government's land titles on the blockchain ledger. Kirby told Reuters that Honduras could leapfrog legacy systems used in the developed world by deploying Factom's blockchain technology, and it would eventually make for more secure mortgages and mineral rights.[64] "Documentation for ownership from patents to houses is extraordinarily paper-based, and there's no reason it should be, other than history. Blockchain works with any transaction or interaction where property rights and timing matters,"[65] said Kausik Rajgopal, who heads up McKinsey's Silicon Valley office and payments practice.

At the end of the day we don't know whether the Honduran government will enforce land titles registered on the blockchain or sustain its use. In previous land registration attempts, the government has backed away from the additional costs of scaling up and including more people. But if the ledger delivers reliable, tamperproof data, then NGOs could get the additional data they need to inform and influence policy decisions and governance. If it eliminates five of the six steps currently required to register land in Honduras, and cuts the length of time from twenty-two days to ten minutes, then those nonmarketed transaction costs drop to nearly zero.[66] And perhaps it would enable journalists and rights advocates to shame large global corporations into not purchasing or building on or sourcing timber or water from land that has been designated for environmental protection or has historically been used by the campesinos or indigenous people without compensating them fairly. We're hopeful!

IMPLEMENTATION CHALLENGES AND LEADERSHIP OPPORTUNITIES

Blockchain technology is obviously not a panacea for the world's economic and financial woes. Technology does not create prosperity; people do. There are obstacles to overcome and opportunities for leadership. The first is technical. According to International Telecommunications Union data, there are still significant gaps in Internet connectivity, either because the telecommunications infrastructure is poor or because service is unaffordable.[67]

The second is literacy. Using smart phones and interacting online requires a workable level of literacy. In the United States, 18 percent of adults over the age of sixteen read below the fifth-grade level, 30 percent have low math literary,[68] and 43 percent of these illiterate adults live in poverty.[69] Literacy is highly uneven in the developing world. In many parts of Africa, literacy hovers around 50 percent, and the problem is even more severe when comparing the genders. For example, in places like Afghanistan, Niger, Sierra Leone, Chad, Mozambique, and other poor nations, the gap between male and female literacy is a staggering 20 percent.[70]

The third is corruption. Blockchain is a powerful tool, but like all technologies, it is not inherently good or bad. People can harness brilliant technologies, from electricity to the radio and through to the Internet, for benevolent or malevolent goals. We need leadership from the institutions in society that can leverage blockchain technology for good, such as aid groups, civil society organizations, companies, and governments, right down to the individuals who are connecting to this vast network. Only when these challenges have been met and overcome will blockchain technology fulfill its potential as an instrument of global prosperity and positive change.

CHAPTER 8

REBUILDING GOVERNMENT AND DEMOCRACY

The Republic of Estonia is a Baltic state with Latvia to the south and Russia to the east. With a population of 1.3 million, it has slightly fewer people than the city of Ottawa.[1] When Estonia regained its independence from the former Soviet Union in 1991, it had an opportunity to completely rethink the role of government and redesign how it would operate, what services it would provide, and how it would achieve its goals through Internet technologies.

Today, Estonia is widely regarded as the world leader in digital government, and its president, Toomas Hendrik Ilves, will be the first to say so: "We're very proud of what we've done here," he told us. "And we hope the rest of the world can learn from our successes."[2]

Estonia ranks second of all countries on the social progress index for personal and political rights, tied with Australia and the United Kingdom.[3] Estonia's leaders have designed their e-government strategy around decentralization, interconnectivity, openness, and cybersecurity. Their goal has been to future-proof infrastructure to accommodate the new. All residents can access information and services online, use their digital identity to conduct business, and update or correct their government records. While much of Estonia's work predates the blockchain, the country introduced a keyless signature infrastructure that integrates beautifully with blockchain technology.

Central to the model of e-Estonia is a digital identity. As of 2012, 90 percent of Estonians had an electronic ID card to access government services and travel within the European Union.[4] The chip embedded in the card holds basic information about the cardholder as well as two certificates—one to authenticate identity and one to provide a digital signature—and a personal identification number (PIN) of their choice.

Estonians use these to vote, review, and edit their automated tax forms online, apply for social security benefits, and access banking services and public transportation. No need for bank cards or Metrocards. Alternatively, Estonians can do the same with mobile-ID on their mobile phones. In 2013, Estonians submitted over 95 percent of taxes electronically and conducted 98 percent of banking transactions online.

Parents and students use Estonia's e-School to track assignments, curriculum, and grades, and to collaborate with teachers. Estonia aggregates in real time diverse health information from various sources into a single record for each citizen, and so these records don't reside on a single database. Each Estonian has exclusive access to his own record and can control which doctors or family members have access to these data online.[5]

Since 2005, citizens have used i-voting for their national elections. Using their ID card or mobile-ID, Estonians can log in and vote from anywhere in the world. In the 2011 parliamentary election, citizens cast almost 25 percent of ballots online, up from 5.5 percent in the previous parliamentary election. The people obviously like and trust the system: the number went up again for the 2014 European Parliament elections in which a third of voters participated over the Internet from ninety-eight different countries. The Estonian cabinet uses a paperless process and makes all draft legislation accessible online. The average length of weekly cabinet meetings has gone from around five hours to under ninety minutes.[6]

Estonia has an electronic land registry that has transformed the real estate market, reducing land transfers from three months to a little over a week.[7] In the last few years, Estonia has launched its e-Residency program, where anyone in the world can apply for a "transnational digital identity" and authentication to access secure services, encrypt, verify, and sign documents digitally. An entrepreneur anywhere in the world can register his or her company online in fewer than twenty minutes and administer the company online. These capabilities contribute to Estonia's image as a digital country.[8]

None of this would work or be acceptable without solid cybersecurity.

As Mike Gault, CEO of Guardtime, noted, "Integrity is the number-one problem in cyberspace and this is what Estonia recognized ten years ago. They built this technology so that everything on government networks could be verified without having to trust humans . . . it is impossible for the government to lie to its citizens."[9]

Estonia's cybersecurity derives from its keyless signature infrastructure (KSI), which verifies any electronic activity mathematically on the blockchain without system administrators, cryptographic keys, or government staff. This capability ensures total transparency and accountability; stakeholders can see who has accessed which information, when, and what they may have done with it. Consequently, the state can demonstrate record integrity and regulatory compliance, and individuals can verify the integrity of their own records without the involvement of a third party. It lowers costs: there are no keys to protect, and no documents to re-sign periodically. According to e-Estonia.com, "With KSI, history cannot be rewritten."[10]

Clearly, blockchain technology applies not only to corporations fixated on profits but also to public institutions focused on prosperity for all, from government, education, and health care to energy grids, transportation systems, and social services. Where to start?

SOMETHING IS ROTTEN IN THE STATE

In his Gettysburg Address in 1863, Abraham Lincoln said that society's greatest goal was a "government of the people, by the people, for the people." Twelve decades later, President Ronald Reagan said in his 1981 Inaugural Address, "Government is not the solution to our problem; government *is* the problem." Many in the nascent blockchain ecosystem agree. In a 2013 survey, over 44 percent of bitcoin users professed to be "libertarian or anarcho-capitalists who favor elimination of the state."[11]

Libertarians of all stripes tend to support bitcoin. It's decentralized and free from government control. It's anonymous and difficult to tax. It resembles gold in its scarcity, and libertarians favor the gold standard. It's a pure market, driven by supply and demand rather than quantitative easing. Not surprising, the first 2016 presidential candidate to endorse bitcoin for campaign payment was Rand Paul.

The libertarian bent has given opponents of digital currencies fodder for dismissing blockchain technologies outright. Jim Edwards, founding editor of *Business Insider UK*, wrote of the libertarian paradise he called Bitcoinistan, a country like Somalia "with as little government interference as possible, in a market free of burdensome laws and taxes." He described the paradise as "a total nightmare . . . characterized by radical instability, chaos, the rise of a boss-class of criminals who assassinate people they don't like, and a mass handover of wealth to a minority even smaller than the one percent that currently lauds it in the United States."[12]

Certainly, we live in a crisis-racked world. "The world has not seen this much tumult for a generation. The once-heralded Arab Spring has given way almost everywhere to conflict and repression," wrote Kenneth Roth, executive director of Human Rights Watch, founded in the 1970s to support citizen groups. "Many governments have responded to the turmoil by downplaying or abandoning human rights," using the Internet to spy on citizens, using drones to drop explosives on civilian populations, and imprisoning protesters at mega public events like the Olympics.[13]

That's the wrong response to turmoil, according to renowned Peruvian economist Hernando de Soto. "The Arab Spring was essentially and still is an entrepreneurial revolution, people who have been expropriated," said de Soto. "Basically, it's a huge rebellion against the status quo," and the status quo is *serial expropriation*—the repeated trampling of citizens' property rights by their governments until they have no choice but to work outside the system to make a living.[14]

So trampling more rights is the worst possible response because it pushes more people—such as journalists, activists, and entrepreneurs—outside the system. During the past twenty years, voter turnout has dropped in most Western democracies, including the United States, the United Kingdom, France, Germany, Italy, Sweden, and Canada. In particular, young people are looking to bring about social change outside the system, certainly not by voting. Most Americans think their Congress is dysfunctional and deeply corrupt. And for good reason: as in many countries, U.S. politicians are beholden to wealthy contributors and interest groups, and many members of Congress go on to become lobbyists. Case in point: 92 percent of Americans want background checks of people buying guns, but the rich and powerful National Rifle Association thwarts any legislation to effect change. So much for a "government of the people, by the people, for the people."

The more citizens don't feel their political institutions reflect their will and support their human rights, the more these institutions overstep their authority, the more citizens question the legitimacy and relevance of the institutions. Political sociologist Seymour Martin Lipset wrote that legitimacy is "the capacity of a political system to engender and maintain the belief that existing political institutions are the most appropriate and proper ones for the society."[15] And increasingly young people look to bring about change through means other than governments and even democracy. The bumper sticker "DON'T VOTE! IT ONLY ENCOURAGES THEM" tells the story.

"For individuals, it might not be desirable for them to be in a searchable, verifiable database of recorded history that governments could potentially use to exploit or subjugate people," de Soto said. "The legislation of most of the countries in the world is so badly done, so unwelcoming, that the cost of coming into the legal system doesn't make sense to poor people. And a country with too many poor and disconnected people causes too many problems."[16]

As legitimacy fades, libertarianism ascends. But it's not the answer to what ails the body politic. In this troubled world, we need strong governments, and ones that are high performance, effective, responsive, and accountable to citizens.

What should governments do? "Build, streamline, and fortify the laws and structures that let capitalism flourish," de Soto wrote in *The Wall Street Journal*. "As anyone who's walked the streets of Lima, Tunis and Cairo knows, capital isn't the problem—it is the solution."[17] So what's the problem? "Getting their people identified," he told us. "There is no way a government can go in and force people inside the system. So I think that governments all over the world right now are willing to turn the system around."[18]

That's where the blockchain comes in. The design principles of the blockchain should drive this transformation as it supports and enables higher levels of the following:

Integrity. To rebuild the public's trust in political institutions, elected officials must behave with integrity. Trust must be intrinsic to the system, encoded in every process, not vested in any single member. Because the blockchain supports radical transparency, it is becoming central to rebuilding trust between stakeholders and their representatives. Ongoing transparency is critical to maintaining this relationship.

Power. Everyone has a right to take part in the government, directly or by voting. Whoever is elected must conduct affairs in the full light of day as a peer among peers. With the Internet, citizens took more responsibility for their communities, learned from and influenced elected officials and vice versa. With blockchain, citizens can go one step further: they can advocate for sealing government action in the public record in an unalterable and incorruptible ledger. Not just checks and balances among the powerful few but broad consensus of the many, for example, to effect background checks on potential gun owners.

Value. Votes must have value. The system must align the incentives of all stakeholders, be accountable to citizens rather than big money, and invest tax dollars wisely. The machinery of government must be high performance, better and cheaper with technology.

Privacy and Other Rights Preserved. No spying on citizens, no arbitrary interference with privacy, family, or home, no attacks upon anyone's honor or reputation. No arbitrary seizure of property—real estate or intellectual property such as the patents of inventors—without compensation. No censorship of news organizations, no interference with efforts to assemble. People can register their copyrights, organize their meetings, and exchange messages privately and anonymously on the blockchain. Beware of any politician who argues for trade-offs between personal privacy and public security. Remember, it's a false dichotomy.

Security. Everyone must have equal protection of the law without discrimination. No arbitrary detentions or arrests. No one person or group of people should live in fear of their own government or law enforcement agencies or be subjected to cruel, inhumane, or degrading treatment from members of those agencies because of their race, religion, or country of origin. Members of police forces can't withhold evidence of undue use of force, and evidence can't go missing. It would all be logged and tracked on the blockchain.

Inclusion. Using the Internet, citizens became more involved, learned from one another. With the blockchain, the system can cost-effectively engage all citizens, recognize everyone as a person before the law, and provide equal access to public services (e.g., health care, education) and social security.

Technology is a powerful tool but it alone cannot achieve the change we need. In the spirit of the saying "The future is not something to be predicted, it's something to be achieved," let's reinvent government for a new era of legitimacy and trust. It's time to stop the tinkering.

HIGH-PERFORMANCE GOVERNMENT SERVICES AND OPERATIONS

The critics of "big government" are right in one sense. When it comes to efficiency, government services and operations have a long way to go. Governments are organized into silos that don't share information. Bureaucracy too often trumps common sense or shared practices. Citizens rarely have one-stop shopping for government services. Every country has countless tales of politicians and bureaucrats squandering taxpayer dollars.

Blockchain can improve client service, increase efficiency, and improve outcomes while enabling both integrity and transparency of government. The potential to improve all facets of government is significant, but some are especially important in the developing world, where governments are establishing new processes and can leapfrog the systems of long, stable, and open governments.

Let's look at two broad areas where we can apply the blockchain: integrated government and the public sector use of the Internet of Things.

Integrated Government

Estonia is cutting administrative inefficiencies and providing integrated services to its residents and businesses by creating an electronic ID card for everyone and using a blockchain-enabled Internet backbone known as the "X-road" to connect across multiple programs and databases in both public and private sectors. Others can do it, too.

Many countries such as Canada, the United Kingdom, and Australia have explicitly rejected the concept of a central population registry and single government ID as a matter of public policy. This decision stems from a concern for personal privacy and an aversion to expanding state power, especially in granting or revoking identities.

However, as Estonia shows, if we hash official documents (passport, birth certificate, marriage license, death certificate, driver's license, health card, land titles, voter ID, business registration, status of tax payments, employment number, school transcripts, etc.) that currently exist in multiple databases into a single blockchain, blockchain-enabled networks could deliver integrated services without going through any central processing. Not only could this model protect privacy, it could enhance it by allowing people to verify the accuracy of their information, and to see who accessed or added to that information (i.e., a permanent information audit).

In fact, in the future it makes sense that each citizen owns her identity information rather than a government. As we explained in chapter 1, just as networks and mass collaboration can eliminate the need for a government to issue currency or for a bank to establish trust, people won't necessarily even need a government-issued identity card. Said Carlos Moreira of the cryptographic security company WISeKey, "Today you need an organization with endowed rights to provide you with an identity, like a bank card, a frequent flier card, or a credit card. But that identity is now yours and the data that comes from its interaction in the world is owned by someone else."[19] On the blockchain, the individual owns the identity. Your "personal avatar" could decide what information is provided to whom under your command. It could also make choices about integrating data. However, rather than everything you do with government being integrated in some massive government database, the integration is achieved by the virtual you—owned and controlled by you.

Better integration would support life events such as marriage. Melanie Swan, founder of the Institute for Blockchain Studies, explained: "The blockchain—with its structure that accommodates secure identities, multiple contracts, and asset management—makes it ideal for situations such as marriage because it means a couple can tie their wedding contract to a shared savings account, and to a childcare contract, land deed, and any other relevant documents for a secure future together."[20] Some have suggested that the blockchain could become a public documents registry outside any government sanction or involvement. The world's first blockchain-recorded wedding took place at Walt Disney World, Florida, in August 2014. Smart prenuptial contracts, anybody?

Beyond integrated services, governments could register and manage documents with transparency and reliability. Consider the staff time spent in issuing, verifying, updating, renewing, and replacing people's official government records. In addition to ensuring document veracity, blockchain-enabled registration through peer-to-peer networks would support self-service, where people verify a document through the network, not through a registrar, as well as personalized service—when you generate an official document, it automatically contains your relevant information and access rights to that information, and tracks who accesses and uses it in the document metadata.

For example, the U.K. government is investigating the use of the blockchain in maintaining numerous records, especially for ensuring their integrity. Paul Downey, a technical architect with the U.K. Government Digital Service, noted that the perfect register "should be able to prove the data hasn't been tampered with" and should store a history of the changes that have been made, plus "be open to independent scrutiny."[21]

Blockchain-based systems can infuse efficiency and integrity into document registries of all kinds and many other government processes. Let's combine supply chain management with the Internet of Things to tag a new piece of equipment with a smart chip that communicates its provenance, ownership, warranties, or special information. Government procurement offices could track items and automate processes at every step: purchasing, releasing payment, paying sales taxes, renewing a lease, or ordering an upgrade. That's simply better asset management, reducing administrative costs to taxpayers while increasing revenues to governments.[22]

Particularly interesting are national and local opportunities to connect different blockchain networks for greater efficiency across jurisdictions. For example, departments of motor vehicles could connect drivers' databases across state or provincial boundaries to create a virtual database that facilitates confirmation of driver identity, status, and track record. Or in the U.S. health care system, "Suppose the patient, insurance company, doctor and a government payer all had their financial records come together on a single ledger, visible to all, for any given transaction. The potential for transparency would be matched only by the opportunities for new levels of efficiency," said Swan.[23]

The Internet of Public Things

We already wrote about public transportation on the Internet of Things. That's perhaps the easier IoT opportunity for government: record smart devices in a blockchain ledger for life cycle asset management of buildings, work and meeting spaces, vehicle fleets, computers, and other equipment. As with bAirbnb, government employees could dynamically match available supply and demand, lowering security, maintenance, and energy costs through automated access, lighting, and temperature controls, and tracking location, repairs, and roadworthiness of government vehicles, as well as the safety of bridges, rails, and tunnels.

Public leaders could also achieve better public outcomes in infrastructure management, energy, waste and water management, environmental monitoring and emergency services, education, and the health sector. In addition to improving efficiency benefits, these blockchain-enabled applications could also improve public safety and health, ease traffic congestion, and reduce energy consumption and waste (e.g., through leaky pipes), to name but a few benefits.

Securing Infrastructure

By intelligently partnering with the private sector and other stakeholders, the government of Estonia has created a public sector infrastructure that enables much greater convenience and access to government, banks, public transit, and other services for its citizenry. In addition to convenience, Estonia also gains competitive advantage in the global economy, attracting business and investment to the country.

Governments already provide services to neighboring jurisdictions (fire and ambulance); outsource to other jurisdictions (data processing); deliver services on behalf of another jurisdiction (federal government processing income taxes on behalf of both the national and provincial/state governments); and sharing services (sharing office buildings).

Estonia's e-Resident service is useful for individuals anywhere in the world who need an official ID to launch a business, particularly online. Estonia is positioning itself to provide services to foreign citizens that other countries are choosing not to provide. While the services available now are fairly limited, there is no limit to other government services ultimately becoming digital from end to end. For example, publicly funded libraries that are free to local residents could offer access to their digital collections to nonresidents and scholars anywhere in the world for a small fee. What other services might lend themselves to similar treatment, especially digital services where data management and integrity are important?

Offering government services beyond national borders often comes with regulatory hurdles. However, we live in an increasingly globalized world where many of our biggest challenges are not exclusive to one jurisdiction. Global problems require new models for problem solving, for working with other stakeholders. Policy that treats borders as porous, combined with blockchain technology such as the Internet of Things, could do more to address big, intractable issues.

EMPOWERING PEOPLE TO SERVE SELVES AND OTHERS

Blockchain-enabled networks make government services more robust and responsive. Self-service, in anything from renewing a permit to getting an official document, will improve how governments operate. By freeing up time, removing the potential for corruption or other artificial barriers, providing self-training modules online, and paying citizens their social security funds on time, governments empower their citizens.

New models, many to be defined, can empower people to collaborate on public policy goals. Through the blockchain, we can strike a new and appropriate balance between government's need for control and accountability for an entire budget, and the need for individuals and groups to control and contribute to portions of that budget. Some jurisdictions have been exploring new models to give individuals (recipients of benefits from multiple government programs) or communities (neighborhoods), or even entire populations (citywide) control of their own personal budgets previously controlled by civil servants.

For example, rather than requiring individuals to apply to many different government programs for various benefits, each with the its own criteria (income, assets, number and age of children, type of housing, level of education, etc.), the government platform could personalize a budget based on identity, stored information, and production and consumption patterns including risk factors such as residence in poor zip code, level of education, and purchase rates of cigarettes, alcohol, and processed foods. The individual could then decide how to use the resources to achieve his or her objectives according to his or her circumstances.

Imagine that—rather than persuading some bureaucrat that your child needs a new winter coat, you can decide on your own! The result is increased personal accountability and empowerment. We could do the same at the community level (portions of budgets related to community-specific services such as parks and community centers) or at a cross-government level (establishing priorities and then spending discretionary budget).

Some jurisdictions are already empowering the least advantaged.[24] The blockchain could accelerate this trend, allowing taxpayers to see where their dollars are flowing, how fellow citizens are using these resources, and whether programs are achieving results (income changes, educational goals reached, housing found, etc.). The platform reduces or even eliminates the need for time-consuming and complex monitoring and report-backs. While the vast scope of the data and how they're tracked through peer-to-peer networks may sound scary and Orwellian, it is actually just the opposite. Rather than all the data and authority resting in the hands of some central authority or anonymous bureaucrat, individuals and communities could act based on verified and trustworthy information. At the same time, the blockchain ledger assures accountability for the use of public funds. We can now achieve two previously seemingly contradictory goals: "more government" through more information and context; and "less government" through providing information and better tools for individual and group decision making and action within that context.

Streaming Open and Trusted Data

Perianne Boring, founder and president of the Chamber of Digital Commerce, champions the idea that distributed ledgers open up government for the better. To her, "Blockchain enables radical transparency because it provides everyone with provable facts. Anyone can view any transaction that has ever happened on the blockchain."[25]

Governments can easily provide data that others can use for public or private good. This differs from so-called Freedom of Information legislation where citizens must request access to important government information. Rather this involves the release of assets—actual data. Governments could release thousands of categories of data in raw format, stripped of personal identifiers: traffic patterns, health monitoring, environmental changes, government property, energy usage, government budget and expenditures, expense accounts. Citizens, companies, NGOs, academics, and others could analyze these data, put them into applications, map them, and otherwise use them for, say, understanding consumer demographic trends, researching patterns in human health, or knowing whether the bus is going to be on time.

As of August 2015, the U.S. government has already published 165,000 data sets and tools on its Open Government Web site.[26] The U.S. government's philosophy that government-held data is public data has made it a pioneer in transparency. Other governments are following suit. As of August 2015, the U.K. government has released 22,000 data sets.[27]

Releasing data through peer-to-peer networks and the blockchain will introduce even greater levels of efficiency, uniformity, utility, and trust. Making data public is an incentive for ensuring data accuracy. People can view and flag data when they find an error or can prove that data have been altered or corrupted.

If you register a complete data set in a blockchain network, then the network can log additions and changes to the data set and can block efforts to tamper with the data. No need for a central administrator. Governments could release more programmatic data to help the public and analysts understand these programs and their impact.

Partnering to Create Public Value

We've already seen how simply making more trusted information available can be used for positive economic and social value, and how individuals and communities can be empowered to improve their own lives. Blockchain-enabled peer-to-peer networks will require us to rethink how we divide responsibility in creating public value. When governments publish raw data, they become a platform on which companies, the civil society, and other government agencies and individuals can self-organize to create services. We have used "pay for success" models for a few years now to engage businesses in solving civic problems. For example, the U.S. Department of Labor funded initiatives that hire ex-offenders and reduce recidivism, and the City of Chicago raised education levels among disadvantaged preschoolers.[28]

This model also encourages innovation and incentivizes achievement of desired results by releasing funds only when these results have been achieved and are measurable. Think about the power of ongoing micropayments to a small not-for-profit group working in a community on sustainable energy initiatives. A government program could link funding to actual declines in consumption. The not-for-profit group could support itself without having to rely on complex paperwork for reimbursement and might even secure financing based on the government's commitment to its participation in the "pay for success" model.

Pegging Smart Social Contracts to Political Reputations

Just as the bitcoin network uses blockchain technology to constantly ensure the integrity of payments, government networks can use blockchains to ensure the integrity of their transactions, records, and important decisions. Officials can't hide "off the book" payments or other government records, including e-mail records, decision logs, and databases. Whereas security often derives from fences, walls, or a perimeter of protection, the blockchain protects against tampering from both inside and outside. Therefore, it keeps "honest people honest."[29]

Transparency is crucial for changing the behavior of an institution. While we of course cannot force these values and behaviors on our public representatives, we can limit their decisions and actions through smart contracts that define their roles and responsibilities as our representatives and then monitor and measure them on the blockchain.

Remember, smart contracts are self-executing agreements stored on the blockchain, which nobody controls and therefore everyone can trust. Political factions such as the Grand Old Party could use them so that candidates like Donald Trump who use their party infrastructure to debate and campaign during primaries couldn't run as independents in a general election. We could apply smart contracts to different government operations (supply chain, external legal services, pay-for-success contracts) and even more complex roles of government and our elected representatives. We do foresee peer-to-peer networks tracking an elected official's commitments and his or her fulfillment of these commitments. Watchdogs already do this through formal and informal peer networks on the Web.

While this approach could not apply to everything we expect from our political leaders, we could use it for all manner of specific commitments and actions. While measurement of eventual outcomes will be more of a challenge (e.g., results achieved with money spent), over time we will build experience and expertise with indicators so that we base our assessments on facts rather than current spin. This is not pie in the sky—in London, a candidate in the 2016 mayoralty campaign is calling already for the use of the blockchain to hold elected officials accountable for public business.[30]

Regulators could implement blockchain processes as a verifiable means to track the commitments of regulated industries in real time, assessing whether they are following through on promises made (e.g., investments in sustainable energy) or complying with regulated practices (e.g., on-time delivery, safety targets). While publication of key performance indicators and results on public Web sites is increasingly common, the blockchain would enable these processes to be automated and guaranteed accurate when applied to measurable results.

The data generated by these processes ensure that the public is constantly aware of who is behaving with integrity. How often did he show up at meetings? How did he vote? Did he abide by his commitments to do such-and-such? Who donated to her political campaign? Who violated the terms of her smart contract? Elected officials and those regulated must honor their commitments or explain why they haven't. It also provides feedback to the electorate on whether their demands as constituents are reasonable and fair, not reactionary. Voters often want more services but lower taxes, or more factories but not in their backyard, or lower prices but higher wages. As such, open data provides a means of understanding trade-offs and increasing the accountability of all participants.

THE SECOND ERA OF DEMOCRACY

While representative democracy is complex and varies globally, one thing remains constant: passive citizenry. To date the discussion has focused on how blockchain technology can help create the conditions for fair, secure, and convenient voting to occur. To be sure, we have big opportunities. Online voting based on the blockchain would enable citizens to give their views more often. But attempting to replace representative democracy would be a mistake. "Motions put to a vote are usually well refined distillations of large and complex issues. They result from a long process involving conflicts, contradictions and compromises. To understand a motion and to vote responsibly, citizens need to participate in some form of refining process," Don wrote in *The Digital Economy* more than twenty years ago.[31] However, if we understand the contours of a new model, we can see how blockchain technology can help far beyond voting.

Technology and Democracy: Not a Happy Story

How has technology affected democracy? Surprisingly, the story is mixed at best. Television arguably degraded democratic discussion, turning what Al Gore called "the marketplace of ideas"[32] into a one-way dialogue. Add equally toxic cable news—where talking heads win ratings by attacking opponents rather than discussing ideas—and you've got stupefying battles of extremes. As the fictitious news anchor Howard Beale in the film *Network* said, "I'm as mad as hell and I'm not going to take this anymore!"

So far the Internet hasn't changed democracy for the better. If anything, with increased surveillance and infringement of privacy on the pretext of national security, democratic governments are behaving more like authoritarian regimes. We'd like to focus on three particular challenges.

1. Fragmenting Public Discourse

Al Gore hoped the digital age would reverse the tide of negativity eroding our basic institutions. "The greatest source of hope for reestablishing a vigorous and accessible marketplace for ideas is the Internet."[33] He was not alone. We have long argued that as the Web extends in usage, resources, and connectivity, increased access to factual information would improve the quality of public discourse.

However, the opposite seems to be occurring: balkanization of perspectives and exploitation of the new tools by ideologues who organize into battalions. Today as content production becomes more distributed and the sources of information and opinion proliferate, anyone can present a certain view and attract a like-minded audience, which may be small but may also be zealous.

The new communications and data analysis tools have also allowed ideologically driven groups to hijack social and political debates. Both liberals and conservatives are using them to create echo chambers that undermine the potential for compromise, let alone consensus.

2. Scaling Ignorance on the World Wide Web

Just as people can't tell a person from a dog on the Internet, they can't always discern the truth. Conspiracy theorists popularize their antievidence views in days or even hours,[34] most recently those around the crash of Malaysia Airlines Flight MH370. Consider that three in ten Americans now believe that human beings have existed since the beginning of time.[35] And despite overwhelming scientific evidence that carbon emissions threaten life on Earth, those with short-term vested interests have effectively denigrated the science and blocked intelligent discussion, let alone action plans. Those using the Web to further ignorance and denialism are outgunning scientists and rationalists. Internationally repressive countries from Iran to North Korea are creating private, restricted versions of the Internet for their citizens, making the Web an ever more powerful tool to trump rationalism with ideology.

3. Complicating Policy and Implementation

In the predigital era, enacting and enforcing policy was less complicated. Policy specialists and presidential advisers had strong command of the issues. Today they can barely keep pace with defining the problems, let alone crafting the solutions or explaining them to the public. So bad is the problem that President Obama signed into law the Plain Writing Act of 2010 requiring federal agencies to use language the public can understand.[36]

Today, many unforeseen issues arise between elections. No government can credibly assert that it has a voter mandate to take specific actions on all pertinent issues. Moreover, governments lack sufficient in-house policy expertise on many issues. So even if a government commissions an opinion poll to discern the public's view, the polling process doesn't tap into the wisdom and insight that a nation's citizens can collectively offer.

Putting Democracy on the Blockchain

All of these problems suggest a new model of democracy that emphasizes public discourse and citizen engagement. Let's not confuse civic engagement with notions of so-called *direct democracy* where we all watch the evening news and vote on public hangings via our mobile devices or interactive televisions. Citizens don't have the time, interest, or expertise to become informed on all issues. We want reasoned opinion, not just any opinion. We still need legislative assemblies to debate, refine, and resolve issues.

But surely a more collaborative model of democracy—perhaps one that rewards participation such as the mining function—could encourage citizens' engagement and learning about issues, while at the same time invigorating the public sector with the keen reasoning the nation can collectively offer. Could we create a culture where people are turned on by the democratic process, not turned off by their representatives' abuse of their seats?

Why has this not occurred to date? The main problem is not a technological one. Most politicians, regardless of political stripe, seem to care more about winning elections than about solving the crisis of legitimacy through citizen engagement.

Let's start with the basics. The most fundamental process for representative democracy is the election. Voting is the right (and in some countries, like Belgium, the responsibility) of all eligible citizens in a democracy. Yet around the world elections are deeply flawed. Corrupt officials tamper with results or outright rig them. Voting can be suppressed using everything from lack of access to bribery and intimidation. Manipulating elections is a complex business but it's done almost everywhere. Could blockchain technologies help improve the voting process?

For all our technological advances, the mechanics of voting in elections have remained largely unchanged for hundreds of years. In many parts of the world, to cast your vote you go to a polling station, identify yourself, mark a paper ballot, put it in a secured box, and wait for human beings to count the ballots.

Electronic voting (e-voting) is a term for voting with the aid of any electronic system. E-voting has, in many cases, proved as unreliable as manually tallied votes. E-voting today suffers from three problems: attacks on the software and hardware, mistakes or bugs in the coding, and human error. In 2004, a voting machine aiding in the general election in North Carolina was accidentally set to store only 3,000 votes. It irrevocably lost 4,438 votes in a race that was decied by a difference of only 2,287.[37]

BLOCKCHAIN VOTING

How might voting on the blockchain work? Imagine the board of elections creating digital "wallets" for each candidate or choice, with approved voters allocated one token or coin each for each open seat. Citizens vote anonymously through their personal avatar by sending their "coin" to the wallet of their chosen candidate. The blockchain records and confirms the transaction. Whoever ends up with the most coins wins.

Some have tried to solve the problem of trust by using end-to-end auditable voting systems. Votes are typically made via kiosk, which produces a cryptographically authenticated paper record of the ballot but allows votes to be counted electronically.

CommitCoin uses cryptographic proof-of-work systems to prove a message was sent at a certain date. The inventors, Jeremy Clark and Aleks Essex, say we can use it to prove the integrity of election data before the event, as a means of "carbon dating commitments," providing a baseline to counteract fraud and error.[38]

End-to-End E-Voting Systems

Citizens are making advances all the time. In 2015, academics from the National and Kapodistrian University of Athens published a paper introducing DEMOS, a new end-to-end (E2E) e-voting system verifiable in the standard model, without reliance on setup assumptions or access to a "randomness beacon."[39] It uses a distributed public ledger like the blockchain to create a digital ballot box that citizens can use to vote from anywhere in the world.

An E2E verifiable election detects election authorities who try to misrepresent outcomes. Voters cast ballots in exchange for receipts that allow them to verify that (a) their vote was cast as intended; (b) it was recorded as it was cast; and (c) it was counted as it was recorded. An external third party could verify the election results. Voters still must accept setup assumptions and take a leap of faith on the election results.

With DEMOS, the voting system generates a series of randomized numbers.[40] Voters get two sets of numbers, or keys: one corresponding to them, and one to their preferred candidate. After the encrypted vote is cast, it's sent across multiple servers. Results are published to a bulletin board publicly displaying all the information related to the election.

Neutral Voting Blocs

In Australia, an organization called the Neutral Voting Bloc (NVB) is using voting on the blockchain to revolutionize democracy in another way entirely. They have a unique approach to government, and it's optimistic: "We believe the best way to fix politics is to participate ourselves."[41]

Founder Max Kaye describes NVB as a "political app" where interested citizens can register their opinions on policy issues by "voting" on the blockchain. Once the time is up, the final tally instructs elected officials on how to vote in governmental proceedings. When asked, why are you using the blockchain? Max Kaye replied, "Because we intend to facilitate a variety of parties, some of them will necessarily disagree strongly. To maintain integrity we need each party to be able to independently verify the voting record, and each vote." Furthermore, Kaye suggests there are anticensorship properties, and immutability. He said, "The only electronic structure on the planet I'm aware of that can do this is the bitcoin network. (Although there are other blockchains, they are not immutable enough because their hash-rates are so low)."[42]

Protecting the Voters

Voter intimidation can take a violent turn. In Zimbabwe, the party opposing Robert Mugabe withdrew from the election when coercion from supporting militias had become too lethal. The elections were carried out anyway—Mugabe won. While technological advances always come with people who exploit them to their own advantage, some are beginning to say that blockchain technology could eradicate corruption in places such as Asia.

In July 2014, during one of the most contested presidential elections in Indonesian history, an anonymous group of seven hundred hackers created an organization called Kawal Pemilu, or "Protect the Vote." Its mission was to publicly tally election ballots online to let voters verify results at each polling station. The principles of decentralization, transparency, and individual anonymity combined to ward off malicious cyberattacks and ensure a fairer election.[43]

"Do corrupt governments want to keep themselves honest?"[44] asked Anson Zeall, CEO of CoinPip, a company specializing in sending fiat currencies across international boundaries using the blockchain. He questions whether everyone embraces advances in voting, and whether politicians actually want fair elections. To others, e-voting seems like an unnecessary or hasty leap forward. We argue that many of these issues belong to the realm of implementation, not design.

The redesign of our electoral and political systems will likely influence more fundamental issues with voting in democratic elections. Compare voter ID fraud with other more insidious factors. A comprehensive investigation of voter ID fraud in the United States in 2014 found thirty-one incidents, including prosecutions and credible allegations, in federal, state, and municipal elections—since 2000.[45] In that time, more than one billion ballots were cast in general and primary elections alone.

In the four states with the harshest ID laws, more than three thousand votes were positively rejected for lack of proper ID.[46] This doesn't include those who didn't bother to try at all—and that is a much bigger problem. While their model of democracy is heralded around the world, most Americans don't vote in elections, citing reasons like "nothing ever gets done," "politics is so corrupt," and "there is no difference between the choices."[47] We expect blockchain technology to have some innovative approaches to these problems, too.

With time and development, blockchain technology might be the impetus that allows e-voting to transform democratic elections and institutions by effectively and reliably bringing them into voters' hands.

ALTERNATIVE MODELS OF POLITICS AND JUSTICE

If the blockchain could enable a more efficient, responsive government and improve how democracy is administered through new voting procedures, could it also catalyze new political processes as well?

For some supporters of next-generation government, the ultimate aim of electoral reform is to enable a system of "liquid democracy." Eduardo Robles Elvira, CTO of Agora Voting, is one such fan. He describes liquid democracy as combining the best parts of direct democracy (like the sort practiced in ancient Athens) with today's representative democracies, which ask very little of their electorates.

Liquid democracy, also called *delegative democracy*, allows citizens the ultimate in customization and personalization of the democratic experience. In Robles Elvira's words, in a liquid democracy "you can choose your level of participation at any point in time."[48] Your input is always welcome, but not required to keep the country running.

Voters can delegate voting authority to multiple representatives delineated across an array of topics.[49] Referenda are then held frequently and categorized by topic, indicating which proxy (if any) should be prompted to cast their vote on the issue. This enables a system in which voters can select many trusted experts or advisers to vote on their behalf. Underlying this ideology is the belief that no one person (or party) has the full, right answer to every question. In representative democracies this axiom is often both assumed and ignored.

Robles Elvira is working with governments to build "a highly distributed, unique log of events really good at solving distributed denial-of-service (DDOS) attacks." Blockchain technology enables this. He said, "It is very difficult to create a secure, distributed system and the blockchain allows us to do this . . . and it's not just that it's distributed, but that it's distributed in a secure way. This is really important and can be useful for a lot of applications; e-voting is just one of them." His company, Agora Voting, provides the technological infrastructure needed to conduct auditable, transparent, and verifiable e-elections. "With top-notch cryptographic technology, humans become the weakest link in the security chain."[50]

Spain's antiausterity party Podemos (translated to "We Can") uses Agora Voting to hold its primary elections. With the party's commitment to participatory democracy came a commitment to transparency, an ideological shift in Spain and elsewhere consistent with the one underpinning distributed technologies.

Robles Elvira sees some limitations, too. To maximize security and anonymity, a user currently needs access to the whole blockchain, a behemoth of a file. Size makes it difficult to access (especially on a mobile) and decidedly user-unfriendly. Still, the technology is always evolving and designs are ever improving. "We are at the beginning of e-voting," said Robles Elvira.[51] The technology is pliable. Undoubtedly, its best applications are yet to come.

Dispute Resolution

Some legal disputes are best left out of the courts. We've seen how smart contracts can enable decentralized, independent, autonomous adjudication of commercial disputes. Smart contracts are indifferent to notions of fairness or justice, however, and unable to reconcile conflicting versions of facts. Even more revolutionary for adjudication than a verifiable record of evidence, the blockchain can be the platform for P2P dispute resolution. In this model, a jury of hundreds or thousands of your peers could weigh in to effectively, as Pamela Morgan of Empowered Law said, "crowdsource justice."[52]

Random-Sample Elections

Another democratic model enabled by blockchain-style governance is random-sample elections. Voters selected at random would receive a ballot in the mail and directions to Web sites with candidate information and statements from interested parties. Anyone may request a ballot, but it will not be counted, and will appear indistinguishable to its valid counterparts to all but the requesting voter. These can be sold to a vote buyer, but they would never know if that vote counted. As these votes are believed more likely to be sold than their countable counterparts, it makes coercion impractically costly. David Chaum, inventor of the concept, said random-sample polling could produce more representative and reliable results than elections today regularly achieve.[53]

Prediction Markets

The company Augur is using the blockchain to aggregate many small wagers about future events into powerful predictive models. With the right application, it could help create collaborative democracy. Governments could use prediction markets to engage citizens in helping better understand future scenarios, enabling governments to make better policy choices.

Ethereum's Vitalik Buterin discusses an alternative model of political life called *futarchy*.[54] Conceived by economist Robin Hanson, its tenets can be neatly summarized as "vote for values, but bet on beliefs." Citizens elect their democratic representatives in a two-stage process: First, pick some metric to determine their country's success (like literacy or unemployment rate). Then, use prediction markets to select government policies designed to optimize the elected metric.

Augur's style of prediction making could engage citizens in making small choices that contribute to national discussions of policy, eventually shaping the future of their own democracy.

Blockchain Judiciary

The blockchain can also transform our judiciary. Combining the concepts of transparency, crowdsourcing, and online citizen participation—over a blockchain—we can envision reintroducing concepts of ancient Athenian democracy into the twenty-first century.[55] CrowdJury[56] looks to transform the justice system by putting several judicial processes online, using both crowdsourcing and the blockchain, including filing a charge or complaint, gathering and vetting of evidence, engaging citizens in open trials online and as online jurors, and issuing a verdict. Think transparent processes with crowdsourced discovery, crowdsourced analysis, and crowdsourced decision making and presto—you get an accurate outcome in a much shorter time frame and at vastly reduced cost.

The process[57] starts with the reporting online of a suspected civil or criminal wrongdoing (e.g., a public official suspected of receiving bribes) and inviting potential witnesses to provide evidence, and combining information from multiple sources. The original complaint or claim, as well as all the evidence, would be cryptographically stored via the blockchain to ensure that it remains on record and is not tampered with.

Once it is filed, relatively small (nine to twelve people) groups of volunteers self-selected based on required expertise would analyze the facts and determine whether there is validity to go to trial. At trial, there would be two possible paths. First, the named "wrongdoer" pleads guilty and proposes restoration (which may or may not be accepted by a jury) or the complaint proceeds to online trial with a massive jury. Just as in Athens, where any citizen over thirty could apply for jury for any given period (but not for a specific case), individuals will apply for juries with final selection by a randomization device, just as Athenian jurors were selected by a *kleroterion* in the fourth century BC.[58] As a result, there is no bias in the distribution of jurors to specific cases. The trial and all the evidence are broadcast online in an open-court-like model. Anyone can "attend" and ask questions of the defendant, but only jurors vote for the verdict via an online vote.

Let's start with conflict adjudication in low-value disputes and resolving issues in global communities across jurisdictional lines, for example social networks. The U.K. Civil Justice Council recently looked at online models worldwide to recommend online dispute resolution.[59] Most of the early models depend upon use of judges or other expert adjudicators at some stages in the online process. Other processes in place rely upon other online participants to call out and address inappropriate behavior online such as defamatory feedback (e.g., eBay's subsidiary in the Netherlands Marktplaats' Independent Feedback Review) or cheating at online games (e.g., like Valve's Overwatch, which allows qualified members of the community to review reports of disruptive behavior and apply temporary bans, if appropriate.)[60]

This is a far cry from mob justice. It's the "wisdom of the crowd" applied to many more judicial processes with beneficial results.

ENGAGING CITIZENS TO SOLVE BIG PROBLEMS

Most people who believe in science understand that human carbon emissions are warming the atmosphere. This climate change is dangerous to ours and many other life-forms on the planet. Governments, companies, and NGOs working on reducing carbon tend to agree that so-called carbon trading is an environmentally effective and economically sensible approach to lowering emissions.

One policy is called "cap and trade." A regulatory body sets a "cap" or limit on carbon emissions and lowers it over time to reduce the amount of pollutants released into the atmosphere. The "trade" represents a market for carbon allowances, helping companies and other organizations to comply with their allocated limit. According to the Environmental Defense Fund, "The less they emit, the less they pay, so it is in their economic incentive to pollute less."[61]

Today the European Union's most developed nations have cap-and-trade exchanges. California, Ontario, and Quebec agreed on the Montreal Protocol, advocating for a global exchange. Officials at nation, state, city, and enterprise levels could allocate cap-and-trade credits to balance set allowances. At the same time, blockchain-based reputation systems could rate the kilowatt-hours of energy providers to the grid according to standards of sustainable greenhouse gas reduction. For example, the system could tag energy sourced from coal with higher cap debits and renewables like solar as credits. Blockchain can help automate the cap-and-trade system on an industrial scale. Efficient pricing algorithms compute credits and debits in real time, and green organizations capture and track their carbon credits on the ledger and roll them into an exchange.

What if we created a cap-and-trade system for people? Surely we need more than our institutions to change their behavior! Personal carbon trading would work through the Internet of Things. Sensors, detectors, and instrumentation would measure your water heater, dishwasher, and household thermostat in real time and inform you of your carbon credits balance. At the same time, you could earn credits by acting in practical, sustainable ways. If you added an array of solar panels to your roof, you would earn credits by returning excess energy to the grid.

Could this create new sources of annual income for people? After all, the poor and homeless are low carbon users. By biking to work, you could save credits that your hot water heater could spend: "Hey, dishwasher—my personal cap-and-trade watch indicates we can afford to run on full wash and thirty-minute dry cycle." Water sensors in the washer could manage water usage based on an acceptable level of particulate density, dampness sensors in the dryer could turn the dryer off when the clothes have reached an acceptable level of dryness, and the building's HVAC system could harness the excess heat.

WIELDING TOOLS OF TWENTY-FIRST-CENTURY DEMOCRACY

As a global, distributed, and programmable ledger that is secure, designed for privacy, and enriched with incentive systems, blockchain technology lends itself to the development of new democratic tools such as:

Digital Brainstorming: Bringing together policy officials and citizens to have real-time, moderated, online brainstorming sessions to identify new policy issues or needs. Consensus is then achieved through one-token, one-vote systems that can help achieve thoughtful discussion and make it harder for disrupters, trolls, and saboteurs to cause damage.

Challenges: Online contests with a panel of judges. Think of pre-blockchain models like the Goldcorp Challenge (mentioned earlier), the X Prize, or the numerous innovation challenges conducted by many Western governments. The goal of challenges is to engage citizens in innovation and the creation of public value.

Online Citizen Juries and Panels: Citizens chosen at random serve as policy jurors or advisers on a topic. The jury uses the Internet to share information, ask questions, discuss issues, and hear evidence. Blockchain reputation systems help questioners to know the background and reputation of the jury and panel members. Decisions and records are recorded on the blockchain.

Deliberative Polling: This gives citizens the resources to learn about and reflect upon the issues in a collaborative and deliberative fashion. This would combine small group discussions on the Internet with scientific, random sampling to contribute more informed public input in policy making than instant polling can provide.

Scenario Planning: Building scenarios with simulation and modeling software to project future policy needs and to understand the long-term consequences of decisions. Politicians, bureaucrats, and citizens could assess the potential impacts on a range of factors, ranging from health, to the environment, to the economy.

Prediction Markets: As we explained in the case of Augur, there are countless opportunities to use prediction markets for trading the outcome of events. Governments can use them to gain insight into many substantive questions: When will the bridge actually be built? What will be the unemployment level in twelve months? Will there be a National Party prime minister after the next election—an actual question from an iPredict market in New Zealand.

Blockchain technologies could supercharge all of these tools. To begin, contributions from citizens could be private, opening up the possibilities of engagement. This is bad for repressive governments but good for democracy as it makes it more difficult for government authorities to censor, suppress, and track down opposition. At the same time, as described earlier in the case of Blockapedia, blockchain-based reputation systems could enhance the quality of discussions, reduce the number of trolls and saboteurs, and ensure that all comments are accurately and indelibly recorded. When there is compensation for winners or other contributors, settlements could be much more granular and instant through digital currencies. Various smart contracts could be constructed with citizens and groups to better clarify the role of everyone in the process.

Melanie Swan, founder of the Institute for Blockchain Studies, argues that blockchain technology might have a maturing impact on how society approaches topics like governance, independence, and civic duty. "It might seem harder to let go of centralized authority in matters of government and economics as opposed to culture and information, but there is no reason that social maturity could not develop similarly in this context."[62]

Clearly, the next-generation Internet provides profound new opportunities. The main challenges are not technological. One cautionary example: Obama's 2008 campaign created an expansive Internet platform, MyBarackObama.com, that gave supporters tools to organize themselves, create communities, raise money, and induce people not only to vote but to get involved in the Obama campaign. What emerged was an unprecedented force: thirteen million supporters connected to one another over the Internet, and self-organized to build thirty-five thousand communities of people with common interests. When young people chanted "Yes We Can," it wasn't just a slogan of hope; it was an affirmation of collective power.

However, in 2012 the Obama campaign shifted from citizen engagement to "big data," replacing "Yes We Can" with "We Know You." It used data to swing voters and target supporters for funds. The campaign won the election, but relegated citizens to consumption of its messages. The big data strategy had fewer risks than a strategy of self-organizing communities.

During both his terms the president did take important steps to engage citizens, primarily through "Challenges," which are elaborate contests for innovative ideas. But in his critical second campaign, Obama failed to engage citizens and missed a historic opportunity to strengthen government legitimacy. In the end, even President Obama, who had been called "The First Internet President," took the expedient route to power, using social media to broadcast messages and raise funds through data-enriched targeted advertising online.

If not the Internet President, then who?

There is a role for everyone in moving government and democracy onto blockchains. First, there are unlimited opportunities for eliminating redundancy and wasted time, voting and participating in new democratic processes, serving as a juror, earning energy credits, paying taxes and receiving public services, and seeing where one's tax dollars are going and how representatives are voting. Elected representatives need to step up and show leadership in designing and implementing smart contracts. If you have integrity, why not encourage the creation of blockchain reputation systems? "Voters have short memories,"[63] said Andreas Antonopoulos. Create better transparency, whether you are a judge, attorney, police officer, or parliamentarian. Civil servants and government employees could use sensors and cameras to track public assets and inventory on the blockchain, prioritize infrastructure repairs, and allocate resources. If you're a young person, don't give up on democracy. It may be broken but it's fixable. Focus on campaign financing as a starting place for blockchain transparency, as big money is currently the most fundamental problem. If you're a government contractor, use smart contracts to clean up graft and waste and evidence your superior performance. The possibilities abound.

Clearly there will be a struggle to bring about change, but citizens of the world, unite! You have everything to gain through the blockchain!

CHAPTER 9

FREEING CULTURE ON THE BLOCKCHAIN: MUSIC TO OUR EARS

It wasn't your typical yearling's birthday party. The celebration took place at the Round House an hour outside London, in a massive barn complete with sound-reactive LED tree, a bouncy castle, and a buffet fit for Henry VIII. The crowd was eclectic: a "contact" juggler, two dozen toddlers, their parents, neighbors, musicians, and a handful of blockchain developers. There was Vinay Gupta, a Scottish-Indian engineer best known for inventing the hexayurt, a small disaster relief shelter. Gupta is now explainer-in-chief when it comes to communicating blockchain technology to the masses. There was also Paul Pacifico, CEO of the Featured Artists Coalition. After a career in banking, Pacifico is now fighting for the rights of musicians. And, of course, there was our host, Imogen Heap, an accomplished composer and musician, voted "inspirational artist of the year" by readers of *Music Week*,[1] and the mother of one-year-old Scout.

"I want to know that the stuff I'm making could be worth something to Scout someday," Heap told us. She was expressing her deep concerns about the music industry. "It's so fragmented; there's so little leadership, and there's so much negativity around the business side of it," she said. "Everything is topsy-turvy. It's all upside down. The artists are at the end of the food chain. It just doesn't make sense. Music is everywhere, all the time. It's on our phones, it's in our taxis, it's everywhere. But the artists are getting less and less."[2]

And therein lies the rub. The Internet is a marvelous muse, both a medium of creativity and a channel for free speech. There's no shortage of ideas for what talented artists, designers, and coders—and their many fans—can do with one another on the World Wide Web. Nor is there a dearth of ways to make money from all this creative collaboration. Creative industries such as music publishing and recording have been tapping new revenue streams like digital downloads and streaming audio. The problem is that, with each new intermediary, the artists get a smaller cut and have little say in the matter. David Byrne of Talking Heads fame summed up the situation in an op-ed piece: "It seems to me that the whole model is unsustainable as a means of supporting creative work of any kind. Not just music. The inevitable result would seem to be that the Internet will suck the creative content out of the whole world until nothing is left."[3]

This chapter looks at ways that blockchain technologies are putting artists at the center of the model so they can not only "have their cake," that is, exercise their freedom of expression, but also "eat it, too," maximizing the value of their moral and material interests in their intellectual property. In other words, to restore their rights. No more big, greedy intermediary, no big government censors. Here, we survey the cultural landscape—art, journalism, and education—where basic human rights and livelihoods hang in the balance.

FAIR TRADE MUSIC: FROM STREAMING MUSIC TO METERING RIGHTS

"If Scout ever did end up being a musician, how on earth would she make money? She wouldn't be able to," said Imogen Heap of her daughter's music career, were it dependent on the current music industry model. "We need something really simple and core, something trustworthy, for people to feel that music is something they can do for a living."[4] Paul Pacifico agreed: "We want a music industry that reflects the cultural, technological, social, and commercial sense of our times and allows for a sustainable and viable future for creators and consumers alike."[5] Heap has teamed up with Pacifico, Vinay Gupta, and others to create this new music ecosystem.

If there were a prediction market for innovations, we'd bet on Team Heap. In 2009, she became the first woman to win a Grammy solo for engineering her own album, *Ellipse.* She took all her Twitter followers to the award ceremony by wearing what has become known as the "Twitter dress." Her outfit, designed by Moritz Waldemeyer, featured an LED zipper that streamed her fans' tweets around her shoulders. In 2013, Heap kick-started the nonprofit Mi.Mu to invent a musical glove system. It combines mapping software with motion detection sensors so that performers can control lights, music, and video with user-customized gestures. The invention won top prize at the 2015 Berlin Awards for WearableIT/FashionTech. The gloves are quickly catching on. Pop star Ariana Grande posted this message on YouTube with her video cover of Heap's "Hide and Seek": "want to thank my idol @imogenheap for allowing me to use the Mimu gloves on my first ever world tour."[6] If anyone doubts Heap's ability to rally a community around new technology, think again.

"We really know what we want," Heap said. "We're not a bunch of airheads who like to smoke pot in our living rooms and make music. We're hardworking entrepreneurs."[7] Heap views blockchain technology as a new platform for creators of intellectual property to get fair value for it. Smart contracts in particular could eliminate the magnitude of industry complexity, simplifying a mission-critical role of music labels in this ecosystem.

Rube Goldberg Strikes Again: Complexity in the Music Business

To paraphrase Talking Heads, how did we get here? How do we work this?[8] It starts with a basic problem for artists—that they signed contracts drafted for the vinyl age, when huge analog production and distribution costs stood between recording artists and their potential consumers. Heap told us, "When I first found a record label, I think I managed to get something like fifteen percent. My last record deal a few years ago, I got maybe nineteen percent. If people are lucky, they might get more now."[9] Artists may have assigned their rights to a label for the full term of copyright. In the States, that's either ninety-five years or the life of the artist plus seventy years. Imagine all the unforeseen innovations that a contract would have to cover to keep the deal fair for artists and their heirs.

Initially, the labels were small, radio was king, the record store was queen, and artist and repertoire personnel not only scouted for new talent but also oversaw their artistic development. In the last twenty-five years, the industry has consolidated from thousands of labels down to three global superpowers—Sony Music Entertainment, Vivendi's Universal Music, and Warner Music Group—and a few hundred indie labels. These three majors have a combined 15 percent stake in Spotify, the most popular and lucrative streaming music service.[10] So they will get an extra cash infusion if and when Spotify goes public. Apple has become the world's largest music retailer, and Live Nation the world's largest live entertainment company.

So control of music copyright is concentrated in the few. The labels and the tour promotion companies have started asking for *360-degree deals* from artists. That means getting a cut of all the revenues an artist generates—publishing rights to the underlying composition, usage rights to the sound recording, performance rights when the artist goes on tour, potentially even merchandise and sponsorship rights—regardless of whether they invested in the cultivation of those rights.

With consolidation comes systems integration, and that's never easy. Each conglomerate has its own process of accounting and its own version of a contract and a royalty statement, making side-by-side comparisons a challenge. "We have a big problem in that the industry is very fragmented. With all of its different platforms, it's a bit of a nightmare," Heap said.[11] These systems must accommodate innovation in production, format, distribution, and context of usage. But rarely is an element immediately outmoded, and so every party must maintain two or more models concurrently, the most obvious two being the physical and the digital.

To add to the complexity, there are many members of the supply chain, not just the publishers and the *performance rights organizations* (PROs)—organizations that monitor public performances of music and collect performance royalties such as the not-for-profit American Society of Composers, Authors and Publishers (ASCAP), not-for-profit Broadcast Music, Inc. (BMI), and the enterprise formerly known as the Society of European Stage Authors and Composers (SESAC)—but also the producers and the studios, the venues, the concert tour organizers and promoters, the wholesalers, the distributors, and the agents, each with its own contract, accounting, and reporting system. They take their cut and pass along the remainder to the artists' managers and agents. Whatever's left goes to the artists themselves, per the terms of their contracts. That's right—the artist is the last to be paid. It could be six to eighteen months before the first royalty check arrives, depending on the timing of the release and the label's accounting cycle.

Finally, an entirely new layer of intermediaries—technology companies like YouTube or Spotify—inserted themselves into the supply chain between artists and labels, slicing the artists' piece of the pie even thinner. Let's look at streaming music. Spotify pays on average between $0.006 and $0.0084 per stream to rights holders, usually the labels.[12] The calculation of this initial payment may seem transparent at first. Spotify's site states that it pays 70 percent of its advertising and subscription revenues to rights holders. But we reviewed its forty-one-page "Digital Audio/Video Distribution Agreement" with Sony USA Inc., and the payout of some $42.5 million in nonrecoupable advances to Sony artists is anything but clear. In fact, the first paragraph of the agreement calls for confidentiality. It appears that neither Spotify nor Sony can inform Sony's artists of the impact of this agreement on artists' revenues. Rich Bengloff, president of the American Association of Independent Music, said that, in his experience, the labels don't usually share money not directly tied to usage.[13] Industry analyst Mark Mulligan said, "Artists are going to feel pain for at least another four to five years, just as they did in the first four to five years after iTunes launched."[14]

So what value do the labels add? Certainly, they attempt to manage this complexity, police piracy, and enforce rights. For example, Universal Music Publishing Group had dedicated a third of its workforce to royalty and copyright administration in local markets around the world.[15] Universal recently deployed an artist's portal that allows artists to peek at the status of their royalties and to request advances against future revenues for no fee. The portal also provides "insights into Spotify usage: how many times a song streams, what kinds of people are streaming it, what else is on those listeners' playlists, and how specific songs resonate with certain audiences." Universal has devoted sixteen staffers to upgrading the portal and interpreting data for artists.[16] The labels also have huge teams of lawyers and lobbyists. They can launch new artists globally, demanding their boilerplate terms, marketing through local foreign media, distributing their music in foreign markets, licensing rights to foreign publishers, supporting international tours, and aggregating all the revenues. The *cost of policing* royalties has increased with the complexity of the business—

that is a cost that directly affects artists everywhere, because it functions like a tax.

Smart contracts on the blockchain can eliminate the magnitude of this complexity, replacing a mission-critical role of music labels in this ecosystem. According to Imogen Heap, "If you're a computer program, a piece of software, a database . . . these issues disappear, as it's just maths half the time. This bit goes to this person . . . and it doesn't take a year or two to reach the artist, writer, performer. . . . It's instant because it's automated and verified. On top of this, culture-shifting new music distribution services gather really useful data from artists' fans, which could massively help us be more efficient if the artists themselves could get to it."[17] That's the future of music on the blockchain.

Emergence of a New Music Business Model

The combination of blockchain-based platforms and smart contracts—plus the artistic community's standards of inclusion, integrity, and transparency in deal making, privacy, security, respect of rights, and fair exchange of value—could enable artists and their collaborators to form a new music ecosystem.

"Wouldn't it be nice if I could just decide how I'd like my music to be shared or experienced?" Heap asked. "To simply upload a piece of music, for example, and all its related content to one place online, for all to tap into and derive from. Usage rights, ownership, the equivalent of today's liner notes. Video, latest biography," and all other parties—not just record labels, music publishers, and tour promoters but also corporations looking for jingles, TV producers looking for soundtracks, mobile service providers looking for ringtones, and the many fans looking to do fan videos—could decide whether to agree to Heap's terms of use. "Wouldn't it be amazing to feel the presence of the artist, that if they make that decision about their music, it's really felt in a real physical sense, even from one day to the next?" she said. "I may decide, hey, it's my birthday today, all my music is for free or . . . if you're under sixteen or over sixty, it's on me! Or to divert all payments due to me to a relief fund, with just a few alterations of wording in the smart contract."[18]

That's the goal of designing an artist-centric model on the blockchain, not a model centered on music labels or tech distributors. Artists could produce music and be paid fairly for the value they create, and lovers of music could consume, share, remix, and otherwise enjoy what they love and pay a fair value. This model wouldn't exclude labels or digital distributors, but they would be equal rather than dominant members of the ecosystem.

The new music industry is not a pipedream. In October 2015, Heap launched her first experiment by releasing her song "Tiny Human," and all related data—the instrumental version, seven stereo stems, front cover image, music video, liner notes on musicians, gear, credits, lyrics, acknowledgments, and useful links, and the story behind the song—on the Internet.[19] These details would increase her *discoverability* on the Internet, that is, to help potential collaborators to find her.

Heap invited fans, developers, and services to upload her song to their various platforms and to share their work as well. She granted them nonexclusive rights to create an Imogen Heap artist profile, provided that they gave her the login details and permissions after uploading her files to their systems. If they expected revenue streams, then she asked them to provide payment models, percentages, and amounts so that she could factor those details into her analysis of the experiment. Finally, she welcomed donations to her bitcoin address and promised to direct half the proceeds to her charitable foundation, Mycelia, her name for this new ecosystem. Usage data and participant behavior would inform the next stage of development on the blockchain.

Various companies are working on its design and collaborating with Heap and other forward-thinking musicians. This new ecosystem has a number of features that the existing industry lacks:

Value Templates to construct deals that respect the artist as entrepreneur and equal partner in any venture, integral to value creation. Good-bye archaic paper contracts that perpetuate unfair beginnings. "No more clawing back royalty percentages," said Heap.

Inclusive Royalties that divide revenues fairly according to each person's contribution to the creative process, not only composers and performers, but also the other artists and engineers. Everyone could have an upside in a smash hit, not just the labels and the distributors.

Transparent Ledgers distributed on the blockchain so that everyone could see how much revenue a song was generating, the timing and magnitude of the revenue streams, and who was getting what percentage. No more archaic and proprietary paper-based accounting systems to hide behind. Separate tags for the nature of revenues, from work-made-for-hire to royalty revenues. Easy accounting, easy auditing, easy payment of taxes.

Micrometering, Micromonetizing functionality to stream the revenues, not just the music. If the music were metered, where consumers made micro- or picopayments per play, the royalty payments could be streamed immediately to the artists and contributors. No more payment delays, no more biannual or quarterly royalty checks, no more cryptic royalty statements. No more hand-to-mouth artists! Blockchain theorist Andreas Antonopoulos gave this example: "Streamium in Argentina is a streaming video service. It allows video producers to charge thousandths of a penny for downloading, say, two hundred milliseconds of live streaming video. It uses multisig[nature], time-locked transactions, atomicity, and sum integrity to implement its scheme. Producers deliver only the video that has actually been paid for, and consumers pay for only the video actually consumed. They automatically renegotiate that contract five times a second. If either of them drops out at any point in time, the contract ends and they cash in the most advantageous transaction between them."[20]

Rich Databases that can interface with one another and associate the core copyrighted material—lyrics, composition, and recordings—with all their metadata, the liner notes, the artwork and photographs, the individual tracks, the rights that the composer and the performer are willing to license, the terms of licensing, contact information, and so forth—in the digital ledger for all to see. No more incomplete databases of rights. Rights availability at your fingertips! Rights holders would be easy to find.

Usage Data Analytics in the hands of artists at last, to attract the right advertisers and sponsors, organize tours, plan promotions and crowdfunding resources and future creative collaborations with other artists. The model could capture "so much of this lost data around the world, like where are your fans, how old are they, what are they interested in," said Heap. "With that information, we could really tailor-make tours, we could connect up with brands and initiatives that resonate with us, or promote artists, products, or charities we love and support. I'm not talking names and e-mail addresses type of info, a bit more zoomed out but still vitally useful data. We could cross-reference it with other bands' data for all kinds of interesting uses for fans and artists alike!"[21]

Digital Rights Management, that is, a means of managing digital rights, not the anticonsumer DRM software wrapper that was all about restricting usage. We're talking about the deployment of smart contracts that actually manage rights and maximize the value of publishing, recording, performance, merchandise, and all other rights. This includes *terms of third-party engagement* for record labels and distribution services: labels and distributors could decide whether to opt into an artist's terms of use and expectations for service. If artists don't want advertising to interrupt the music experience, then they can forbid it. If they want a certain cut of ad revenue, then they can insist on it. If they want one of these large firms to handle licensing, distribution, and copyright enforcement in certain territories such as China, then they can do that. They can also set term limits. If the firms don't deliver a certain level of revenue, then the contracts can automatically terminate. Artists also need *automated subsidiary rights management*, wherever possible or desired, where prospective licensees either accept or reject the artist's terms of use and payment. The contract itself enforces each deal and can notify the artist of any breaches and terminations.

Auction/Dynamic Pricing Mechanisms to experiment with promotions and versioning of content, even peg subsidiary rights royalty percentages to the demand of a song. For example, if consumer downloads of a song spike, then an advertiser who licensed that song for a commercial automatically has to pay more when the commercial runs.

Reputation System that culls data from a bitcoin address's transaction history and social media, to create a reputation score for that address. Artists will be able to establish their own credibility as well as that of prospective partners in deal making, whether among artists as collaborators or between artists and consumers, labels, merchants, advertisers, sponsors, licensees, and so forth. Using multisig smart contracts, artists could refrain from doing deals with entities that don't meet certain reputational standards or don't have necessary funding in their accounts.

The key point of this new fair music industry is that the artists are at the center of their own ecosystem, not at the edges of many others. "I see a place for Spotify and YouTube. I see a place for curation, and I see a place for user-generated content," said Heap. "I see a place for record companies because we still need people to sift through the hundreds of millions of hours of music, or billions of bits of music and art all over the planet being created every day."[22] With software templates, they can engage creative collaborators, the big music labels, the big distributors, and the many smaller intermediaries as they see fit on the blockchain.

The Self-Launched Artist: Signs of the New Music Paradigm

One of Imogen Heap's friends, Zoë Keating, Canadian-born cellist and composer, has always controlled her own music. She owns all her publishing rights and the masters to her recordings. She carefully orchestrates her own marketing, sales, licensing, and distribution strategy. Given all the complexity mentioned above, we're hugely impressed. "An artist like me couldn't exist without technology. I can just record music in my basement and release it on the Internet," Keating told *The Guardian*. For her, the Internet has leveled the playing field for independent artists, but her experience with the big online music distributors has not differed greatly from Imogen Heap's with the traditional labels. "This is not just an excuse for services to replicate the payment landscapes of the past. It's not an excuse to take advantage of those without power," said Keating. "Corporations do have a responsibility not just to their shareholders but to the world at large, and to artists."[23]

Keating was alluding to the new contract that Google's YouTube presented to her. It was wrapped in nondisclosure. For several years, she's distributed her music on YouTube and monetized third-party uploads of her material using Content ID, a program that automatically alerts rights holders to instances of potential copyright infringement. Keating wasn't concerned about piracy, file sharing, or royalties. To her, commercial streaming was a means of promotion, reaching new audiences, and analyzing usage data. The music aggregators and the hit makers were the ones who made significant money by offering complete catalogs through on-demand services. Not her. The largest share of her revenues had always come from hard-core fans who'd pay from twenty dollars to a hundred for a new album. She would release new work on Bandcamp first, then upload it to iTunes, and finally make selections available elsewhere—YouTube, Spotify, Pandora. That *windowing strategy*—making content available exclusively in a particular channel for a period of time—had proven itself effective for her and her hard-core fans. She could thank her existing supporters and cultivate new relationships.

YouTube was launching a new subscription service, Music Key, where users would pay a fee to avoid advertising. If Keating wanted to continue monetizing her work through YouTube, then she would have to agree to YouTube's terms: include her entire catalog, and stop windowing elsewhere. It was all or nothing. She knew that the independent labels weren't happy either about the new licensing terms, but they were more upset by the financial repercussions. Keating wanted to maintain control over her music, on her terms.

She saw the potential of the bitcoin blockchain as a technology that could ensure that goal, starting with its transparency. "I just believe in transparency in everything," she told *Forbes*. "How can we build a future ecosystem without knowing how the current one actually works?"[24] For example, on YouTube, Keating estimates that there are fifteen thousand videos—dance performances, films, TV shows, art projects, gaming sessions—that use her music as soundtracks without her authorization. She should be able to leverage all that enthusiasm for her work, but only YouTube knows exactly how popular her music is. Nielsen's SoundScan is only one facet of a multidimensional picture.

Like Heap, Keating wants to register copyright and leverage copyright metadata on the blockchain. That way, people could more easily track her down as the copyright holder. She could then track derivative works through the blockchain. A distributed ledger of music metadata could track not only who created what, but who else was materially involved. She imagines visualizing usage and relationships, calculating the real value of a song for dynamic pricing, and enabling ongoing micropayments to collaborators and investors without third-party black boxes like ASCAP or BMI.[25]

Again, we're not saying that there is no role for labels or technology companies, and that artists could just make it on their own in a purely peer-to-peer ecosystem. Rather we're talking about a new kind of music ecosystem centered on the artists, where they control their own fate and receive fair compensation for the value they create. Blockchain technology will not create a new standard for how artists get compensated. Instead, it will liberate them to choose and customize an infinite array of solutions that work for their specific needs and beliefs. They can give it away for free, or micromonetize everything—but it's their choice, not the label's or the distributor's.

Other Elements of the New Music Ecosystem

Basic Copyright Registration

There are two fundamental dimensions to music copyright. The first is the worldwide right to the underlying composition—the musical notes and the lyrics—in all forms and languages, and that's usually held by the composer-lyricist. The music and the lyrics can be copyrighted separately. The composer-lyricist makes royalties whenever someone records or performs the song, buys the sheet music, adapts it into another genre (e.g., elevator Muzak), translates it into a foreign language, or includes the music in an anthology or textbook. The second is the worldwide right to the sound recording, a performance captured and preserved in some medium like a digital file or music video. The recording is usually copyrighted by the performer or band members, who earn royalties when the recording is played on the radio, television, or the Internet; synced with TV shows, commercials, or video games; streamed, downloaded, or purchased in some hard medium such as vinyl, CD, or DVD.

Zoë Keating's level of autonomy was what motivated Toronto's industrial rock band 22Hertz to turn to the blockchain. In Canada, registering copyright for one song costs fifty loonies, and the certificate contains only the title of the work. The band's founder, Ralf Muller, didn't think that'd be helpful in court, if anyone ever used the lyrics or melody. So he decided to go the hashing route, by creating a hash of the whole song using something called an OP_RETURN feature—and encoded it into the blockchain instead. If anyone ever used his words or music, he could simply demonstrate his ownership by pointing to the transaction on the blockchain, doing another hash of the song, and comparing the second hash with the hash on the blockchain. They would be identical. "Once you encode a hash in the OP_RETURN and block upon block get written on top, it is basically impossible to go back and change anything. This, to me, is incredible." When asked why the band's online store was accepting bitcoin and offering discounts to bitcoin users, Muller responded emphatically, "I'm not into 'business as usual.'"[26]

Digital Content Management System

Nor is Colu, a digital content management platform based on bitcoin blockchain technology. It provides developers and enterprises with tools for accessing and managing digital assets including copyright, event tickets, and gift cards—much of what a truly distributed music industry would need. Colu has partnered with music technology leader Revelator to build a rights management application programming interface (API). The goal is to realize what Imogen Heap and Zoë Keating were imagining—the demystification of rights ownership, digital distribution, and actual usage. The API will also provide incumbents a means of providing the much-called-for transparency and efficiency. "We are very excited about the potential for Colu's platform to simplify the management of music rights, starting with those associated with songwriters and their compositions," said Bruno Guez, founder and CEO of Revelator. "Colu has made the complex technology of the blockchain accessible for integrations into platforms like ours, and we're looking forward to exploring all the ways it can improve service to our clients."[27]

The New Artists and Repertoire (A&R)

Finally, a key aspect of any creative industry is talent scouting and coaching. Musicians naturally embrace mentoring and play an "artists and repertoire" role in such competitions as *The Voice.* The blockchain supports this type of A&R with usage algorithms. Consider PeerTracks. According to its landing page, it is "the ultimate one stop music platform" for both music lovers and artists. PeerTracks attaches a smart contract to every song that an artist uploads, and the contract automatically divvies up the revenues according to whatever deal the performer made with the lyricist, the composer, and other members of the band. Artists create their own tokens, bearing their name and likeness similar to a virtual baseball card. Tokens are collectibles. The artist sets the number of available tokens. So there could be limited editions, so to speak. The concept is simple: create a store of value, the valuation of which correlates to the artist's popularity.[28]

Users get full access, on demand, to the entire PeerTracks music catalog for free without advertising interruptions. They can save songs and playlists for offline use and download any song or album from the catalog. Unlike with Spotify or iTunes, users can also purchase tokens of artists and trade these tokens like baseball cards. As an artist's star rises, the value of the tokens rises, and so users could potentially benefit financially from supporting artists before they become famous. Loving an artist translates into VIP treatment, perquisites, and freebies from artists. This incentive turns people who would be passive listeners on Spotify into active promoters and builds a long-term, engaged fan base. PeerTracks intends to pay artists more for streams and downloads—specifically 95 percent of revenues—and to pay them instantaneously on the blockchain. Artists can set their own price for music downloads and merchandise. PeerTracks claims that "swarms of profit driven talent scouts/curators looking for the next hot star/token" will hear a new artist's song because PeerTracks users will vote them to visibility.[29]

ARTLERY FOR ART LOVERS: CONNECTING ARTISTS AND PATRONS

The traditional art market is notoriously exclusive and opaque. A relatively small number of artists and collectors represent an incredibly large percentage of the market, and there are very few, narrow, and sometimes circuitous paths for emerging artists to enter the art world. Even so, the openness and generally unregulated nature of the art market encourages experimenting with new concepts and new media, democratizing the art market on one hand, and democratizing capital markets on the other, with both hands on the transformative and disruptive power of the bitcoin blockchain.

Artlery describes itself as a network of artists who have agreed to share some of their earnings with patrons and peers who engage socially with their works.[30] Artlery's goal is to mint an art-as-asset-backed-currency on the blockchain by engaging art lovers as partial owners and stakeholders of the art with which they interact. Its approach is to provide the right incentives for all sides of the market—artists, patrons, curators, and venues such as galleries, museums, studios, and fairs—rather than provide perverse incentives for one party at the expense of the others. To foster patronage and build reputation for an artist, Artlery stages initial public offerings (IPOs) of digital pieces of that artist's work. Artlery's app enables artists such as JaZoN Frings, David Perea, Keith Hollander, Benton C Bainbridge, and the Bazaar Teens to replicate their physical works digitally, break them into many pieces like a picture puzzle, and apportion them to patrons based on each patron's level of appreciation from within the Artlery app. During a work's IPO period, patrons can accumulate interest up to the specified percentage of the piece that the artist initially gifted to the community. As the platform matures, Artlery plans to allow for the transfer and sale of the accumulated interests in the works.

At the 2015 Stanford Blockchain Summit sponsored by Artlery, Don decided to back a work by Anselm Skogstad, titled *EUR/USD 3081*, an artistic rendering of a euro currency note magnified and printed on a fifty-eight-by-forty-four-inch Dibond aluminum composite.

Buying Art Through the Bitcoin Blockchain: How It Works

To purchase the piece, Don opened his bitcoin wallet app. He used it to create a message that specified the amount of bitcoin representing the purchase price of the piece, designated Artlery's public key as the recipient of that bitcoin, and used his private key to "sign" or authenticate the message. Don double-checked all the fields because, unlike traditional payment methods, there was no reversing this bitcoin transaction. Then he broadcast the message not to his Canadian bank but to the entire network of computers running the full bitcoin blockchain.

Some people refer to these computers as *nodes*, where some nodes are donating their processing power to solve the math problem associated with creating a block. As we've explained, the bitcoin community refers to them as *miners* and to their problem-solving work as *mining*, as in gold mining. It's an awkward analogy because it conjures images of experts whose talent might confer some competitive advantage over noobs (newbies). It doesn't. Each miner is running the software like a utility function in the background, and the software is doing all the computations. Serious miners configure their machines to optimize their processing power, minimize their energy consumption, and leverage high-speed network connectivity. Beyond that, there's really no human aptitude necessary and no human interference tolerated.

Not all nodes are mining. In fact, the vast majority of nodes on the bitcoin network are simply performing bitcoin rule verification of received data before routing the data to peer connections. The network verified the two bits of data—that Don controlled the amount of bitcoin specified and authorized the transaction—and recognized Don's message as a transaction. The miners then race to convert unordered and unrecorded transactions into transactions ordered and recorded in a block of data. Each block had to include the digest or hash of the previous block of transactions, as well as a random number known as a *nonce*. To win the race, a computer must create a hash of the block; this hash must have a certain, but arbitrary, number of zeros at the beginning. It's unpredictable which nonce will produce a hash with the correct number of zeros, and so the computers have to try different nonces until they stumble upon the right value. It's really like winning the lottery because there's no skill involved. However, a human being can increase her chance of winning the lottery by buying a state-of-the-art computer processor that specializes in solving bitcoin's math problems; buying more tickets, that is, running multiple high-powered nodes; or, as human beings often do, pooling her node with other nodes—like colleagues at the office—and agreeing to split the pot if one of their nodes wins. So winning is a matter of luck, processing power, and the size of one's mining pool.

Hash rate is a measure of the total processing power of the bitcoin network. The higher the total aggregate hash rate of the entire network, the more difficult it is to find the right nonce. When a miner finds a hash with the correct number of zeros, it shares its proof of work with all other miners on the network. This is the other big scientific breakthrough in distributed computing: using proof of work to achieve network consensus. It's known as the Byzantine Generals' Problem. The other miners signal their acceptance of the block by focusing on assembling the next block, which has to include the hash of the newly made block. Just as Don's public and private keys are unique to him, the hash of each block is unique: it works like a cryptographic fingerprint that makes all the transactions in the block verifiable. No two block fingerprints are the same. The winning miner receives a set quantity of new bitcoins as a reward—the software itself mints and allocates the new coins—and the hashed block is appended to the chain.

So, within ten minutes of Don's broadcasting his message, he and

Artlery received a confirmation that Don's bitcoin transaction created what is referred to as "unspent transaction output," meaning that Artlery can spend it by doing what Don just did—broadcasting a message that specifies the amount to send and the address of the recipient, and authorizing the transaction with Artlery's public key. If the artist and the patrons knew both Don's and Artlery's public keys, then they could see that the deal between them went through and could see the amount of the transaction. That's why we call it a *public ledger*—all transactions are transparent—and pseudonymous, in that we can see the parties' addresses, though we can't see the names of the persons behind them. Every subsequent block served as further confirmation of their transaction.

Profile of the Next-Gen Art Patron: Redefining Money

Now Don owns a percentage interest in the rights of an artistic rendering of a euro. When the physical work sells, the artist, the venue, Don, and his virtual patrons all receive a portion of the sale according to their level of participation. In other words, patron participation matters. Patrons who interact with the artist and the work, who share their appreciation in social networks, who spark others to engage with both artist and art, and who essentially serve to promote the artist's brand, receive more than does a passive patron who viewed it once online and bought a stake. We're not sure whether writing about it in this book counts directly toward Don's participation points. Artlery wants signals of appreciation—in the form of positive references to artists and their works—to correlate with the appreciation in value of the work itself, and so perhaps future platform releases will take examples like ours into account. Artlery is initially concentrating on gifting a percentage of each piece's sale. Future releases of the platform will allow patrons to purchase ownership stakes in artwork directly, perhaps sharing portions of subscription royalty revenue or copyright licensing of the work.

By involving multiple parties directly, including patrons, in the process, and engaging them as stakeholders, Artlery is focusing more eyes on accounting. The blockchain as a public, distributed ledger ensures open, accurate, and timely processing of transactions. As payouts expand beyond first sale, secondary sale, and subsidiary rights such as prints and merchandising, individual artists will never be acting alone. They will have a community of patron stakeholders behind them to negotiate and enforce their contractual rights.

Artlery utilizes the bitcoin blockchain in several ways. First, it registers the provenance of the art as metadata on the blockchain through a partnership and API integration with another bitcoin start-up, Ascribe.io, and it uploads the payout table so that all stakeholders are paid immediately according to their asset shares with immediate transparency for all parties. It is exploring various techniques for encoding this information, such as a bitcoin script inside transactions. While its initial target market is fine art, Artlery has significant traction in other copyright industries such as music, books, and movies, which it will target through the release of its own API.

PRIVACY, FREE SPEECH, AND FREE PRESS ON THE BLOCKCHAIN

Personal privacy, free speech, and free press are essential to an open, free, and prosperous society. On one hand, citizens must be able to communicate privately and anonymously. On the other hand, they must be able to speak freely and securely without fear of repercussion. Online censorship, the hacking of large institutions and civil society, and Edward Snowden's revelations of mass and targeted surveillance and data fracking have driven citizens of well-established democracies to seek anonymity and encryption technologies. These tools enable them to disguise their identities and scramble their messages in transit and in storage so that only authorized persons may access them.

Here's the rub—encryption technologies are either not legal for individual use or not readily available in those countries whose citizens need them most. The Wassenaar Arrangement, a multilateral export control regime that technologically advanced nations agreed to, governs the export of "dual use" products, that is, those that could be deployed for both good and evil. Wassenaar's original goal was to keep high-tech products out of the hands of dictators in North Korea, Libya, Iran, and Iraq. Anonymity and encryption technologies such as public key infrastructure were considered dual use.

Today, in countries such as Russia and China, both individuals and corporations—including foreign firms—must seek authorization to use them. In countries where their use is discretionary, governments—even the Obama administration—have asked tech firms to include "backdoor access," that is, a secret means of bypassing the normal authentication process (e.g., logging on with a password or other security code) and gaining remote access to a computer and its data *without authorization or detection*. It is far more insidious than Big Brother, because at least everybody knew that Big Brother was watching.[31] Here, tech firms aren't supposed to tell users that there's a back door. No doubt hackers look for them, find them, and use them, too.

"The trend lines regarding security and privacy online are deeply worrying," wrote David Kaye, Special Rapporteur of the Office of the United Nations High Commissioner for Human Rights. "Encrypted and anonymous communications may frustrate law enforcement and counterterrorism officials, and they complicate surveillance, but State authorities have not generally identified situations—even in general terms, given the potential need for confidentiality—where a restriction has been necessary to achieve a legitimate goal."[32] He went on to say that law enforcement and counterterrorism agencies have come to downplay good old-fashioned detective work and deterrence measures, including transnational cooperation.[33]

Not surprisingly, on global measures of the preservation of political and personal rights—that is, privacy and the freedoms of speech, assembly, and the press, and the tolerance of other religions, immigrants, political refugees, and homosexuals—Russia ranks 114th, and China second to last, at 160th.[34] For what it's worth, the United States is no epitome: it ranks 28th.

Blocking Web sites without a court order has become customary in such countries, and many censors have figured out how to thwart virtual private network software used to prevent censorship.[35] According to Reporters Without Borders, Russia has been curtailing freedoms of expression and information and blocking increasingly more sites since Vladimir Putin's return to the presidency in 2012, among them Wikipedia.[36] China has mastered the art of the targeted data blackout, censoring search terms related to Hong Kong's "Occupy Central" prodemocracy movement and the twenty-fifth anniversary of the Tiananmen Square protests on Weibo, China's Twitter clone. It managed to block nearly 90 percent of all Google services. Imprisonment is also popular in these countries when people post content deemed questionable by the government online. Following the Chinese stock market collapse in July 2015, authorities arrested more than one hundred people for using social media to spread rumors that "caused panic, misled the public and resulted in disorders in stock market or society."[37]

Governments that wish to repress the voices of citizens everywhere and have captured technologies like the Internet to silence dissidents and block outside media will find blockchain technology significantly more challenging for several reasons. First, citizens and journalists could use public key infrastructure to encrypt information and conceal their identities from would-be censors and attackers. Second, where governments discourage and deprive good and honest journalism of funding, journalists could raise funds on the blockchain, casting a wider net for investors sympathetic to their cause, especially investors who preferred to remain anonymous. Finally, governments could not destroy or alter information recorded on the blockchain; therefore, we could use it to hold governments and other powerful institutions accountable for their actions.

Consider crowdfunding journalists on the blockchain. If we released them from the financial grip of state-controlled media, they could cover politics freely while preserving the anonymity of donors. Veteran Chinese journalists could try one of the distributed peer-to-peer crowdfunding platforms such as Koinify, Lighthouse, or Swarm that use PKI to protect the identities of sender and recipient better than Internet-only systems. Another great blockchain tool is the free mobile app GetGems, which both guards and monetizes instant messaging through bitcoin. Users can send all sorts of files securely, with GetGems functioning like private e-mail, not just SMS.[38] These apps are just the beginning of what is possible.

Another solution is a distributed platform for filing stories in an immutable ledger that makes the ledger unique, such as what Factom aims to accomplish in the developing world. Reporters could purchase entry credits—rights to create entries on Factom's ledger. As with the bitcoin ledger, everyone would get the same copy, and anyone could add to it but no one could alter entries once they were filed. Factom has a commit/reveal commitment scheme that serves as an anticensorship mechanism: servers in China, for example, couldn't prevent the filing of an otherwise valid entry because of its content. If the reporter had attached an entry credit to the filing, then it would get recorded. A government could identify certain entries as offensive but couldn't delete or block them as the Chinese government has done on Wikipedia. If an official court were to order a change in the ledger, an officer of the court could make a new entry to reflect the ruling, but the history would remain for all to see.[39]

A third solution is distributed peer-to-peer microblogging that doesn't go through centralized servers. Stephen Pair, CEO of BitPay, described how to reinvent Twitter or Facebook so that users controlled their own data. "Instead of having just one company like Facebook, you might have many companies tying into this common database [the blockchain] and participating in building their own unique user experiences. Some of those companies might ask you for or might require certain information to be shared with them so that they could monetize that. But as a user, you would have full control over what information you're sharing with that company."[40] There is Twister, a Twitter clone in terms of feel and functionality developed in 2013 by Miguel Freitas, a hacker and research engineer at PUC-Rio University in Rio de Janeiro, Brazil. Twister leverages the free software implementations of bitcoin and BitTorrent protocols and deploys cryptography end to end so that no government can spy on users' communications.[41]

GETTING THE WORD OUT: THE CRITICAL ROLE OF EDUCATION

Joichi Ito is among an elite group of widely successful entrepreneurs—from Bill Gates and Steve Jobs to Biz Stone and Mark Zuckerberg—who dropped out of college to invent something new in the digital economy.[42] It is a hallmark of our entrepreneurial culture that one's pursuit of an idea, to go deep and understand its nuances as Ito likes to say, drives a visionary out of the classroom and into business. Henry Ford and Walt Disney pursued their passions without college degrees. And so it is one of those paradoxes that the Massachusetts Institute of Technology would choose Ito to direct its legendary Media Lab, at the epicenter of all things digital and relevant to culture.

The timing was perfect. "Digital currency is something that I've been interested in even before the Media Lab. . . . I ran one of the really early digital test servers from the DigiCash days in the 1990s. One of the first books I ever wrote was in Japanese, called *Digital Cash*, that I coauthored with someone from the Bank of Japan. So this has been an area of interest for me for a long time and predates a lot of the other stuff I've been doing."[43]

When he got to the Media Lab, various academics were dabbling in some aspect of bitcoin relative to their core discipline—consensus models, cryptography, computer security, distributed systems, and economics—but no one specifically focused on it. He didn't see faculty doing fundamental research around bitcoin, even though MIT students had launched the MIT Bitcoin Project to give $100 of bitcoin to undergraduates.

Ito had a sense of urgency similar to Imogen Heap's in spreading the word and forming teams around legal, technical, and creative challenges. Blockchain technologies were moving much faster than Internet technologies had, but without much academic involvement. The core developers of the bitcoin protocols were recovering from reputational hits: the Bitcoin Foundation was bankrupt, and board member Mark Karpeles was arrested in Japan for embezzlement through his Mt. Gox exchange. Ito moved quickly. He launched the Digital Currency Initiative (DCI) at the Media Lab and hired former White House adviser Brian Forde to direct it. He brought three of the bitcoin core developers into the DCI to provide them with stability and resources so they could focus on the code.

He thought that creating an academic network of universities interested in supporting bitcoin was important, and that's under way. "We're setting up courses, we're trying to organize research, but we're still very early-stage," he said. "We've just got the core funding together to support the program, and we're just trying to drum up interest from the faculty and the students in the space." More broadly, he wants the MIT Media Lab to reinvent higher education so that people like him won't drop out and will see the value of a diverse place like the Lab. It's an opportunity to pilot the future of academia.[44]

Melanie Swan, a leading blockchain theorist and academic, was more specific about where to educate students about the blockchain, and it's not in traditional universities. *It's on the blockchain.* "It's really a complete revolution in how we do everything. Academia is not the right place to do academic thinking about very new things like the blockchain," she said. For example, rather than submitting research to scholarly journals for publication and waiting from six to eighteen months for a rejection or publication, a scholar could post the paper immediately as Satoshi Nakamoto did to a limited audience of peers, receive reviews in real time, and establish the needed credibility to publish to a larger audience. Reviewers could vote reviews up or down as redditors do on Reddit so that the scholar would know which ones to take to heart. The paper might even be available for free, but other scientists could subscribe to deeper analysis or threaded discussion with the author. The scholar could make her raw data available or share it with other scientists as part of a smart contract. If there was a commercial opportunity flowing from the paper, she could protect the rights in advance, taking into consideration who funded the research and any claims they might make to the discovery.

Swan is the founder of the Institute for Blockchain Studies. "There's the start of a development of an educational infrastructure to support learning about these technologies. Obviously, all the meetups, user groups, and hackathons are tremendously useful," she said. "Every strategy and accounting consultancy has a blockchain practice group now, and there are education institutions like Blockchain University."[45] Swan herself teaches a blockchain workshop at Singularity University.

She talked of an education system where a college student would become what she called an "educational sommelier," pairing interests or needed skills with accredited courses, potentially massive open online courses (MOOCs). "The benefit of MOOC is decentralized education. So I can take the top machine learning class from Andrew Ng at Stanford University via Coursera. I can take the top other courses at MIT." So students could fund their own personal development programs anywhere in the world and receive accreditation. She explained: "Just as when I go take the GRE or the GMAT or the LSAT, I show up with my ID, it confirms locally that I am who I say I am, I take that test," and that local confirmation "could easily be part of the MOOC infrastructure."

Swan has been working on how to do MOOC accreditation and tackle student debt on the blockchain. The blockchain provides three elements toward this goal: (1) a trustable proof of truth mechanism, an oracle, to confirm that the students who signed up for the Coursera classes actually completed them, took the tests, and mastered the material; (2) a payment mechanism; and (3) smart contracts that could constitute learning plans. Consider smart contracts for literacy. "Why don't we target financial aid toward personal development? Like Kiva, but Kiva for literacy," Swan said, except that everything would be super transparent and participants would be accountable. Donors could sponsor individual children, put money toward learning goals, and pay out according to achievement. "Say I wanted to fund a schoolchild in Kenya's literacy program. Every week this child would need to provide proof of completion of a reading module. Perhaps it's all automated through an online test where the blockchain confirmed the child's identity and recorded progress before disbursing the next week's worth of funding into what we might call the child's 'smart wallet for learning' so that the child could continue payment for school without interference. Money toward a girl's education couldn't be diverted to her brother's schooling," she said.[46]

CULTURE ON THE BLOCKCHAIN AND YOU

After two world wars in a single generation, global leaders admitted that political and economic agreements could not—would never—maintain long-term world peace. Those conditions changed, sometimes frequently, sometimes drastically so. Peace had to be rooted in something richer, more universal, in the shared moral values and intellectual freedoms of society. In 1945, three dozen nations convened to form an educational body of sorts that would model a culture of peace. It became known as the UN Educational, Scientific, and Cultural Organization (UNESCO). Its mission in the world today is "to create the conditions for dialogue among civilizations, cultures, and peoples."[47]

Through the lens of blockchain technologies, musicians, artists, journalists, and educators are seeing the contours of a world that protects, cherishes, and rewards their efforts fairly. All of us should care. We are a species that survives by its ideas, not by its instincts. We all benefit when creative industries thrive and when the creatives themselves can make a living. Moreover, these are the bellwethers of our economy—they reveal faster than nearly any other industry how both producers and consumers will adopt and then adapt a technology to their lives. Musicians have long been among the first to exploit innovations for the benefit of a great many others, too often at their own expense. These dedicated members of our society inspire us, and every business executive, government official, and other organizational leader has much to learn from them about the new era of the digital age.

PART III

PROMISE AND PERIL

CHAPTER 10

OVERCOMING SHOWSTOPPERS: TEN IMPLEMENTATION CHALLENGES

Lev Sergeyevich Termen was a gifted musician, but he preferred playing with physics. Born into Russian aristocracy before the turn of the twentieth century, Termen joined the Bolsheviks in dismantling the tsarist autocracy. One of his early missions was to create a device that could measure the electrical conductivity and capacity of various gases. He tried gas-filled lamps, he tried a high-frequency oscillator, and he even tried hypnosis.[1] The oscillator ended up working well, and so Termen's boss encouraged him to seek other applications for it. Two apps would become legendary. The more whimsical of the two started out as two metal terminals with nothing between them, like a lamp without the glass. Termen discovered that, if he infused this void with gas, he could gauge the gas's electrical properties. His design was brilliant: he substituted headphones for dials so that he could take acoustic rather than visual readings, monitoring the pitch of the signal that each gas produced. It was way ahead of its time, the stuff of Dr. Emmett Brown's garage in *Back to the Future*.

Devotees of TED talks and students of technological history already know the end of this story: Termen stumbled upon a means of making music out of thin air. Whenever he put his hands near the metal terminals, the pitch of the signal changed. He learned that he could manipulate the pitch by the precise position and motion of his hands. He called his device the "etherphone," known today as the theremin, an anglicized version of his name. The other app was a larger-scale version of this apparatus, one that was sensitive to movement within a radius of several meters. It was the first motion detector—sentry of the ether. He demonstrated both of these instruments at the Kremlin, playing his etherphone with abandon for Comrade Lenin. While Lenin delighted in the etherphone, he put the motion detector immediately to work in watching over the Soviet stashes of gold. If anyone crossed the electromagnetic line around the gold, they'd set off a silent alarm. Big Brother suddenly had electric eyes.

The moral of the story is simple: Termen's devices brought both light and darkness to the world. In a poignant talk, "Our Comrade the Electron," Maciej Ceglowski pointed out these two themes in all of Termen's inventions: as soon as they gave shape to airy nothing, they were usurped by dark forces. Lenin even co-opted electricity in his propaganda, equating communism with Soviet power plus the electrification of the country.[2] But it was Stalin who rounded up Termen and his peers, threw them into the Kolyma gulag, and forced them to invent instruments of tyranny.

We've heard bitcoin used with similar grandiosity in campaigns of all stripes. Like every revolutionary technology, the bitcoin blockchain has its upside and its downside. In the previous chapters, we've walked you through the many promises of this technology. This chapter shines a spotlight on ten showstoppers—problems and perils. Forgive us if some of these are technically complicated. We think it imprudent to supersimplify these issues: we need a certain level of detail for precision.

As well, after reading this section you may be tempted to dismiss these blockchain innovators because they face serious obstacles. We encourage you to consider whether these are either "reasons the blockchain is a bad idea" or "implementation challenges to overcome." We think it's the latter, and we'd like innovators to view these as important problems to solve creatively as we transition to the second era of the Internet. For each challenge, we propose some solutions. In the final chapter, we present our thinking on what we can do overall to ensure the fulfillment of the blockchain's promise.

1. THE TECHNOLOGY IS NOT READY FOR PRIME TIME

As of this writing, most people have only a vague understanding of bitcoin the cryptocurrency, and very few have heard of blockchain the technology. You the reader are among the forward-thinking few. Bitcoin conjures images ranging from a pyramid scheme and a money Laundromat to a financial E-ZPass on the economic highway for value. Either way, the infrastructure isn't ready for prime time, so goes the argument.

The challenge is multifaceted. The first facet borrows from science fiction author William Gibson, that the future is here; its *infrastructure* is just unevenly distributed. Had Greek citizens known about bitcoin during their country's economic crash in 2015, they still would've been hard-pressed to locate a bitcoin exchange or a bitcoin ATM anywhere in Athens. They wouldn't have been able to transfer their drachmas into bitcoins to hedge against the plummeting fiat currency. Computer scientist Nick Szabo and information security expert Andreas Antonopoulos both argued that robust infrastructure matters and can't be bootstrapped during catastrophes. Antonopoulos said that Greece's blockchain infrastructure was lacking at the time of the crisis, and there was insufficient bitcoin liquidity for an entire population to move its troubled fiat currency into it.

On the other hand, the bitcoin blockchain isn't ready for Greece either. That's the second facet: it falls short on security controls for such a massive bump in usage. "The system lacks the *transactional capacity* to on-board ten million people. That would represent almost a tenfold increase in user base overnight," said Antonopoulos. "Remember what happened when AOL dumped 2.3 million e-mail accounts onto the Internet? We quickly discovered that the Internet wasn't ready, in terms of spam protection and Net etiquette, to absorb 2.3 million noobs who didn't have the culture. That's not good for an immature technology."[3] The blockchain would be susceptible to capacity problems, system failures, unanticipated bugs, and perhaps most damaging, the huge disappointment of technically unsophisticated users, none of which it needs at the moment.

That relates to the third facet of this showstopper, its *inaccessibility* to the average person. There's not enough wallet support, and many interfaces are user-*un*friendly, requiring a high tolerance for alphanumeric code and geekspeak. Most bitcoin addresses are simply strings of between twenty-six and thirty-five characters beginning with a one or a three, quite tedious to type. As Tyler Winklevoss said, "When you go to Google.com you don't type in a string of numbers. You don't type in an IP address. You type in a name, a word that you can remember. And the same goes with the bitcoin addresses. Bitcoin addresses shouldn't be exposed to the average user.

Little things like that make a difference."[4] So there's much work to be done in basic user interface and experience.

Critics have also raised concerns about long-term *illiquidity* because bitcoin is finite in quantity—21 million by 2140—and mined at a diminishing rate. It's a rules-based monetary policy intended to prevent inflation triggered by arbitrary and discretionary monetary policies, a phenomenon commonplace for many fiat currencies. Satoshi wrote, "It's more typical of a precious metal. Instead of the supply changing to keep the value the same, the supply is predetermined and the value changes. As the number of users grows, the value per coin increases. It has the potential for a positive feedback loop; as users increase, the value goes up, which could attract more users to take advantage of the increasing value."[5]

At the margin, coins stored in lost wallets or sent to addresses whose owners have lost their private keys are not recoverable; they just sit dormant on the blockchain, and so there will be fewer than 21 million in circulation. Early adopters have tended to hold on to bitcoin as they hold on to gold, hoping that its value will increase in the long run, and therefore treating bitcoin as an asset rather than as a medium of exchange. According to economic theorists, low or no inflation motivates holders to hoard rather than spend their bitcoin. Still, if more trusted bitcoin exchanges facilitate consumers' movement in and out of bitcoin, then the frequency and volume of trading could increase. If more merchants accept bitcoin as a medium of payment, then people who've been sitting on bitcoins may start to use their store for purchases, thereby freeing up more bitcoins. If merchants begin to issue bitcoin-denominated gift cards, then more people should be exposed to cryptocurrencies and become more comfortable transacting in bitcoin. And so, hypothetically, people will have fewer reasons to hoard bitcoin. Advocates of the bitcoin protocol argue that, because bitcoins are divisible to eight decimal places—the smallest unit is called a Satoshi, worth one hundredth of a millionth of a bitcoin—the smallest denominations will buy more if demand for bitcoin increases. There's also the possibility of tweaking the protocols to allow for greater divisibility, say, picopayments (trillionths of a bitcoin) and to remine stranded bitcoin after a period of dormancy.

A fifth dimension is *high latency*: for the bitcoin blockchain network, the process of clearing and settling transactions takes about ten minutes, which is far faster end to end than most payment mechanisms. But clearing transactions at the point of sale instantaneously is not the issue; the real problem is that ten minutes is simply too long for the Internet of Things where devices need to interact continuously. Core developer Gavin Andresen said solving for a trillion connected objects is "a different design space from bitcoin," a space where low latency is more critical and fraud is less of an issue or where parties could establish an acceptable level of trust without the bitcoin network. Ten minutes is also too long for financial transactions where timing matters to get an asset at a particular price, and where latency exposes traders to time-based arbitrage weaknesses such as market timing attacks.[6] The immediate solution for entrepreneurs has been to fork the bitcoin code base, that is, to modify the source code by tweaking a few parameters, and to launch a new blockchain with an altcoin in place of bitcoin as incentive to participate. Litecoin is a popular altcoin with a block time of 2.5 minutes, and Ripple and Ethereum are entirely reengineered blockchain platforms that have latency of seconds, not minutes.

A sixth dimension is *behavioral change* in a deeper sense than Netiquette. Today, many people count on their bank or credit card company, even talking with a real person, when they make an accounting error, forget their passwords, or lose their wallets or checkbooks. Most people with bank accounts aren't in the habit of backing up their money on a flash drive or a second device, securing their passwords so that they needn't rely on a service provider's password reset function, or keeping these backups in separate locations so that, if they lose their computer and all other possessions in a house fire, they don't lose their money. Without this discipline, they might as well stuff their mattress with cash. With greater freedom—better privacy, stronger security, and autonomy from third-party cost structures and system failures—comes greater responsibility. For those consumers who don't trust themselves to keep safe backups of their private keys, third-party storage providers could provide backup service.

A seventh dimension is *societal change*. Money is still a social construct representing what a society values. It is endogenous to that society, it manifests because of human relationships, and it adapts to evolving human needs. "You can't take the social out of money," said Izabella Kaminska of the *Financial Times*. "A lot of these protocols attempt to do that by creating an absolutist and very objectified system. It just doesn't reflect the world as it is." She pointed to the euro system as an example of how one size—one set of protocols—doesn't fit all countries.[7] She echoed what Antonopoulos said about the very human need for societies to forgive and forget in order to move on. "There's a very long tradition in finance of obliterating records, because we as a society believe that it's wrong to persecute or discriminate against individuals for something they did ten or fifteen years ago. We have this whole debt jubilee-esque mentality because we think people should be given another chance. Creating a system that never forgets is slightly sociopathic," she said.[8]

That leads us to the eighth dimension, the *lack of legal recourse* in a world of irrevocable transactions and unvoidable smart contracts. According to legal scholars Primavera De Filippi and Aaron Wright, "People are, indeed, free to decide the particular set of rules to which they want to abide, but—after the choice has been made—can no longer deviate from these rules, to the extent that smart contracts are automatically enforced by the underlying code of the technology, regardless of the will of the parties."[9] This very high degree of certainty—mathematical certainty—as to the outcome of a transaction or a smart contract is unprecedented in society. It delivers greater efficiencies and effectively eliminates nonperformance risk because we have no choice of breach, no choice of damages. But that's also a downside. It allows no room for human beings. To Josh Fairfield of Washington and Lee University School of Law, that means "more messiness, not less. We're going to see more fights. 'You didn't actually renovate my house, I want my money back.' We're going to see more human messiness, but more human messiness doesn't mean the technology is bad."[10]

But will people actually take the counterparty to court? De Filippi estimated that, in the analog world, 80 percent of contract breaches aren't enforced because they're too costly to pursue in court, too expensive to go into proceedings. Why should those numbers improve in a blockchain world? When the code indicates that the contract has been fully executed rather than breached, except one party is dissatisfied with the outcome, will the dissatisfied party actually pursue a lawsuit? Will the courts recognize the case? Will the small business owner back away from the corporate legal team of Dewey, Cheatham, and Howe or—with his modest resources—even be able to identify his anonymous counterparty, so that he could file a lawsuit in the first place?

2. THE ENERGY CONSUMED IS UNSUSTAINABLE

In these primordial days of the bitcoin blockchain, the proof-of-work method described in chapter 2 has been critical to building people's trust. Years from now, we will look back and appreciate the genius of its deployment, from minting and allocating new bitcoins to assigning identity and preventing double spending. Pretty remarkable. And pretty unsustainable, according to critics of cryptocurrencies that use proof of work to keep the network safe and pseudonymous.

Hashing, the process of running pending transactions through the secure hash algorithm 256 (SHA-256) to validate them and solve a block, burns a lot of electricity. Some people in the blockchain ecosystem are making back-of-the-envelope calculations that become memes in the community. Estimates liken the bitcoin network's energy consumption to the power used by nearly seven hundred average American homes at the low end of the spectrum and to the energy consumed by the island of Cyprus at the high end.[11] That's more than 4.409 billion kilowatt-hours,[12] a Godzilla-sized carbon footprint, and it's by design. It's what secures the network and keeps nodes honest.

In early 2015, *The New Republic* reported that the combined processing power of the bitcoin network was hundreds of times greater than the aggregate output of the world's top five hundred supercomputers. "Processing and protecting the more than $3 billion worth of bitcoins in circulation requires more than $100 million in electricity each year, generating a volume of carbon emissions to match." The article's author, Nathan Schneider, wrote what has been on our minds ever since: "All that computing power, which could be curing cancer or exploring the stars, is locked up in machines that do nothing but process bitcoin-type transactions."[13]

As citizens who care about our planet, we should all be concerned. There are two issues, one around the electricity used to run the machines and another around the energy used to cool them so that they don't fail. Here's a rule of thumb: for every dollar a computer burns up in electricity, it needs fifty cents to cool down.[14] The acute drought in California has raised serious concerns over using precious water to cool data centers and bitcoin mining operations.

As the value of bitcoin increases, the competition for mining new bitcoin increases. As more computing power is directed at mining, the computational problem that miners need to solve becomes more difficult. One measure of the total processing power of the bitcoin network is the hash rate. Gavin Andresen explains: "Let's say we have millions of transactions per block, each paying an average of a dollar transaction fee. Miners would be paid millions of dollars per block, and they would spend a little less than that in electricity to do that work. That's how the proof-of-work economics work out. It really is the price of bitcoin and however much reward is in a block that drives how much hashing is done."[15] The hash rate has been increasing considerably over the last two years, rising forty-five-fold in less than a year. And the trend is toward using more energy, not less.

"The cost for having no central authority is the cost of that energy," said Eric Jennings, CEO of Filament, an industrial wireless sensor network.[16] That's one side of the argument. *The energy is what it is*, and it's comparable to the cost incurred in securing fiat currency. "All forms of money have a relationship to energy," said Stephen Pair of BitPay. He revisited the gold analogy. "Gold atoms are rare on earth because an intense amount of energy is needed to form them." Gold is precious because of its physical properties, and those properties derive from energy. Pair mused that artificially manufacturing gold would require nuclear fusion.[17]

From one perspective, all this electricity consumption makes sense. Erik Voorhees, founder of the coin exchange ShapeShift, said critics were unfair in calling the energy spent on bitcoin mining a waste. "The electricity is being burned for a purpose. There is a real service being provided, the securing of these payments." He urged critics to compare it with the energy burned by the current financial system. Think of the big vaults, the bunkerlike architecture with majestic Grecian facades, HVAC systems pushing frigid air into bright lobbies, competing branches on every corner, and ATMs in between. "The next time you see a Brink's armored truck pumping black soot into the air, compare that to the burning of electricity in bitcoining. It is not quite clear which is worse," Voorhees said.[18]

The second energy-related issue is computer architecture itself. For backward compatibility with slower-changing legacy systems, your laptop or PC is likely a type of complex instruction set computer (CISC) that can run a wide range of math apps that the average person will never ever use. When engineers realized that they'd seriously overshot the market, they created the reduced instruction set computer (RISC). Your mobile device is likely an advanced RISC machine (ARM). What miners realized was that they could also harness their graphics processing unit to increase processing speed. Because modern GPUs have thousands of computing cores on each chip, they are ideal for computations that can be done in parallel, such as the hashing done in bitcoin mining, There were some trade-offs, and estimating the machine's energy consumption got slightly more complicated, but for the most part GPUs could do the work.[19]

"If I can design a RISC computer to be oh-so-superfast and massively, near insanely parallel to try the billions of kazillions of codes simultaneously with little or no electricity, I will make money out of thin air,"[20] said Bob Tapscott, Don's CIO brother. That's what the BitFury Group has done: built a massively parallel bitcoin solver with application specific integrated circuits (ASICs) that are energy efficient and designed solely to mine bitcoins. Its founder and CEO, Valery Vavilov, argued the view that machines and mining operations overall will continue to get more energy efficient and environmentally friendly. Some of that depends on relocating to cold climates where energy is cheap and preferably renewable, such as hydro or geothermal, and where either Mother Nature handles the cooling or manufacturers figure out an efficient way to capture the heat. BitFury, for example, has two data centers—one in Iceland and another in the country of Georgia—with plans for additional centers in North America, and it acquired the Hong Kong–based start-up Allied Control, which specializes in immersion cooling technology.[21] And so BitFury is working to reduce the ecological impact of the bitcoin infrastructure.

Even if these initiatives limit mining's carbon footprint, we still have the rapid consumption and disposal of these continually upgraded devices. Miners who want to make a career of it must continually upgrade and specialize their systems. Most mining equipment has a useful life span of three to six months.[22] Bob Tapscott likened firms such as BitFury to those Yukon shopkeepers during the great gold rush: they made their real fortune by selling better and better shovels to the miners.[23] We found one miner's description of his Cointerra TerraMiner IV bitcoin with an ASIC chip that was so energy intensive that his home's electrical system couldn't handle it. "I am selling three units because my house is old and has substandard wiring. I do not want a fire." The starting bid was five thousand dollars.[24] Vendors such as MRI of Australia are applying new approaches to recycling, first disassembling rather than shredding all these computing components, and then managing resulting waste streams. Such creative processes are enabling them to reclaim precious metals and reuse up to 98 percent of product by weight.[25] Unfortunately, hardware recycling is still not widely available to most consumers.

For bitcoin's core developers, the concern is legitimate and worth solving: "If bitcoin really does become a global team network, I think we will need to slowly move away from proof of work as the only way it's secure," said Andresen. "In the very long run, maybe we will move away from proof of work as the way the network is secured, and we'll combine it with something else."[26]

That's what several altchains have done: explored alternative consensus algorithms such as proof of stake for securing the network while retaining decentralization. The open source nature of the bitcoin protocol makes it technically easy to do. Remember, the purpose of consensus algorithms is to distribute the right to decide what the state of the blockchain is to a decentralized set of users. To the mind of Vitalik Buterin, the visionary behind Ethereum, there are only three securely decentralized sets of users, and each set corresponds to a set of consensus algorithms: *owners of computing power*, with standard proof-of-work algorithm; *stakeholders*, with various proof-of-stake algorithms in wallet software; and *members of a social network*, with a "federated style" consensus algorithm.[27] Note that only one of those consensus mechanisms includes the word *power*. Ethereum version 2.0 will be built on a proof-of-stake model, whereas Ripple uses a federated model, a small controlled group akin to something like SWIFT, the global provider of secure financial messaging, where authorized groups reach consensus on the state of the blockchain.[28]

Those systems don't burn electricity as the bitcoin blockchain does. Bram Cohen, founder of Tor, has introduced a fourth way to address the energy waste, what he calls "proof of disk," where *owners of disk storage space*—people who have committed a chunk of computer memory to maintaining a network and performing network functions—defines the economic set of users. Of these alternatives to proof of work, Blockstream's Austin Hill cautioned against using alternative methods for securing consensus. "Experimentation with your proof-of-work algorithm is dangerous, and it's a new area of computer science."[29] It adds an additional dimension to innovation: not only must developers worry about whether their new features and functions will work in their own right, but they must also check how the choice of consensus algorithm keeps them secure and distributed to the most appropriate economic set.

Overall, the expression "If there is a will, there is a way" applies. The smartest technologists on the planet are working on creative solutions to the energy problem, with more efficient devices and use of renewable energy. Further, as computers become inexorably smarter, they will undoubtedly provide their own solutions. Bitcoin angel investor Roger Ver, nicknamed Bitcoin Jesus, said, "Say the smartest human has an IQ up close to 200. Imagine artificial intelligences with an IQ of 250, or say 500, or 5,000 or 5 million. There will be solutions, if we humans want them."[30]

3. GOVERNMENTS WILL STIFLE OR TWIST IT

To libertarians and anarchists, Satoshi Nakamoto wrote, "You will not find a solution to political problems in cryptography."[31] They would have to look elsewhere for a cure-all to big government. Satoshi viewed his experiment as a gain in a new territory of freedom, not a total upheaval. Where governments had succeeded in beheading centrally controlled networks like Napster, pure peer-to-peer networks like Tor were able to persist. Could the bitcoin blockchain network hold its own against mighty central authorities?

That might be the biggest unknown. What will legislators, regulators, and adjudicators around the world make of blockchain technologies? "The courts are going to get it wrong. They've already started to get it wrong, applying intellectual property rules to anything that is intangible. They think that physicality is the dividing line between virtual property and intellectual property, and it's not," said Josh Fairfield. "There's no intellectual property element, there's no part of a bitcoin that is intellectual property, there's no creative spark for copyright, there's no patentable idea, there's no patent, there's no trademark."[32] According to Stephen Pair of BitPay, "The biggest threat to bitcoin is that it becomes so heavily regulated at some point that a competitor that's more private and more anonymous shows up and everybody switches to that."[33] One thing's for sure: "Whatever the particular policy issue is, if you don't understand the technology and you don't understand the implications, you're setting yourself up for failure," said Jerry Brito of the bitcoin policy think tank Coin Center. "If you don't understand it, you can introduce law and policy that's going to harm the development of the technology. We just want you to understand what you're doing."[34]

So their challenge is formidable. They must oversee the unforeseeable.

On the one hand, they must avoid stifling innovation by overreacting to worst cases—human trafficking, illicit drug trade, gunrunning, child pornography, terrorism, tax evasion, and counterfeiting, for instance. On the other hand, they must not twist new but unproven applications such as blockchain-based platforms for identity management to restrict civil liberties. There must be a stable approach to regulation, legislation, and the international negotiation of treaties to minimize regulatory uncertainty, so that investors will continue to support the technology's global development.

Jurisdiction already matters when it comes to using bitcoin. Some governments have banned it or banned state banks from exchanging it, as China has done. Brito said, "In a typically Chinese way, it's not illegal, but it could be at any moment and everybody knows it."[35] China is allowing a serious professional mining community to flourish and those mining pools have become quite influential in debates over upgrades to the bitcoin protocol. What happens to blockchain security if China suddenly bans mining, too? Other jurisdictions have moved to define bitcoin narrowly, as the U.S. Internal Revenue Service has done. The IRS has labeled bitcoin as an asset for calculating taxes on the appreciation of value.

Legal frameworks also matter. Legal scholars De Filippi and Wright don't think the current one can handle the questions raised by smart property deployed globally at scale. Smart contracts both define and manage ownership rights. Their code makes no assumptions about the assignment of rights, and code can't arbitrarily seize, divest, or transfer these rights. For example, if during the process of land registration, government officials assigned the ownership of a parcel of land to someone who isn't the legal owner of that parcel, that person would have absolute sovereignty over the parcel, and the legal owner couldn't simply reverse the assignment.

Josh Fairfield focuses more on process: "The common law isn't affecting technology law; the common law *is* technology law. The common law is the process of adapting human systems to technological change . . . the real fight is how do we take old rules meant for old technology and adapt them rapidly and competently," so that they are recognizable when we start using them but iterated so that they're state of the art when the technology really hits.[36]

Last but not least, and this should be no surprise, identity matters bigtime—or at least how we construct it on the blockchain matters. "People have a very simplistic view of identity," said Andreas Antonopoulos. "I am actually terrified of the implications of digital identity because I think people will take shortcuts. . . . If we transfer identity to the digital world where views are inflexible, we actually end up with a construct that does not resemble the social construct of identity, but is a terrifying, fascist copy of it."[37]

Combine a precisely coded version of personhood with a precisely coded version of society, and you get the stuff of science fiction novels and Arnold Schwarzenegger movies. Legal scholars De Filippi and Wright conjured images of "self-enforcing contracts, *walled gardens* or *trusted systems*, owned and managed by a sophisticated network of decentralized organizations that dictate what people can or cannot do, without any kind of constitutional safeguards or constraints." In other words, a machine-driven totalitarian regime.

Artificial intelligence expert Steve Omohundro threw this phrase at us: the *dictator's learning curve*, or how cave dwellers end up with space age technology. Think about all the AI labs out there staffed by the world's smartest PhDs with access to the world's most powerful computers. PhDs might fork the bitcoin code or write a smart contract that controls a drone's delivery of a package, where bitcoin is held in escrow until that exact moment when the package arrives. Let's say these PhDs post that software as open source code to the Internet, because that's what they do to move their ideas forward; they share ideas. So now ISIS doesn't need an AI lab, it doesn't need a software development team. It just needs to substitute a grenade for the package. That's the dictator's learning curve, and it's not steep. But don't blame the code or the culture of sharing. It's not necessarily what we do with the code; it's what we don't realize we're doing with it—the unintended consequences of a friction-free world.

4. POWERFUL INCUMBENTS OF THE OLD PARADIGM WILL USURP IT

Many of our concerns about the first generation of the Internet have come true. Powerful corporations have captured much of the technology and are using it in their vast private empires to extract most of the value. They have closed off opportunity and privatized much of our digital experience. We use proprietary stores to acquire and use new apps on our phones, tablets, and now watches. Search engines and marketing departments alike interrupt our content with advertising. Big companies that promote and prosper from consumer transparency are notoriously secretive about their activities, plans, technology infrastructures, and information assets. To be sure, some companies have opened up voluntarily, but many others have merely reacted to the sunlight of whistleblowers and investigative journalism. Such disclosures are dwarfed by efforts to hide operations and conceal information.

Simply put, they haven't been good stewards of the public trust.

Case in point: the banking industry. "Banks are traditionally secret keepers," according to Kaminska of the *Financial Times*. She explained that banks make good judgments about whom to lend to and how to process payments when they have good access to private information, and they get that information by promising to keep the secret. The more secrets they hold, the greater the information asymmetry and the greater their advantages, but those advantages have harmful systemic implications.[38] So what's to prevent huge corporations or powerful nation-states from capturing blockchain technologies for their own narrow interests? "Any consensus mechanism that you have is going to be susceptible to marketing—where powerful interests spend money trying to convince people to do a certain thing," said Pair of BitPay.[39]

To be clear, we are not suggesting that corporations and governments should leave this technology alone. After all, blockchain technology is emerging as an important global resource that could enable new capabilities. Moreover, society needs governments to deliver services for their citizenry and corporations to create jobs and wealth. But that's different from capturing a disruptive technology and its largesse in ways that limit its greater benefits to society.

Also consider what the core developers and blockchain companies are already doing to secure their networks, anticipating and responding quickly to worst-case scenarios. For example, in 2014, thieves stole eight million VeriCoins, a proof-of-stake cryptocurrency, from the MintPal exchange. Within days of the attack, VeriCoin developers released new code that forked the VeriCoin blockchain prior to the hack—in a sense, they rolled back time—and collaborated with exchanges to make sure it was adopted.[40] Similarly, "if money and power do try to capture the network, the miners would stop them by going to the real version of bitcoin and initiating a fork,"[41] according to Keonne Rodriguez, product lead at Blockchain.

What's to prevent China from aiming all its state processing assets and all its mining pools at the bitcoin blockchain to stage a 51 percent attack or at minimum destabilize the process? Let's say some wealthy despot has decided that bitcoin, like the Internet before it, has become so influential that it is eroding his power. This despot could seize all the mining power within reach and purchase the rest from countries that still tolerate his bad behavior, to put him over the 50 percent hash rate threshold. He could then decide which transactions to include in blocks and which to reject. With controlling interest, he could also decide whether to fork the code and introduce a few prohibitions, maybe blacklisting addresses associated with gambling or free speech. So do honest nodes adopt this centrally controlled fork or do they fork over to a new code? Andrew Vegetabile, director of the Litecoin Association, said there was no escape from such a scenario because the despot controlled 51 percent of the network. And he needn't represent a government; he could be one of the world's wealthiest people or an executive of a highly profitable company with substantial purchasing power.[42]

A third scenario is that the incumbents will defend their territory, lobbying to make sure existing regulations for well-established firms apply to small start-ups, and suing any start-up that survives the regulatory inquisition. This litigate-not-innovate strategy may buy them time to sort out a strategy. Or it may simply drain the incumbent of whatever real value it contains. Think of those twin tyrants—legacy systems and active inertia. Academics have well documented the effects of lock-in and switching costs and have identified the challenges of postmerger systems integration. Organizations with huge technology investments in their installed base may be more likely to throw more money at their old system, sharpening their knives for the pistol fight rather than conducting strategic experiments on the blockchain.

5. THE INCENTIVES ARE INADEQUATE FOR DISTRIBUTED MASS COLLABORATION

Miners do have an incentive to maintain the bitcoin infrastructure because, if the network fails, all the unconverted bitcoin they'd earned (or could earn) through mining would be lost or worthless or otherwise at risk. Before we dig further into incentives, let's be clear about the service that miners provide: it is *not* transaction validation. Every full node can validate transactions. Rather, miners preserve the distribution of power—the power to decide which transactions to include in each block, the power to mint coins, the power to vote on the truth.

SO YOU WANNA BE A BITCOIN MINER?

As part of our research, we recruited Bob Tapscott—former bank CIO, management consultant at large, and Don's brother—to download the entire bitcoin blockchain stack and ledger in early 2015. The experiment was instructive in terms of the elapsed time, the effort required, the energy consumed, and the (lack of) payoff for hobby mining of bitcoin.

Bob dedicated his spare four-thread, two-core Windows PC to the task. Downloading took a full three days and consumed on average about 20 percent of the available processing power. Mining uses slightly more than 200 MB of memory and 10 percent of the CPU to stay current.

Although Bob's computer was hardly optimized for mining bitcoin, he entered it into a mining pool. In a 137-hour session, it mined 152.8 microbitcoin (μBTC), roughly three and a half U.S. cents at the time. But at ten cents per kilowatt-hour, Bob's computer used about fourteen cents of electricity. Bob concluded, "The days of mining bitcoins from your PC are now over."

So any design change to the original bitcoin protocol, whether through an altcoin or an upgrade, must keep in mind appropriate economic incentives to sustain miner decentralization, so that the network gets good value from miners in exchange for the large sums of bitcoin. Bitcoin core developer Peter Todd likened this task to designing a robot that can buy milk at the grocery store. "If that robot doesn't have a nose, before long store owners are going to realize it can't tell the difference between unspoiled and spoiled milk, and you're going to get ripped off paying for a bunch of spoiled milk."[43] To Todd, that means that smaller miners in geographically dispersed locations should be able to compete nose to nose with larger miners that are geographically centralized, that is, large mining pools in Iceland or China.

The question is whether that's possible. Because the number of new bitcoins minted halves every four years, what will happen when the reward drops to zero? The mining cycle depends on the market price of bitcoin. When the price drops, some bitcoin miners park their supply, but they continue to play the lottery until the price increases. Other miners can't afford to park and play; they just dry-dock their mining rigs or divert their processing power to another altchain that might be more profitable. Still others join mining pools, pooling their computing power with nodes with the hope of increasing their odds and at least getting some fraction of the winnings rather than nothing at all. And then there's the industrial bitcoin mining complex. Valery Vavilov of BitFury estimated that his mining operation would have at least 200 megawatts' capacity by the end of 2016.

One answer is charging fees. Satoshi wrote, "There will be transaction fees, so [mining] nodes will have an incentive to receive and include all the transactions they can. Nodes will eventually be compensated by transaction fees alone when the total coins created hits the predetermined ceiling."[44] So once all bitcoins have been minted, a fee structure will likely emerge. Think in terms of billions of nanopayments. Because each block has a fixed maximum size, there is a limit to how many transactions a miner can include. Therefore, miners will add transactions with the highest fees first, leaving those with low or zero fees to fight for whatever space might be left over. If your transaction fee is high enough, you can expect a miner to include it in the next block; but if the network is busy and your fee is too low, it might take two, three, or more blocks before a miner eventually records in the blockchain.

What does that mean for people who can't afford fees now? Won't levying fees lower the blockchain's advantage over traditional payment methods? According to venture capitalist Pascal Bouvier, the "fees reflect the marginal cost of verifying a transaction." Without fees to incentivize miners, as the block reward keeps halving, the hash rate would likely drop. If the hash rate drops, network security declines.[45]

That leads us back to the 51 percent attack, where a huge mining pool or a cartel of large mining pools controlled 51 percent of the hash rate. With that much firepower, they would constitute a majority vote of miners and could hijack block generation and force their version of the truth on the bitcoin network. They wouldn't necessarily get rich. Far from it. All they could do is to reverse their own transactions within a previous block, rather like a credit card chargeback. Let's say the attackers bought some big-ticket item from the same merchant, waited until it shipped, then attacked the network to get their money back. That wouldn't mean tacking its own block to the end of the blockchain. That would mean going back and redoing the block that contained all their purchases as well as all subsequent blocks, even as the network continues to generate new blocks. When the cartel's branch became longer, it would become the new valid one. Satoshi bet on that being wildly more costly than mining new coins.

Where 51 percent attacks on proof-of-work models stem from concentrated mining power, attacks on proof-of-stake models come from concentrated coin control, and coin exchanges are typically the biggest stakeholders. In some jurisdictions, exchanges must be licensed and are under regulatory scrutiny. They also have reputation at stake, and so they have multiple incentives to protect the value of their brand and the value of the coins held in account wallets. However, with more coins in circulation, a greater diversity of value, and more strategic assets registered on PoW and PoS blockchains, an attacker may not care about any of these costs.

6. THE BLOCKCHAIN IS A JOB KILLER

At the 2015 World Economic Forum annual meeting in Davos, Switzerland, a panel of technology executives from Microsoft, Facebook, and Vodaphone discussed the impact of technology on jobs. All agreed that, although technological innovations may disrupt labor markets temporarily, overall they generate new and incrementally more jobs. "Why should this time be any different?" said Eric Schmidt, executive chair of Google.

The displacement of workers through automation is nothing new. Consider the Internet's impact on travel agents and music retailers. Uber and Airbnb have created income for drivers with extra time and home owners with spare rooms, but neither provides health insurance or other employee benefits, and both are displacing better-paid jobs in the travel and hospitality industries.

The blockchain is an extraordinary platform for radical automation, where computer code rather than humans do the work, managing assets and people. What happens when autonomous vehicles replace Uber drivers? Or digital currencies obviate Western Union's five hundred thousand points of sale around the world.[46] Or when a shared blockchain platform for financial services eliminates tens of thousands of accounting and IT systems management jobs? While there will be many new business and employment opportunities created through the IoT, will it drive further unemployment, especially in the relatively unskilled market for relatively routine tasks?

In the developing world, the blockchain and cryptocurrencies could enable entrepreneurs to raise capital, protect assets and intellectual property, and create jobs even in the poorest communities. Hundreds of millions could become microshareholders in new corporations and participate in economic exchange. The technology could radically improve the delivery and deployment of aid, increase government transparency, reduce corruption, and set the conditions for good government—a precondition for jobs in many parts of the world.

Even in the developed world the effects are not determinable. A global platform that drops transaction costs, in particular the costs of establishing trusted commerce and wealth creation, could result in more participants.

Even if this technology enables us to do more with fewer human resources, we still have no case to fear, delay, or halt its march. Ultimately, what matters is not whether new capabilities exist but the extent to which societies turn these into social benefit. If machines are creating so much wealth, then maybe it's time for a new social contract that redefines human work and how much time we should all spend making a living.

7. GOVERNING THE PROTOCOLS IS LIKE HERDING CATS

How should we steward this new resource to fulfill its potential? Unlike the Internet, the bitcoin community does not yet have formal oversight bodies such as ICANN, the Internet Engineering Task Force, or the World Wide Web Consortium to anticipate development needs and guide their resolution—*and the community prefers it that way.* That presents uncertainty. People who want to keep the blockchain decentralized, open, and secure can't agree on a way forward. If we don't address governance, then the movement could collapse on itself as it disintegrates into warring factions.

There are countless issues. Bitcoin core developers Gavin Andresen and Mike Hearn have been advocating for an increase in block size from one megabyte of data to as large as twenty megabytes. Bitcoin is not "a token for rich people to trade back and forth. . . . It is a payment network," said Andresen.[47] They argue that if bitcoin is ever to compete seriously as a global payment mechanism, then it has to prepare for mainstream adoption. It couldn't grind to a halt one day when transaction flow suddenly surpassed blockchain capacity. Fees would skyrocket for people who didn't want to wait months or years for their transactions to settle. Or perhaps some central power would step in, in the interest of consumer protection, and process the overflow. In August 2015, they went ahead and launched Bitcoin XT, a fork of the blockchain that allows for eight-megabyte blocks. It is still a controversial compromise.

Opponents argue that people shouldn't be using bitcoin to buy their venti lattes at Starbucks. "Some developers want every single person in the world to be running a fully validating node that sees every single transaction and has absolutely no trust on anybody else," said Andresen. "The volunteer contributors who have been actually making the software work for the last few years are worried that they personally may not be able to handle larger blocks if transaction volume ramps up. . . . I don't have a whole lot of sympathy for that."[48] In other words, if the bitcoin blockchain is to scale and remain secure, then we can't have it both ways. Some nodes will run full protocols and process more transactions into increasingly larger blocks, and others will run simplified payment verification models and trust that 51 percent of full nodes get it right.

The biggest pushback against Bitcoin XT came from the mining pools in China. Serious bitcoin miners, like hard-core online gamers, need not only seriously powerful computers to find a correct hash but also seriously high-speed bandwidth to broadcast it quickly across the network. China is an exception to Nielsen's law of Internet bandwidth: bandwidth doesn't increase by 50 percent each year. If the block size increase is too large, it would put low-bandwidth Chinese miners at a disadvantage compared with miners in other parts of the world. Receiving new blocks to build upon would take longer; and when they did find a new block, they would take longer to send it out to the rest of the network. These delays would ultimately result in the network's rejecting some of their blocks. They would lose out to miners with more bandwidth whose blocks propagated faster.

"Trying to bootstrap or change a network protocol is just a monumental task," said Austin Hill. "You just don't want to be making changes ad hoc or very fast on an ecosystem that's managing anywhere from three to ten billion dollars' worth of people's wealth and assets."[49] At the end of the day, said Andresen, "That governance model is driven very much by what code the people actually want to run, what standards people want to implement in the equipment they sell." He said that bitcoin, like the Internet, will "have a similar messy, chaotic governance process that will eventually come down to what codes the people choose to run."[50]

Again, we're not talking about regulating but about stewarding this resource for viability and success. Governance includes setting standards, advocating and adopting sensible policies, developing knowledge about the technology's potential, performing watchdog functions, and actually building out the global infrastructure. We discuss a multistakeholder governance model in the next chapter.

8. DISTRIBUTED AUTONOMOUS AGENTS WILL FORM SKYNET

There are highly distributed enterprises with a range of good and bad actors. Anonymous, a distributed affinity group of volunteers, consists of corporate saboteurs, whistleblowers, and watchdogs. With the blockchain, Anonymous could crowdsource bitcoin and hold these funds in a wallet. Let's say a group of French shareholders would like to give that money to a few assassins who would track down and kill off the unaccounted-for terrorists responsible for the Paris massacre. They'd need thousands of signatories to reach consensus and release the funds. In this scenario, who legally controls those funds? Who is responsible for the outcome of that transaction? If you've contributed one ten-thousandth of a vote, what is your legal liability?[51]

If vending machines are programmed to order the most profitable products, will they discover a supplier for illegal goods or drugs? (Hey, the candy machine is selling Ecstasy!) How should the law handle an autonomous vehicle that accidentally kills a human being? For *Wired* magazine, two hackers demonstrated how to hijack the control systems of a Jeep Cherokee on the highway. Chrysler responded by recalling 1.4 million vehicles and alarming drivers, manufacturers, and policymakers alike.[52] Could terrorists figure out how to hack smart devices so that they performed unwanted actions with devastating consequences?

There are other challenges with distributed models of the enterprise. How does society govern these entities? How can owners keep ultimate control? How do we prevent hostile takeovers of personless businesses? Let's say we own a decentralized Web hosting company where each of the servers has a say in company management. A human hacker or some malware could pretend to be a million servers and outvote the legitimate servers in the network. When such takeovers of traditional companies occur, the results can vary. With a DAE, the results will most likely be disastrous. Once that malevolent entity controls our distributed Web hosting company, it could cash out. Or it could release the private data from other servers or hold the data hostage until we human owners paid a ransom.

Once machines have intelligence and the ability to learn, how quickly will they become autonomous? Will military drones and robots, for example, decide to turn on civilians? According to researchers in AI, we're only years, not decades, away from the realization of such weapons. In July 2015, a large group of scientists and researchers, including Stephen Hawking, Elon Musk, and Steve Wozniak, issued an open letter calling for a ban on the development of autonomous offensive weapons beyond meaningful human control.[53]

"The nightmare headline for me is, '100,000 Refrigerators Attack Bank of America,'" said Vint Cerf, widely regarded as the father of the Internet. "That is going to take some serious thinking not only about basic security and privacy technology, but also how to configure and upgrade devices at scale," he added, noting that no one wants to spend their entire weekend typing IP addresses for each and every household device.[54]

We do not recommend broad regulation of DAEs and the IoT or regulatory approvals. We do recommend that managers and entrepreneurs who are developing apps identify any significant public impacts—good, bad, or neutral—and alter source code and designs. We think they should consult with those likely affected by their creations to minimize risks in advance, identify alternative paths forward, and build support.

9. BIG BROTHER IS (STILL) WATCHING YOU

"There will be many attempts to control the network," said Keonne Rodriguez of Blockchain. "Big companies and governments will be devoted to breaking down privacy. The National Security Agency must be actively analyzing data coming through the blockchain" even now.[55] While blockchains ensure a degree of anonymity, they also provide a degree of openness. If past behavior is any indication of future intent, then we should expect corporations known for spying and countries known for waging cyberwarfare to redouble their efforts because value is involved—money, patents, access to mineral rights, the titles to land and national treasures. It's as if we've placed a big bull's-eye on top of the Internet. The good news is that everyone will be able to see the shenanigans. Some may be highly motivated to "out" spying, because they bet on the likelihood of a particular regime's attacking the blockchain in a prediction market.

What happens to privacy when the physical world starts collecting, communicating, and analyzing infinite data that could dog an individual forever? In a 2014 presentation at Webstock, Maciej Ceglowski ranted about Google's acquisition of Nest, a maker of luxury thermostats with sensors that collect data about rooms. His old thermostat didn't come with a privacy policy. This smart thermostat could report back to Google, maybe even eat his leftover pizza like a sketchy roommate.[56] Many of us are already uncomfortable with a social media environment that tracks our whereabouts and barrages us with personalized marketing messages wherever we go. In the blockchain world, we'll have better control over such, but will we be vigilant enough to manage our media diet?

None of these privacy challenges are true showstoppers. Continued Ceglowski: "The good news is, it's a design problem! We can build an Internet that's distributed, resilient, irritating to governments everywhere, and free in the best sense of the word," as we wanted it to be in the 1990s. Ann Cavoukian of the Privacy and Big Data Institute outlined seven principles for design that are "good for business, good for government, good for the public." The first is critical: make privacy the default setting. Reject false dichotomies that pit privacy against security; every IT system, every business practice, and all infrastructure should have full functionality. Leaders need to prevent rather than react to violations, maintain transparency in all operations, and subject their organizations to independent verification. Brands will earn people's trust by respecting user privacy, keeping users at the center of design, and ensuring end-to-end security of their data, destroying it when no longer needed. She said, "It really is a win-win proposition, rejecting zero-sum and embracing positive-sum."[57]

Said Ceglowski, "But it will take effort and determination. It will mean scrapping permanent mass surveillance as a business model, which is going to hurt. It will mean pushing laws through a sclerotic legal system. There will have to be some nagging. But if we don't design this Internet, if we just continue to build it out, then eventually it will attract some remarkable, visionary people. And we're not going to like them, and it's not going to matter."[58]

10. CRIMINALS WILL USE IT

In its early days, naysayers often condemned bitcoin as a tool for laundering money or buying illicit goods. Critics argued that, because the technology is decentralized, lightning quick, and peer to peer, criminals would exploit it. Chances are, you've heard of Silk Road, the dark Web marketplace for illegal drugs. At its peak in October 2013, Silk Road had 13,756 listings priced in bitcoin. Products were delivered by mail with a guide to avoiding detection by authorities. When the FBI seized the site, the price of bitcoin plummeted and digital currencies became synonymous with crime. It was bitcoin's darkest hour.

But there is nothing unique to bitcoin or blockchain technology that makes it more effective for criminals than other technologies. Authorities in general believe that digital currencies could help law enforcement by providing a record of suspicious activities, maybe even solving a multitude of cybercrimes, from financial services to the Internet of Things. Marc Goodman, author of *Future Crimes*, argued recently, "There's never been a computer system that's proven unhackable."[59] Opportunities for crime have scaled with technology. "The ability of one to affect many is scaling exponentially—and it's scaling for good and it's scaling for evil."[60] So this falls under the category of human beings wanting to harm other human beings. Criminals will use the latest technology to do it.

However, bitcoin and blockchain technology could discourage criminal use. First, even criminals must publish all their bitcoin transactions in the blockchain, and so law enforcement can track payments in bitcoin more easily than cash, still the dominant payment medium for criminals. The old Watergate adage, "follow the money" to find the crook, is actually more doable on the blockchain than with other payment methods. Bitcoin's pseudonymous nature has regulators dubbing bitcoins "prosecution futures" because they can be tracked and reconciled more easily than cash.

After each mass shooting in America, U.S. representatives whose constituents and campaign funders are card-carrying members of the National Rifle Association are quick to say, "Don't blame guns for all the gun violence in America!" It would be very rich indeed if these same people banned blockchain technology because of the crimes some people might commit on it. Technology does not have agency. It does not want for anything or have an inclination one way or the other. Money is a technology, after all. When someone robs a bank, we don't blame the money that sits in the vault for the robbery. The fact that criminals use bitcoin speaks more to the lack of strong governance, regulation, advocacy, and education than to its underlying virtues.

REASONS BLOCKCHAIN WILL FAIL OR IMPLEMENTATION CHALLENGES?

So the obstacles are formidable. Looming in the distance is quantum computing, the cryptographer's Y2K problem. It combines quantum mechanics and theoretical computation to solve problems—such as cryptographic algorithms—vastly faster than today's computers. Said Steve Omohundro, "Quantum computers, in theory, can factor very large numbers very rapidly and efficiently, and most of the public key cryptography systems are based on tasks like that. And so if they turn out to be real, then the whole cryptography infrastructure of the world is going to have to change dramatically."[61] The debate over technological innovation and progress is an ancient one: Is the tool good or bad? Does it advance the human condition or degrade it? As satirist James Branch Cabell observed, "The optimist proclaims that we live in the best of all possible worlds. The pessimist fears this is true."[62]

As the story of Lev Termen shows, individuals and organizations can use innovations for good and for evil, and that has been true across a broad range of technologies, from electricity to the Internet. Yochai Benkler, author of the seminal work *The Wealth of Networks*, told us, "Technology is not systematically biased in favor of inequality and structure of employment; that is a function of social, political, and cultural battles." While technology can change business and society dramatically and swiftly, Benkler believes it is "not in a deterministic way, one way or the other."[63]

In balance, the arc of technological history has been a positive one. Consider the many advances in food and medicine, from R&D to treatment and prevention: technology has made for greater human equity, productive capability, and social progress.

There is nothing to suggest blockchain couldn't fall into the same trap as the Internet did. It may be resistant to centralization and control; but if the economic or political rewards are great enough, powerful forces will try to capture it. Leaders of this new distributed paradigm will need to stake their claim and initiate a wave of economic and institutional innovation in order to ensure that everyone has the opportunity. This time, let's fulfill the promise. Which brings us to the issue of making all this happen.

CHAPTER 11

LEADERSHIP FOR THE NEXT ERA

Prolific is an adjective that should precede all titles used to describe twenty-one-year-old Vitalik Buterin, the Russian-born Canadian founder of Ethereum. (*Prolific* founder, that is.) Ask his legion of followers about Ethereum, and they'll tell you it's a "blockchain-based, arbitrary-state, Turing-complete scripting platform."[1] It has attracted IBM, Samsung, UBS, Microsoft, and the Chinese auto giant Wanxiang, and an army of the smartest software developers in the world, all of whom think that Ethereum may be the "planetary scale computer" that changes everything.[2]

When Buterin explained "arbitrary-state, Turing-complete" to us, we got a glimpse of his mind. Listening to music is very different from reading a book or calculating the day's revenues and expenses, and yet you can do all three on your smart phone, because your smart phone's operating system is Turing complete. That means that it can accommodate any other language that is Turing complete. So innovators can build just about any digital app imaginable on Ethereum—apps that perform very dissimilar tasks, from smart contracts and computational resource marketplaces to complex financial instruments and distributed governance models.

Buterin is a polyglot. He speaks English, Russian, French, Cantonese (which he learned in two months on vacation), ancient Latin, ancient Greek, BASIC, C++, Pascal, and Java, to name a few.[3] "I specialize in generalism," he said. He is also a polymath, and a modest one at that. "I had all these different interests, and somehow bitcoin seemed like a perfect convergence. It has its math. It has its computer science. It has its cryptography. It has its economics. It has its political and social philosophy. It was this community that I was immediately drawn into," he said. "I found it really empowering." He went through the online forums, looked for ways to own some bitcoin, and discovered a guy who was starting up a bitcoin blog. "It was called *Bitcoin Weekly*, and he was offering people five bitcoins to write articles for him. That was around four dollars at the time," Buterin said. "I wrote a few articles. I earned twenty bitcoins. I spent half of them on a T-shirt. Going through that whole process, it felt almost like working with the fundamental building blocks of society."[4]

All this from a man who, nearly five years earlier, had dismissed bitcoin. "Around February 2011, my dad mentioned to me, 'Have you heard of bitcoin? It's this currency that exists only on the Internet and it's not backed by any government.' I immediately thought, 'Yes, this thing has no intrinsic value, there's *no way* it's going to work.'" Like many teenagers, Buterin "spent ridiculous amounts of time on the Internet," reading about different ideas that were heterodox, out of the mainstream. Ask him which economists he likes, and he rattles off Tyler Cowen, Alex Tabarrok, Robin Hanson, and Bryan Caplan. He can speak on the works of game theorist Thomas Schelling and behavioral economists Daniel Kahneman and Dan Ariely. "It's actually surprisingly useful, how much you can learn for yourself by debating ideas like politics with other people on forums. It's a surprising educational experience all by itself," he said. Bitcoin kept coming up.

By the end of that year, Buterin was spending ten to twenty hours a week writing for another publication, *Bitcoin Magazine*. "When I was about eight months into university, I realized that it had taken over my entire life, and I might as well let it take over my entire life. Waterloo was a really good university and I really liked the program. My dropping out was definitely not a case of the university sucking. It was more a matter of, 'That was fun, and this is more fun.' It was a once-in-a lifetime opportunity and I just basically couldn't let it go." He was only seventeen years old.

Buterin created Ethereum as an open source project when he realized that blockchains could go far beyond currency and that programmers needed a more flexible platform than the bitcoin blockchain provided. Ethereum enables radical openness and radical privacy on the network. He views these not as a contradiction but as "a sort of Hegelian synthesis," a dialectic between the two that results in "volunteered transparency."

Ethereum, like so many technologies throughout history, could dislocate jobs. Buterin believes this is a natural phenomenon common to many technologies and suggests a novel solution: "Within a half century, we will have abandoned the model that you should have to put in eight hours of labor every day to be allowed to survive and have a decent life."[5] However, when it comes to blockchain, he's not convinced that massive job losses are inevitable. Ethereum could create new opportunities for value creation and entrepreneurship. "Whereas most technologies tend to automate workers on the periphery doing menial tasks, blockchains automate away the center," he said. "Instead of putting the taxi driver out of a job, blockchain puts Uber out of a job and lets the taxi drivers work with the customer directly." Blockchain doesn't eliminate jobs so much as it changes the definition of work. Who will suffer from this great upheaval? "I suspect and hope the casualties will be lawyers earning half a million dollars a year more than anyone else."[6] So Buterin knows his Shakespeare: "The first thing we do, let's kill all the lawyers."[7]

Ethereum has another apparent contradiction. It is unabashedly individualistic and private and yet it depends upon a large, distributed community acting openly in collective self-interest. Indeed, Ethereum's design neatly captures both his enduring faith that individuals will do the right thing when equipped with the right tools, and his healthy skepticism of the motives of large and powerful institutions in society. While Buterin's critique of the problems of contemporary society is grave, his tone is clearly one of hope. "While there are many things that are unjust, I increasingly find myself accepting the world as is, and thinking of the future in terms of opportunities." When he learned that $3,500 would enable someone to combat malaria the rest of her life, he didn't bemoan the lack of donations from individuals, governments, and corporations. He thought, "Oh wow, you can save a life for only $3,500? That's a really good return on investment! I should donate some right now."[8] Ethereum is his tool to effect positive change in the world. "I see myself more as part of the general trend of improving technology so that we can make things better for society."

Buterin is a natural-born leader, in that he pulls people along with his ideas and his vision. He's the chief architect, chief achiever of consensus in the Ethereum community, and chief cultivator of a broader community of brilliant developers who have strong opinions about anything technical. What if he succeeds?

WHO WILL LEAD A REVOLUTION?

In 1992, MIT computer scientist David Clark said, "We reject kings, presidents, and voting. We believe in rough consensus and running code."[9] That was the mantra for stewards of the first generation of the Internet. It was voiced at a time when most people could scarcely imagine how the Internet would become a new medium of human communications, one that would arguably surpass previous media in its importance for society and daily life. Clark's words embodied a philosophy for the leadership and governance of a global resource that was radically different from the norm, yet one that engendered a remarkably effective governance ecosystem.

Since the end of World War II, state-based institutions have governed important global resources. Two of the most powerful—the International Monetary Fund and the World Trade Organization—were born at the Bretton Woods Conference in 1944. The United Nations and other groups under its umbrella, such as the World Health Organization, received a wide berth to exercise their monopoly on global problem solving. These organizations were hierarchical by design, because hierarchies were the dominant paradigm during the first half of a war-torn century. But these industrial-scale solutions are ill suited to the challenges of the digital era. The rise of the Internet marked a significant departure from the traditional culture of governance.

In 1992, most Internet traffic was e-mail. The graphical browser that enabled Tim Berners-Lee's extraordinary World Wide Web was two years away. Most people weren't connected and didn't understand the technology. Many of the important institutions that would come to steward this important global resource were either embryonic or nonexistent. Barely four years old was the Internet Engineering Task Force, an international community that handles many aspects of Internet governance. The International Corporation for Assigned Names and Numbers (ICANN), which delivers essential services such as domain names, was *six years away* from existence; and Vint Cerf and Bob Kahn were just recruiting people for what would ultimately become the Internet Society.

The second generation of the Internet enjoys much of the same spirit and enthusiasm for openness and aversion to hierarchies, manifested in the ethos of Satoshi, Voorhees, Antonopoulos, Szabo, and Ver. Open source is a great organizing principle but it's not a modus operandi for moving forward. As much as open source has transformed many institutions in society, we still need coordination, organization, and leadership. Open source projects like Wikipedia and Linux, despite their meritocratic principles, still have benevolent dictators in Jimmy Wales and Linus Torvalds.

To his credit, Satoshi Nakamoto aligned stakeholder incentives by coding principles of distributed power, networked integrity, indisputable value, stakeholder rights (including privacy, security, and ownership), and inclusiveness into the technology. As a result, the technology has been able to thrive in the early years, blossoming into the ecosystem we know today. Still, this deistic hands-off approach is starting to show signs of strain. As with all disruptive technologies, there are competing views in the blockchain ecosystem. Even the core blockchain contingent has split into different cryptocamps, each advocating a separate agenda.

Brian Forde, the former White House insider and blockchain advocate who now heads MIT's Digital Currency Initiative, said, "If you look at the block-size debate, is it really a debate about block size? In the media, it's a debate about block size, but I think what we're seeing is that it's also a debate on governance."[10] What kind of governance, and more specifically, what kind of leadership is needed? Indeed, Mike Hearn, a prominent bitcoin core developer, caused quite a stir in January 2015, when he wrote a farewell letter to the industry foretelling bitcoin's imminent demise. In it, he outlined a few pressing challenges facing the industry; namely, that important technical standards questions had gone unanswered and that there was discord and confusion in the ranks of the community. Hearn's conclusion was that these challenges would cause bitcoin to fail. We disagree. Indeed, what Hearn intended as a damning critique of bitcoin's shortcomings became, in our eyes, one of the most eloquent treatises on the importance of multistakeholder governance, based on transparency, merit, and collaboration. Code alone is just a tool. For this technology to reach its next stage and fulfill its long-term promise, humans must lead. We now need all constituents—all stakeholders in the network—to come together and address some mission-critical issues.

We've already outlined some of the showstoppers. They are significant. But they are challenges to this revolution's success, not reasons to oppose it. To date, many issues are still unsolved and many questions unanswered, with little collective movement to resolve them. How will the technology scale, and can we scale it without destroying the physical environment? Will powerful forces choke innovation or co-opt it? How will we resolve controversial standards questions without reverting to hierarchy?

How to answer those questions has been the focus of our research over the last two years. We found that, instead of state-based institutions, we need collaborations of civil society, private sector, government, and individual stakeholders in nonstate networks. Call them *global solution networks* (GSNs). These Web-based networks are now proliferating, achieving new forms of cooperation, social change, and even the production of global public value.

One of the most important is the Internet itself—curated, orchestrated, and otherwise governed by a once-unthinkable collection of individuals, civil society organizations, and corporations, with the tacit and sometimes active support of nation-states. But no government, country, corporation, or state-based institution controls the Internet. It works. In doing so, it has proven that diverse stakeholders can effectively steward a global resource by inclusiveness, consensus, and transparency.

The lessons are clear. Good governance of such complex global innovations is not the job of government alone. Nor can we leave it to the private sector: commercial interests are insufficient to ensure that this resource serves society. Rather, we need all stakeholders globally to collaborate and provide leadership.

THE BLOCKCHAIN ECOSYSTEM: YOU CAN'T TELL THE PLAYERS WITHOUT A ROSTER

Although blockchain technology emerged from the open source community, it quickly attracted many stakeholders, each with different backgrounds, interests, and motives. Developers, industry players, venture capitalists, entrepreneurs, governments, and nongovernment organizations have their own perspectives, and each has a role to play. There are early signs that many of the core stakeholders see the need for leadership and are stepping up. Let's review who the players are:

Blockchain Industry Pioneers

Vanguards in the industry, from Erik Voorhees to Roger Ver, believe any form of formal governance, regulation, stewardship, or oversight is not only foolish, but antithetical to the principles of bitcoin.[11] Said Voorhees, "Bitcoin is already very well regulated by mathematics, which are not up to the whims of governments."[12] However, as the industry has expanded, many entrepreneurs are seeing a healthy dialogue with governments, and a focus on governance more broadly, as a good thing. Companies like Coinbase, Circle, and Gemini have joined trade organizations; and some even maintain close relations with emerging governance institutions, such as the Digital Currency Initiative at MIT.

Venture Capitalists

What started as a clique of cryptoinsiders quickly snowballed into Silicon Valley's biggest and brightest VCs, including the venerable Andreessen Horowitz. Now financial services titans are playing venture capitalist: Goldman Sachs, NYSE, Visa, Barclays, UBS, and Deloitte have made direct investments in start-ups or supported incubators that nurture new ventures. Pension funds are entering the fray. OMERS Ventures, the billion-dollar venture arm of one of Canada's largest public sector pensions, made its first investment in 2015. Jim Orlando, who runs that group, is looking for the next killer app that "does for blockchain what the Web browser did for the Internet."[13] Investment has exploded—from two million dollars in 2012 to half a *billion* in the first half of 2015.[14] The excitement is palpable. Tim Draper told us that, if anything, "financiers are underestimating the potential of blockchain."[15] Vocal venture capitalists can advocate for the technology and support nascent governance institutions, such as Coin Center, bankrolled by Andreessen Horowitz. Digital Currency Group, a venture firm founded by Barry Silbert, has appointed academics and other nontraditional advisers to its board to accelerate the development of a better financial system through both investment and advocacy.

Banks and Financial Services

Perhaps in no other industry have we seen a swifter change of opinion. For the longest time, most financial institutions dismissed bitcoin as the speculative tool of gamblers and criminals, and barely even registered blockchain on their radars. Today they are quite literally "all in." Watching this unfold in real time in 2015 was truly incredible. Before 2015, few major financial institutions had announced investments in the sector. Today Commonwealth Bank of Australia, Bank of Montreal, Société Générale, State Street, CIBC, RBC, TD Bank, Mitsubishi UFJ Financial Group, BNY Mellon, Wells Fargo, Mizuho Bank, Nordea, ING, UniCredit, Commerzbank, Macquarie, and *dozens* of others are investing in the technology and wading into the leadership discussion. Most of the world's biggest banks have signed up to the R3 consortium and many more have partnered with the Linux Foundation to launch the Hyperledger Project. Banks should be included in the discussion about leadership, but other stakeholders must remain cautious of powerful incumbents looking to control this technology, just as they had to tread cautiously in the early days of the Internet.

Developers

Developers in the community are split on basic technical issues, and the community is expressing a need for coordination and leadership. Gavin Andresen, the bitcoin core developer at the center of the block-size debate, told us, "I'd prefer to stay in the engine room, keeping the bitcoin engine going"[16] rather than spending every waking moment advocating his position. However, given the lack of clear leadership, Andresen has been inadvertently cast in the spotlight. In the summer of 2015, he told us, "My job over the next six months is to focus on bitcoin's technical life, making sure bitcoin is still around in two or three years for those businesses to happen: micropayments, stock trading, or property transfer, all these other things," which involves a lot of advocating and lobbying. To him, the Internet governance network is a useful starting point. "I always look for role models. The figure role model is the IETF."[17] How the Internet is governed is "kind of chaotic and messy," he said, but it works and it's reliable.

Academia

Academic institutions are funding labs and centers to study this technology and collaborate with colleagues outside their silo. Brian Forde told us, "We started DCI to catalyze some of the great resources we have at MIT to focus on this technology, because we think it's going to be one of the most important technological transformations over the next ten years."[18] Joichi Ito, director of the MIT Media Lab, saw an opportunity for academia to step up: "MIT and the academic layer can be a place where we can do assessments, do research, and be able to talk about things like scalability without any bias or special interests."[19] Jerry Brito, one of the most prominent legal voices in the space—first at the Mercatus Center at George Mason University and now as director of Coin Center, a not-for-profit advocacy group—said, "Governance comes into play where there are serious decisions that need to be made, and you need a process for that to happen."[20] He recommended starting with the Hippocratic oath: first, do no harm. The current bottom-up approach that bitcoin's core developers are using "is showing a little bit of its rough edges right now with the block-size debate. It's going to be very difficult to get any consensus," Brito said. "We want to help develop that forum and foster a self-regulatory organization if it comes to that."[21] Notable universities such as Stanford, Princeton, New York University, and Duke also teach courses on blockchain, bitcoin, and cryptocurrencies.[22]

Governments, Regulators, and Law Enforcement

Governments all over the world are uncoordinated in their approach—some favoring laissez-faire policy, others diving in with new rules and regulations such as the BitLicense in New York. Some regimes are openly hostile, though this is increasingly a fringe response. Likewise, the industry is splitting into factions, those who support the new rules and those who do not. Even those who resist government intervention acknowledge that their enthusiasm to wade into governance debates is a net positive. Adam Draper, a prolific VC in the industry, acknowledged, albeit reluctantly, "Government endorsement creates institutional endorsement, which has value."[23] Central banks globally are each taking different steps to understand this technology. Benjamin Lawsky, former superintendent of financial services for the State of New York, said strong regulations are the first step toward industry growth.[24]

Nongovernment Organizations

The year 2015 proved transformative for the burgeoning constellation of NGOs and civil society organizations focused specifically on this technology. Though Forde's DCI is housed within MIT, we include it here. Other such groups include Brito's Coin Center and Perianne Boring's Chamber of Digital Commerce. These groups are gaining traction in the community.

Users

This means you and me—people who care about identity, security, privacy, our other rights, long-term viability, fair adjudication, or a forum for righting wrongs and fighting criminals who use technology to destroy what we care about. Everyone seems divided on basic taxonomy and categorization: Does blockchain refer to the bitcoin blockchain or the technology in general? Is it big "B" Blockchain or little "b" blockchain? Is it a currency, commodity, or technology? Is it all of these things or none of these things?

Women Leaders in Blockchain

As many have observed, the blockchain movement is overpopulated with men. In technology and engineering, males still outnumber females by a wide margin. However, high-profile women are founding and managing companies in the space: Blythe Masters, CEO of Digital Asset Holdings; Cindy McAdam, president of Xapo; Melanie Shapiro, CEO of Case Wallet; Joyce Kim, executive director of Stellar Development Foundation; Elizabeth Rossiello, CEO and founder of BitPesa; and Pamela Morgan, CEO of Third Key Solutions. Many of them have suggested the industry is very welcoming to all voices, male and female alike. Venture capital in blockchain is also gaining in diversity. Arianna Simpson, former head of business development at BitGo, is now an investor in the sector. Jalak Jobanputra is an investor whose VC fund focuses on decentralized technology.

When it comes to governance and stewardship of this global resource, women have taken the lead.

Primavera De Filippi, faculty associate at the Berkman Center at Harvard and a permanent researcher at the National Center of Scientific Research in Paris, is a tireless advocate of blockchain technology and has emerged as one of academia's clearest and most eloquent voices on governance. She is organizer, instigator, and promoter of dialogue within the ecosystem. With lawyer-turned-entrepreneur Constance Choi, another vocal proponent in the industry, De Filippi has led a series of blockchain workshops at Harvard, MIT, and Stanford, as well as in London, Hong Kong, and Sydney. They have brought together diverse stakeholders from the industry and beyond to debate big issues. Nothing is off limits, and the events often mash up people of different backgrounds, persuasions, and beliefs.

Elizabeth Stark is another emerging star in governance. The Yale Law School professor has taken up the mantle of convener-in-chief for the industry. Like another prominent woman—Dawn Song, MacArthur fellow and computer science professor at Berkeley, and an expert in cybersecurity—Stark comes from a distinctly academic background but has other ambitions. She organized Scaling Bitcoin, convening developers, industry players, thought leaders, government officials, and other stakeholders in Montreal. A "constitutional moment" for the sector, Scaling Bitcoin was credited with clearing logjams in the block-size debate. Today she is also leading as an entrepreneur, collaborating on the development of the Bitcoin Lightning Network to solve the blockchain's scalability issue.

Perianne Boring, a former journalist and TV reporter, is the founder of the Chamber for Digital Commerce, a trade-based association in Washington, D.C. Within a year, CDC has attracted a high-profile board (e.g., Blythe Masters, James Newsome, George Gilder). The movement needed "boots on the ground in Washington to open a dialogue with government," she said. With her background in journalism, Boring focused on messaging, positioning, and polish. Her organization is "open to anyone who is committed to growing this community," she said, and is now a leading voice in policy, advocacy, and knowledge in the burgeoning blockchain governance ecosystem.[25]

This growing chorus of leaders lobbying for governance is as prescient as it is urgent. When we talk about governing blockchain technology, we are not talking about regulation, at least not exclusively. For one, there are serious limitations to using regulations for managing an important global resource. As Joichi Ito said, "You can regulate networks, you can regulate operations, but you can't regulate software."[26] So regulations will be one of several important components. Blockchain is not like the Internet because money is different from information. Blythe Masters, consummate Wall-Street-insider-turned-blockchain-pioneer, expressed her concern: "Newcomers are simply able to do things that regulated institutions are not able to do, but one needs to think very carefully about why those regulations exist, and what purpose they serve, before one can conclude that exposing consumers to unregulated financial activities is a good thing."[27] Ultimately, the debate is not about the kind of society we want but about the opportunities for leaders to steward an important global resource.

A CAUTIONARY TALE OF BLOCKCHAIN REGULATION

Benjamin Lawsky, the former superintendent of financial services for the State of New York (NYDFS), was once the most powerful bank regulator in the United States. To Washington insiders, Lawsky was known for his early morning selfies on his daily jogs around the city. But to the titans of Wall Street, he was a gutsy, ambitious (not to mention overzealous) scrapper who would routinely take the fight to any bank he thought was misbehaving and seek his just deserts.

Appointed by friend and longtime political ally Governor Andrew Cuomo, Lawsky was the first ever to hold the office of top watchdog of the state's chartered banks. In 2012, only one year into the job, he made headlines when NYDFS reached a $340 million settlement with U.K. bank Standard Chartered PLC for its handling of more than $250 million in transactions from Iran, prohibited at the time by U.S. and E.U. sanctions. In the process, NYDFS scooped the Justice Department, which was seeking a similar penalty.[28] To those who thought bank regulations were too lax, he was the new sheriff in town, a fearless leader and reformer of an industry run amok. To the banks, he was quickly becoming Public Enemy Number One. Lawsky was just getting started.

It was mid-2013 Lawsky was at his desk, probably working on another blockbuster case against the big banks, when an economist on his staff knocked on his door to discuss some unusual inquiries. According to a few lawyers on the street, several client firms were transacting in some strange new virtual currency called bitcoin. Lawsky's first reaction was "What the heck is bitcoin?"[29] The economist went on to explain that these companies had customers who were buying, selling, trading, and paying for goods and services with this digital dollar and that the lawyers, ever cautious, wanted to know whether this kind of activity qualified as money transmission, and if so, what to do about it. In New York, money transmissions are typically regulated at the state level; and so the NYDFS, as the state regulator in New York, had a duty to regulate any entity engaged in money transmission. But how? Lawsky hadn't even heard about the technology, and he had a sneaking suspicion this would be a very different kind of challenge.

Almost immediately, Lawsky was confronted with a problem that has become all too commonplace, that disruptive technology does not fit neatly into existing regulatory boxes, a hallmark of the digital age. In his mind, bitcoin didn't fit at all. Bitcoin is global in reach; federal and state governments would be limited in the scope of what they can do to govern and regulate it. Moreover, the technology is peer to peer and decentralized. Regulators make a living monitoring large intermediaries. Their centralized ledgers contain troves of data, ideal for building cases. And in the digital age, officials in government are rarely, if ever, in possession of all the information needed to make decisions in the public interest. Often, they lack resources to govern it effectively and can be ill informed about innovation. Lawsky was coming to terms with something that governments and regulators of digital technologies had wrestled with for twenty years. Thanks to luck, foresight, and a different regulatory framework, the Internet was able to grow and thrive. Cryptocurrencies were another example of how digital technology is wresting control from traditional decision makers, including governments.

Still, Lawsky had a job to do. Upon reviewing the existing statutes, he found them woefully inadequate. The department initially wanted to regulate this technology by enforcing rules written around the time of the Civil War. Those money transmission laws couldn't possibly address any kind of digital technology like the Internet, let alone digital currencies or cybersecurity. "The more I learned, the more interested I got in how powerful this technology is, and I saw all the various applications and platforms that were going to be built, over time," he said. If he "could get regulation right, to make sure the bad stuff we didn't want to see happening in the ecosystem was avoided, and at the same time not have regulation be too overbearing, then we had a real chance of helping a very powerful technology make serious improvements to our system."[30] Lawsky concluded, "Maybe we need a new type of regulatory framework to deal with something that is just qualitatively different?"[31] His proposal, the BitLicense, was the first serious attempt to provide a regulatory lens onto this industry. A controversial piece of law, it revealed how even well-intentioned regulations can produce unintended consequences. When the BitLicense went into effect, there was a mass exodus of companies such as Bitfinex, GoCoin, and Kraken from New York; they cited the prohibitive cost of the license as a main cause. The few that stayed are well-capitalized and more mature businesses.

The benefits, such as improved oversight and consumer protection, are significant. Licensed exchanges, such as Gemini, have gained ground, perhaps because their institutional clientele know they're now as regulated as banks. But with fewer competitors, will the BitLicense stifle innovation and cripple growth? Brito argued that the BitLicense misses the mark by applying old solutions to new problems. He cited the BitLicense rule that if you take custody of consumer funds, you need to get a license. "With something like bitcoin and other digital currencies, you have technologies like multisig [multisignature] that, for the first time, introduce the concept of divided control. So if the three of us each have a key to a multisig address that needs two out of three, who has custody of the funds?"[32] In this case, the concept of custody, once very clear in the law, is now ambiguous.

"My belief is the next five to ten years will be one of the most dynamic and interesting times in history for our financial system," Lawsky said.[33] He resigned from NYDFS to keep working on important issues at the heart of this dynamic environment. "I would enjoy my career if I got to spend my time working in the middle of what I believe is going to be an enormously transformative, dynamic, interesting time . . . you have this world of technology, which is usually largely unregulated, colliding with probably the most regulated system in the world, the financial system. No one really knows what comes of that collision," he said. "It's all going to work out over the next five to ten years and I want to be in the middle of that collision."[34]

THE SENATOR WHO WOULD CHANGE THE WORLD

The Canadian Senate surprised many when, in June 2015, its Committee on Banking, Trade, and Commerce released an unambiguously positive and thoughtful report, "Digital Currency: You Can't Flip This Coin."[35] Incorporating feedback from multiple stakeholders in the blockchain ecosystem, the report detailed why governments should embrace blockchain technology.[36]

"This could be the next Internet," said Doug Black, the Canadian senator from Calgary, Alberta, and a major contributor to the report. "This could be the next TV, the next telephone. We want to signal both within and outside Canada, we support innovation and entrepreneurship."[37] Like Ben Lawsky, Black is a veteran lawyer. He made his career in the country's oil patch, working on behalf of oil and gas producers as a partner at one of the country's most prestigious law firms. Senator Black differs from Mr. Lawsky, however, in his reluctance to rush new regulations out the door. "Government should get out of the way!" Black told us.[38] As members of the Canadian Senate, Senator Black and his colleagues have no formal legislative role, but can move the needle on important issues by issuing guidance or making recommendations to the government. Still, with an average age of sixty-six, the Canadian Senate wouldn't be the odds-on favorite to embrace this cutting-edge technology. But that's exactly what they did.

Reflecting on the process, Black recalled thinking, "How do we create an environment that encourages innovation as opposed to stifles innovation? . . . That's unusual for a government to take that point of view from the get-go." According to Black, governments "tend to be concerned about maintaining control and minimizing risk."[39] While acknowledging the risk any new technology poses to consumers and business alike, Black explained, "There's risk in anything; there's risk in fiat currency. We can manage risk at some level, but let's also create an environment where innovation can be fostered."[40] With this report, Black believes they've hit the mark.

The report makes a number of recommendations, but two stand out. First, the government should start using the blockchain in its interactions with Canadians. Black said, "The blockchain is a more confidential vehicle to protect data"; therefore, "government should be looking to start utilizing this technology, which would be a powerful message."[41] This is a powerful statement: if you want to be the hub for innovation and a pioneer in the sector, put your money where your mouth is, and start innovating yourself.

The second recommendation is perhaps even more surprising: the government should take a light touch on regulation. A number of respected figures in the legal profession who focus on blockchain technology have made this argument. Aaron Wright of Cardozo School of Law, Yeshiva University, advocates for "safe harbor" laws that allow innovators to keep innovating while minimizing government regulations until the technology matures.[42] Josh Fairfield, of Washington and Lee University Law School, said, "We need regulations that act like technology—humble, experimental, and iterative."[43]

CENTRAL BANKS IN A DECENTRALIZED ECONOMY

Finance may be the second-oldest profession, but central banking is a relatively modern phenomenon. The U.S. Federal Reserve (the Fed), the world's most powerful central bank, celebrated its centennial in 2013.[44] Central banks, in their relatively short history, have gone through multiple reincarnations, the last one a big shift from the gold standard to a floating-rate system of fiat currencies. Because digital currencies challenge the role of central banks in an economy, we might expect central bankers to oppose blockchain technology. However, over the years, these bankers have shown a willingness to innovate. The Fed pioneered electronic clearing of funds by championing the Automated Clearing House (ACH) system when all checks were settled and cleared manually. Like central banks elsewhere, the Fed has savored experimentation. It has embraced unorthodox and untested policies, most famously (or infamously) the quantitative easing program in the wake of the 2008 financial crisis, when it used newly minted money to buy financial assets such as government bonds at an unprecedented scale.

Not surprisingly, central bankers have been forward thinking in understanding blockchain technology's importance to their respective economies. There are two reasons for this leadership. First, this technology represents a powerful new tool for improving financial services, potentially disrupting many financial institutions and enhancing the performance of central banks in the global economy.

Second, and this is the big one, blockchain raises existential questions for central banks. How do they perform their role effectively in a global market with one or many cryptocurrencies outside their control? After all, monetary policy is a key lever in a central banker's toolbox to manage the economy, particularly in times of crisis. What happens when that currency is not issued by a government but exists globally as part of a distributed network?

Central bankers everywhere are exploring these questions. Carolyn Wilkins, deputy governor of the Bank of Canada and a central banking veteran, told us, "We are confident in our paradigm right now, but we understand many paradigms have a shelf life: they're going to work well for a number of years and then things are going to start to go wrong. You can fix it at the margin first, but eventually you just need to switch to something else." She believes the blockchain could be that something else. "It's hard not to be fascinated by something so transformative. This technology is being used in ways that have implications for central banking that span all the functions that we have," she said.[45]

Ben Bernanke, former chair of the Fed, said in 2013 that blockchain technology could "promote a faster, more secure, and more efficient payment system."[46] Today, both the Fed and the Bank of England (and likely other central bankers who have not been as vocal) have teams dedicated to this technology.

To understand why central banks are so interested, let's first address what central banks do. Broadly speaking, these august institutions perform three roles. First, they manage monetary policy by setting interest rates and controlling the money supply and in exceptional circumstances by injecting capital directly into the system. Second, they attempt to maintain financial stability. This means they act as the banker for government and for the banks in the financial system; they are the lender of last resort. Finally, central banks often share the responsibility with other government entities of regulating and monitoring the financial system, particularly the activities of banks that deal with savings and loans to average consumers.[47] Invariably, all of these roles are intertwined and codependent.

Let's start with financial stability. "As a central bank, our role is as a liquidity provider of last resort. We do that in Canadian dollars. Therefore, Canadian dollars are important as a source of liquidity for the Canadian financial system," Wilkins said. What if transactions are in another currency like bitcoin? "Our ability to provide lender of last resort services would be limited."[48] The solution? Central banks could simply begin holding reserves in bitcoin, as they do in other currencies, and assets such as gold. They could also require financial institutions to hold reserves at the central bank in these nonstate currencies. These holdings would enable a central bank to perform their monetary role in both fiat and cryptocurrencies. Sounds prudent, right?

When considering financial stability relative to monetary policy, Wilkins said, "The implications [for monetary policy] of electronic money depend on how it's denominated." She suggested in a recent speech that "e-money," as she called it, could be denominated by a government in a national currency or as a cryptocurrency.[49] A digital currency denominated in Canadian dollars would be easy to manage, she said. If anything, it would help a central bank to respond more quickly. Most likely, we will see a combination of the two: central banks will hold and manage alternative blockchain-based currencies as they do foreign reserves and will explore converting fiat currency to so-called e-money through a blockchain-based ledger. This new world will look a lot different.

What about central banks as regulators and watchdogs? They have considerable regulatory power in their respective countries, but they do not operate in silos. They coordinate and collaborate with other central banks and with global institutions like the Financial Stability Board, the Bank for International Settlements, the International Monetary Fund, the World Bank, and others. We need stronger global coordination to address blockchain issues. Today, central bankers are asking important questions. Carolyn Wilkins said, "It's easy to say that regulation should be proportionate to the problem, but what is the problem? And what are the innovations that we want?"[50] These are great questions that we could address more effectively in an inclusive environment.

Bretton Woods is a good model. How about a second meeting of the minds, not conducted in smoky rooms behind closed doors, but in an open forum where various stakeholders, including the private sector, the technology community, and governance institutions could participate? Wilkins said, "The Bank of Canada works with other central banks on understanding this technology and what it means. We've had conferences that invited a variety of central banks and academics and people from the private sector."[51]

Indeed, the story of central banks reveals a bigger issue: governments often lack the know-how to respond in a fast-changing world. Central bankers certainly have views that matter profoundly to this discussion, but they should look to other stakeholders in the network and other central banks globally to share ideas, collaborate on substantive leadership issues, and move the agenda forward.

REGULATION VERSUS GOVERNANCE

To be sure, value and money are different from traditional information. We're talking about savings, a pension, a person's livelihood, her company, her stock portfolio, her economy, and that affects everyone. Don't we need regulation, and fast? Can and should government show restraint in the face of the seismic shifts to come?

Important shifts are revealing the limits of government in an age of accelerating innovation. For example, the 2008 financial crisis showed how the speed and complexity of the global economic system renders traditional centralized rule making and enforcement increasingly ineffective. But stronger regulation isn't the antidote. Governments cannot hope to oversee and regulate every corner of the financial market, technology, or the economy, because there are simply too many actors, innovations, and products. If anything, the experience illustrates that governments can at least force transparency to shed light on behavior and create change. Governments can demand that the actions of banks, for example, be transparent on the Web and let citizens and other parties contribute their own data and observations. Citizens can even help enforce regulations, too, perhaps by changing their buying behavior or, armed with information, by organizing public campaigns that name and shame offenders.

Of course, governments must be key stakeholders and leaders in governance. They must also acknowledge that their role in governing the blockchain will be fundamentally different from their historical role in monetary policy and financial regulation. For millennia, states have had a monopoly on money. What happens when "money" is not issued exclusively by a central authority but instead is (at least in part) created by a distributed global peer-to-peer network?

While generally positive, the U.S. response has seemed at times contradictory. "In the U.S. there is a realization from Congress to the executive branch to different agencies including law enforcement that this technology has serious, legitimate uses," said Jerry Brito.[52] Indeed, the Internet has shown us that, by temperament and institutional design, the United States not only tolerates but welcomes innovations that push the boundaries. It also fences off innovation through regulations—some of which may be misguided and are almost certainly premature.

The risks of regulating prematurely—before firmly grasping the implications—can have profound consequences. During Victorian era England, so-called self-driving locomotives (i.e., automobiles) were mandated by law to be accompanied by a man walking in front waving a red flag to alert bystanders and horses of the coming arrival of this strange contraption. Steve Beauregard, CEO of GoCoin, a leading company in the industry, described the pitfalls of regulating too soon: "When Web pages were first going up, regulators were trying to determine what regulatory regime they should belong under. One idea surfaced requiring people who built and hosted Web sites to get a citizen's band radio license because you're broadcasting. Can you imagine having to have a CB radio license so you could put a Web site up?"[53] Thankfully, this never came to pass.

Let's be clear: regulation differs from governance. Regulation is about laws designed to control behavior. Governance is about stewardship, collaboration, and incentives to act on common interests. But experience suggests governments should approach regulating technologies cautiously, acting as a collaborative peer to other sectors of society, rather than as the heavy hand of the law. They must participate as players in a bottom-up governance ecosystem rather than as enforcers of a top-down regime of control.

Brito of Coin Center argued there is a role for governments, but they should exercise caution. He advocates for a multistakeholder solution, which starts with education: "briefing folks in Congress, at the agencies, in the media, and answering any of their questions or putting them in touch with the people who can intelligently answer their questions."[54]

A NEW FRAMEWORK FOR BLOCKCHAIN GOVERNANCE

Rather than simply regulating, governments can improve the behavior of industries by making them more transparent and boosting civic engagement—not as a substitute for better regulation but as a complement to the existing systems. We believe effective regulation and, by extension, effective governance come from a multistakeholder approach where transparency and public participation are valued more highly and weigh more heavily in decision making. For the first time in human history, nonstate, multistakeholder networks are forming to solve global problems.

In recent decades, two major developments have provided the basis for a new model. First, the advent of the Internet has created the means for stakeholders of all sizes, down to individuals, to communicate, contribute resources, and coordinate action. We no longer need government officials to convene for the rest of us to align our goals and efforts. Second, businesses, academia, NGOs, and other nonstate stakeholders have gained the ability to play an important role in global cooperative efforts. There were no businesses, NGOs, or nonstate stakeholders at the table at Bretton Woods. Today, these stakeholders routinely engage with governments to address issues in all facets of society—from the governance of a global resource like the Internet to addressing global problems like climate change and human trafficking.

The combination of these developments enables the new model. For a growing list of global challenges, self-organizing collaborations can now achieve global cooperation, governance, and problem solving—and make faster, stronger progress than traditional state-based institutions.

In considering the foundation for a blockchain governance network, we pose a number of critical questions and develop a framework for answering them:

- How do we design such a governance network?
- Do we create a new network from scratch or build around an existing institution that already has a constituency that deals with international financial issues?
- What will be the mandate for this network and will it have the power to implement and enforce policy?

- In whose interests will a blockchain governance network act and to whom is it accountable?
- And critically, will nation-states actually cede any authority to a global network?

Overall, the ecosystem that governs the Internet is rich with lessons. That it has become a global resource in so short a time is astounding, in no small part thanks to strong leadership and governance and despite the powerful forces against it.

So who governs the first-generation Internet and how? A vast ecosystem of companies, civil society organizations, software developers, academics, and governments, namely the U.S. government, in an open, distributed, and collaborative manner that we cannot measure by traditional command-and-control hierarchies and frameworks. No governments or group of governments control the Internet or its standards, though several U.S. government agencies once funded it.[55]

In the early days of the Internet, governments showed both restraint and foresight. They showed restraint by limiting regulation and control throughout the Internet's evolution and they showed foresight by allowing the ecosystem to flourish before trying to impose rules and regulations. This multistakeholder network worked for the Internet, but we need to recognize that there will be a greater role for regulation of blockchain technologies. Whereas the Internet democratized information, the blockchain democratizes value and cuts to the core of traditional industries like banking. Clearly there will be a regulatory role to ensure that consumers and citizens are protected. Yet our research suggests that the Internet governance model is a good template.

Questions persist over how much new leadership will come from the old Internet governance community. Vint Cerf, who coinvented the Internet itself and led the creation of the Internet Society and the Internet Engineering Task Force, which has created virtually all the important Internet standards, suggested that a good starting point for blockchain would be to create a BOF (Birds of a Feather) interest group within the IETF.[56] Initially, many organizations involved in Internet governance viewed digital currencies and blockchain technologies as outside their purview, but that is changing. The World Wide Web Consortium, W3C, has made Web payments a priority, and blockchain is central to that discussion.[57] Additionally, the Internet Governance Forum (IGF) has hosted sessions about blockchain and bitcoin, where participants have explored new decentralized governance frameworks enabled by this technology.[58] Boundaries between old and new are fluid, and many leaders in the Internet governance network, such as Pindar Wong, the Internet pioneer, former vice-chair of ICANN and trustee of the Internet Society, have been the most effective leaders in blockchain governance as well.[59]

What does the new governance network look like? There are ten types of GSNs. Each involves some combination of companies, governments, NGOs, academics, developers, and individuals. None of them are controlled by states or state-based institutions like the UN, IMF, World Bank, or the G8. All will play an important role in the leadership and governance of blockchain technology.

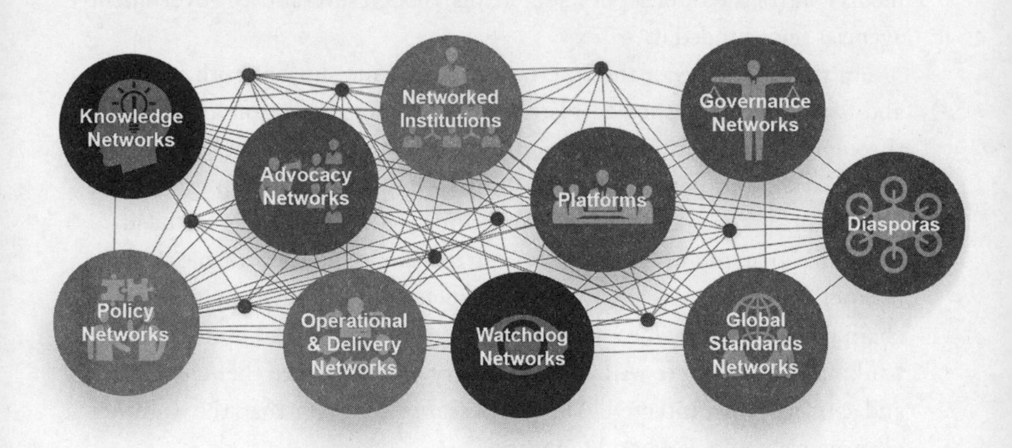

1. Knowledge Networks

The primary function of knowledge networks is to develop new thinking, research, ideas, and policies that can help solve global problems. More informed and savvy users can better protect themselves from fraud and theft and protect their privacy. They can also realize the full value of this disruptive technology, creating opportunities for a greater share in global prosperity and greater financial connectivity.[60] Knowledge networks must foster a culture of openness and inclusion, be transparent, and involve multiple stakeholders.

Blockchain Implications: Knowledge networks are the origination points for disseminating new ideas to other GSNs and the broader world. They are the key to avoiding pitfalls and showstoppers. Knowledge will prepare stakeholders to advocate more effectively, create or cocreate policy, and spread critical information to users. Knowledge sharing instigates a fruitful dialogue with government. According to Jerry Brito of Coin Center, whatever the particular policy issue is, if governments "don't understand the technology and don't understand the implications, they're setting themselves up for failure."[61] Many voice the need to create spaces for ideas and information to be shared and debated. "There should be a forum to present proposals or ideas," Tyler Winklevoss said.[62] MIT's Digital Currency Initiative is a leading knowledge network, trying to unite and excite academics and universities globally. Below the radar, informal meetups, like the San Francisco developer meetup and the New York developer meetup, are also making knowledge a priority. Blockchainworkshops.org is another group that has convened stakeholders to spread knowledge and key lessons. Reddit, the online forum and community, is also a breeding ground for new knowledge in the space.

2. Delivery Networks

This class of networks actually delivers the change it seeks, supplementing or even bypassing the efforts of traditional institutions. For example, ICANN performs an essential role in the Internet governance network, delivering solutions in the form of domain names.

Blockchain Implications: How do we ensure that the incentives are adequate for distributed mass collaboration, making the technology ready for prime time? We will likely have an "ICANN moment" for blockchain, where organizations will form to deliver essential functions. However, whereas ICANN and many other GSN types in the Internet governance network are distinctly American, blockchain leaders should push to make these organizations international. Joichi Ito said, "I do think there's already a big push to make governance non-American and international from the beginning because that's one thing we learned from ICANN, that it's hard to get out from under America once you get started as part of America."[63] The Coalition for Automated Legal Applications (COALA) is a global organization that performs a few key roles: It disseminates knowledge, influences policy, and advocates for blockchain technology, and supports the development and deployment of blockchain-based applications, all critical to overcoming major potential showstoppers.[64]

3. Policy Networks

Sometimes networks create government policy, even though they may consist of nongovernmental players. Policy networks support policy development or create an alternative for policy, whether governments support them or not. The goal of policy networks is not to wrest control of the policy-making process from governments. Instead, their goal is to turn decision making from the traditional hierarchical broadcast model to one of consultation and collaboration.

Blockchain Implications: Today, a nascent policy network is emerging. Coin Center, a not-for-profit policy group in Washington, D.C., focuses on five core verticals: innovation, consumer protection, privacy, licensing, and AML/KYC (anti–money laundering/know your customer). The Chamber of Digital Commerce, a trade organization, focuses on promoting the acceptance and use of digital currencies.[65] The United Kingdom has its own Digital Currency Association, as do Australia and Canada, that speaks for industry. With the hiring of John Collins, a former senior adviser to the U.S. government, Coinbase became the first company to install a permanent policy advocate.[66] Promoting and uniting many strong voices in the policy arena will ensure that blockchain has a better chance of fulfilling its potential. For example, we know mining consumes a lot of energy and that climate change is a big problem. Responsible policy will go a long way toward building a sustainable future, and government can't do it alone.

4. Advocacy Networks

Advocacy networks seek to change the agenda or policies of governments, corporations, and other institutions. The Internet has lowered the cost of collaboration, and today the world is witnessing the dramatic rise of increasingly powerful advocacy networks that are more global, widely distributed, and technologically sophisticated than anything we've seen.

Blockchain Implications: Advocacy networks arise with the disillusionment with traditional political and civic institutions, making them a logical fit for the blockchain community, which is trying to upend how those traditional institutions solve problems. However, in these early days, advocacy networks must work with government as a partner. Advocacy networks are closely tied to policy networks, so it's unsurprising that Coin Center and the Chamber of Digital Commerce are taking the lead in this area. We could also include here COALA, MIT's Digital Currency Initiative, and others. Advocacy is critical to scaling blockchain technology. In the absence of strong advocates who stand up for stakeholders and stakeholder rights, governments and other powerful institutions could try to stifle, twist, or usurp this powerful open network to their exclusive advantage, another dangerous potential showstopper.

5. Watchdog Networks

These networks scrutinize institutions to ensure that they behave appropriately. Topics range from human rights, corruption, and the environment to financial services. In the process, they drive public debate, boost transparency, and ignite movements for change. The role of watchdogs is inherently intertwined with that of advocacy networks and policy networks. Policy networks collaborate with government to shape policy that works. Watchdogs ensure that industry complies with policies and effectively monitors and enforces compliance. Governments that abuse the public trust can also be scrutinized and held accountable.

Blockchain Implications: The Blockchain Alliance is a partnership between law enforcement, NGOs, trade organizations, and the private sector and is the first true advocacy network to form in the space. Coin Center and the Chamber of Digital Commerce, with support from BitFury, Bitfinex, BitGo, Bitnet, Bitstamp, Blockchain, Circle, Coinbase, and others, have partnered with law enforcement agencies such as the U.S. Justice Department, the FBI, the Secret Service, and the Department of Homeland Security. As we highlighted in the previous chapter, blockchain being co-opted by criminals on a widespread scale is a showstopper. These watchdogs have an important advocacy role as well. In the aftermath of the Paris terrorist attacks, some European lawmakers, regulators, and law enforcement blamed bitcoin as the source of terrorism financing. The Blockchain Alliance called for patience: Let's not regulate out of fear,[67] they said. As of this writing, we don't know how effective they were, but surely in their absence things would have turned out worse, with government approaching the job unilaterally. Other than the self-policing role of community members who convene, collaborate, and debate on forums and on Reddit, few other watchdog networks have stepped up. Partnerships with law enforcement are a helpful start, but the blockchain ecosystem needs fully independent organizations, perhaps like traditional watchdogs such as Amnesty International and Human Rights Watch, to monitor governments, corporations, and other large institutions. Otherwise, we risk falling victim to another showstopper: that blockchain becomes a new and powerful surveillance tool used by corrupt and unscrupulous governments.

6. Platforms

The digital age allowed organizations to be much more than closed, siloed institutions; they can also be platforms for value creation, innovation, and global problem solving. Organizations like Change.org empower individuals to initiate campaigns in support of social causes from human rights to climate change. A "petition platform" harnesses the collective force of millions of people and catalyzes their passion into lasting impact. Open data platforms can apply to many issues—from climate change to the blockchain.[68]

Blockchain Implications: As blockchain technology gains in systemic importance, stakeholders must aggregate and scrutinize data. The bitcoin blockchain may be radically open, transparent, and reconcilable, but closed blockchains used in everything from financial services to the Internet of Things might not be. Imagine a platform that allowed regular citizens to aggregate and scrutinize data, proving a strong bulwark against creeping showstoppers of scalability, government encroachment, or unsustainable energy use. They would enable watchdogs and advocates among us to hold institutions and corporations more accountable and drive constructive discussion.

7. Standards Networks

Standards networks are non-state-based organizations that develop technical specifications and standards for virtually anything, including standards for the Internet itself. They determine the standards that form the fundamental building blocks for product development and allow a promising innovation to make the leap to mass adoption. For global standards networks to work, they must engage the expertise of individuals, institutions, civil society organizations, and, most of all, private sector enterprise.

The Internet Engineering Task Force, one of the primary standards bodies for the Internet governance network, excels at incorporating the many views of diverse stakeholders.

Blockchain Implications: Originally, the Bitcoin Foundation funded development of the bitcoin core protocol, the common standards used by the community. However, the near-collapse of the foundation (precipitated by mismanagement and waste) proved the need for networked governance solutions. Recognizing the profound importance of this technology and the need for careful stewardship and nurturing, MIT created the Digital Currency Initiative, which has since bankrolled the bitcoin core developers so they can continue their work. "We stepped in immediately and provided them with positions at the MIT media lab, so they could continue to independently work on supporting the core development of bitcoin," said Brian Forde.[69] For the core developers, their ability to work autonomously was central to the design.

Gavin Andresen is among the core developers working at MIT. He believes leadership is required to move the agenda forward on common standards, such as the much-debated block-size question. "Maybe you can design light socket set waves by committee, but you can't design software standards that way," he suggested. Pointing to the early days of the Web, Andresen said, "The Internet model shows that you can have technologies where consensus does arise, even though there's no one clear leader," but that "you can either have a person or a process that ends in a person. You definitely need one or the other."[70] Consensus mechanisms alone can't support standards developments.

Scalingbitcoin.org is an organization that convenes engineers and academics to address major technical issues, including standards questions. Pindar Wong, who chairs the planning committee for Scalingbitcoin.org (among his many other important leadership roles), has been a key leader in convening key stakeholders and clearing technical logjams in the sector. In financial services, both R3 and the Hyperledger Project are tackling critical standards issues. Invariably, there will have to be standards networks on a variety of things, from the blockchain protocol that forms the basis of the financial services industry of the future, to common standards for privacy and payments in the Internet of Things.

While each of these groups attacks the problem from different angles and with different agendas, each shares a common goal to make this technology ready for prime time—by building infrastructure, developing standards, and making it scalable.

8. Networked Institutions

Some networks provide such a wide range of capabilities that we describe them as "networked institutions." They are not state-based but true multistakeholder networks. The value they generate can range from knowledge, advocacy, and policy to actual delivery of solutions.

Blockchain Implications: The World Economic Forum (WEF), a leading networked institution, has been a vocal proponent of blockchain technology. The blockchain was front and center at Davos in January 2016. Jesse McWaters, financial innovation lead at the WEF, believes blockchain technology is a general-purpose technology, like the Internet, which we can use to make markets radically more efficient and improve access to financial services. The WEF predicted that within a decade, we could store 10 percent of global GDP on blockchains.[71] As an organization, the WEF has championed and advanced big issues, such as income inequality, climate change, and even remittances. Other networked institutions, from the smallest groups to the biggest foundations in the world, such as the Clinton Foundation and the Bill and Melinda Gates Foundation, would be wise to champion this technology to advance such big issues as financial inclusion and health care delivery. Networked institutions often have a role to play in influencing government policy making, making them a critical link and strategic partner in overcoming a number of major showstoppers.

9. Diasporas

Diasporas are global communities formed by people dispersed from their ancestral lands and united by culture and identity with their homeland. Thanks to the Internet, these people and affiliated organizations can collaborate in multistakeholder networks. One of the functions of many of today's diasporas is to address and help solve common global problems.

Blockchain Implications: Diasporas are critical to blockchain's future. For one, blockchain makes the process of sending remittances simple and affordable. Far from being a job killer, blockchain actually creates time and resources for these people to pursue other wage-earning opportunities or entrepreneurial endeavors. While a few companies have originated in places such as the Philippines and Kenya, diasporas must do more to accelerate knowledge, adoption, and acceptance of blockchain payment methods. Today, the majority of companies targeting the opportunity, such as Abra and Paycase, are U.S., U.K., Canada, or China based.

10. Governance Networks

The blockchain governance network will combine all the features and attributes of the nine other GSN types. Ultimately, a blockchain governance network should strive to be inclusive and welcome participation from all relevant stakeholder groups. The network should be a meritocracy, meaning that the community would champion viable proposals regardless of the rank and status of the proposer. The network should be transparent, releasing all of its data, documentation, and meeting minutes for public scrutiny. Finally, decisions should be reached, as much as possible, by consensus in order to gain legitimacy for the outcomes.

A NEW AGENDA FOR THE NEXT DIGITAL AGE

A blockchain governance network is critical to stewardship of this global resource. But how can we ensure that this next generation of the Internet fulfills its promise?

The next era of the digital age is delivering unlimited possibilities, significant dangers, unknown roadblocks, formidable challenges, and a future that is far from certain. Technology, especially the distributed kind, creates opportunities for everyone, but inexorably humans determine the outcome. In the words of Constance Choi, "This technology holds both promise and peril. It's how we wield it."[72] As this chapter has discussed, there is a role for everyone to play in achieving the new promise of the digital age.

In previous epochal transitions, societies took action to implement new understandings, laws, and institutions. These transformations of civilization took time, usually centuries, and were often punctuated by strife or even revolutions.

Today the situation is different. Change is happening infinitely faster. More important, Moore's law indicates that the *rate of change* is accelerating exponentially. We're moving to the proverbial "second half of the chessboard" where exponential growth upon exponential growth creates the incomprehensible.[73] The upshot is that our regulatory and policy infrastructures are woefully inadequate and adapting too slowly or not at all to the requirements of the digital age. The disruptions of today are moving so fast they are getting beyond the capacity of individuals and institutions to comprehend them, let alone manage their impact. Our democratic institutions and instruments were designed for the industrial age—in fact they originated precisely in the transformation from agrarian feudal societies into industrial capitalist states.

How can we accelerate the human transformation required to keep pace with accelerating technological innovation and disruption? How can we avoid massive social dislocation, or worse? Lest we be accused of being technology determinists or utopians, may we propose that it's time for a new social contract for the digital age. Governments, the private sector, the civil society, and individuals need to collaborate to forge new common understandings.

As we enter this second generation of the Internet, it's time for a Manifesto for the Digital Age. Call it a Declaration of Interdependence. Digital age citizens have *Rights*—access to digital infrastructures, literacy, media literacy, lifelong learning, and renewed freedom of speech online without the fear of surveillance.

The digital economy and society should be governed according to *Principles*. Surely, those who work should share in the wealth they create. If computers can do the work, then the workweek, not our standard of living, should be reduced. In fact, Satoshi's implicit design principles for the blockchain revolution should serve us well—we need institutions that act with integrity, security, privacy, inclusion, rights protection, and distributed power. Let's work to distribute opportunity and prosperity at the point of origin, rather than simply redistributing wealth after it's been created by traditional class structures.

Blockchain technology may reduce the costs and size of government, but we'll still need new *Laws* in many areas. There are technological and business model solutions to the challenges of intellectual property and rights ownership. So we should be rewriting or trashing old laws that stifle innovation through overprotection of patents. Better antitrust action must stem the trend toward monopolies so that no one overpays for, say, basic

Internet or financial services. Eighty percent of Americans have no choice when it comes to Internet service providers, which might help explain why bandwidth is one of the slowest and most expensive in the developed world. Criminal fixers who manipulate everything from foreign exchange to diesel emissions should be prosecuted and punished appropriately.

We'll need *Institutional Transformation* across the board. Central banks will need to change their role in currency management and monetary policy and collaborate multilaterally with more stakeholders in the economy and society. We need schools and universities with student-focused, customized collaborative mastery of information on the blockchain, freeing up students and teachers alike to participate in small group discussion and projects. We need a universal patient record on the blockchain, to ensure collaborative health when we can manage our own wellness outside of the system. When we enter the health care system, we should not suffer because of ignorance-inspired drug interactions or medicine not based on evidence. Politicians will need to adapt to a transparent world where smart contracts ensure their accountability to electorates. How do we manage the disruption after digital currencies upend the $500 billion remittances market?

Blockchain technology can enable new *Physical Infrastructures* requiring new partnerships and understandings among stakeholders. What happens to the millions of Uber drivers when SUber wipes out their jobs? What can cities do to ensure that in 2025 citizens think positively about intelligent transportation systems? How do we effectively move to a distributed blockchain-enabled electrical power grid where home owners are contributors rather than just customers of electricity? How will we find the leadership to implement a blockchain-enabled personal carbon trading system?

THE TRUST PROTOCOL AND YOU

Will the law of paradigms kick into effect—that leaders of the old have the greatest difficulty embracing the new? Consider the leaders who endorsed Don's 1994 book *The Digital Economy*: the CEOs of Nortel Networks, MCI, Nynex, Ameritech, and GE Information Services, all of which are gone. At least he didn't include the CEOs from Kodak, Borders, Blockbuster, or Circuit City. (Another cautionary note for the kind jacket endorsers of *Blockchain Revolution*.)

Why didn't Rupert Murdoch create *The Huffington Post*? Why didn't AT&T launch Skype, or Visa create PayPal? CNN could have built Twitter, as it is all about the sound bite, no? GM or Hertz could have launched Uber, and Marriott, Airbnb. Gannett could have created Craigslist or Kijiji. eBay would have been a natural play for the Yellow Pages. Microsoft had the resources to create Google or any number of business models based on the Internet rather than the personal computer. Why didn't NBC invent YouTube? Sony could have preempted Apple's iTunes. Where was Kodak when it was time for Instagram or Pinterest to be invented? What if *People* or *Newsweek* had come up with *BuzzFeed* or *Mashable*?

As we wrote at the beginning of this tome, "It appears that once again the technological genie has been unleashed from its bottle . . . now at our service for another kick at the can—to transform the economic power grid and the old order of human affairs. If we will it." Like the first generation of the Internet, the Blockchain Revolution promises to upend business models and transform industries. But that is just the start. Blockchain technology is pushing us inexorably into a new era, predicated on openness, merit, decentralization, and global participation.

We expect a period of volatility, speculation, and misuse. We also expect a strong and steady barreling forward, a plowing aside of the sacred cows on the tracks. No one knows yet what impact this train will have on financial services. Is Ben Lawsky right—that the industry could be unrecognizable in five to ten years? Tim Draper said, "Bitcoin is to the dollar as the Internet is to paper."[74] Could it be that blockchain's most ardent supporters are actually *underestimating* the long-term potential here? Will the blockchain be the biggest boon to industry efficiency and value since the invention of double-entry accounting or the joint-stock corporation? Hernando de Soto said blockchain holds the potential to bring five billion people into the global economy, change the relationship between the state and citizens (for the better), and become a powerful new platform for global prosperity and a guarantor of individual rights. To him, "the whole idea of peace through law, the whole idea of one family humankind is that we reach agreement on common standards. We should consider how the Universal Declaration of Human Rights could be better served with blockchain."[75] How can we achieve this better future?

Most of the people leading the revolution are still unknowns, except for veterans like Netscape's progenitor Marc Andreessen. You've likely never heard of most of the people quoted in this book. Then again, who'd heard of Iranian immigrant Pierre Omidyar or Wall Street programmer Jeff Bezos in 1994? Much depends on how the leaders of the industry get on board. Is a blockchain alternative to Facebook or Twitter really achievable or will the incumbents respond by addressing user concerns about data ownership and privacy? Doesn't matter. Consumers win either way. Will Visa wither or will it change its business model to embrace the power of blockchain? How will Apple respond to an artist-centered music industry? What will tin-pot dictators think about a decentralized Internet that they can't turn off or control? Can the blockchain make technology accessible to the world's two billion unbanked people?

The failure rate of start-ups is high, and so we expect a good number of our case studies to fall by the wayside, not because blockchain technology is a bad idea, but because—for each one of our examples—there are many competing start-ups. All of them can't survive. We believe those that follow Satoshi's principles have a better shot than those that don't.

These are exciting and perilous times. As a business leader, use *Blockchain Revolution* as your playbook, sure, but realize also that the rules of the game themselves are changing. Think about your business, your industry, and your job: How will I be affected and what can be done? Do not fall into the trap brought about by many paradigm shifts throughout history. Today's leaders cannot afford to be tomorrow's losers. Too much is at stake and we need your help. Please join us.

AFTERWORD TO THE PAPERBACK EDITION

Don Tapscott and Alex Tapscott

A NEW SOCIAL CONTRACT FOR THE DIGITAL AGE

Permit us some final words about a modest but related topic—the future of civilization in the digital age.

Since we wrote *Blockchain Revolution*, the world was hit in the face by the one-two punch of Brexit and Donald Trump.

In the book, we discussed the danger of the "prosperity paradox," where wealth is growing, yet prosperity for most is stagnant or declining. Sure enough, it did turn out that people everywhere were "mad as hell and not going to take it anymore." As such, they have become vulnerable to populism and xenophobia, even scapegoating any minority—any group that represents a difference, be it ethnic, racial, gender, or religious—for their circumstances or even the consequences of their own decisions.

Centrist parties are in rapid decline, and extremist, right-wing parties are on the rise, from Hungary and Poland to France and Germany. In other countries, particularly in southern Europe where memories of fascism and dictatorship are still vivid, the left is ascendant.

Perhaps as unthinkable as the success of Donald Trump is that of Bernie Sanders, an avowed socialist who almost won the Democratic presidential nomination. Some think he could have beaten Trump in the general election. The unfolding story is one of growing discontent with the deepening economic crisis and the establishment that created it with impunity.

The world has seen this story before, in the run-up to the Second World War, but the analogy is imperfect. Among other things, the rate of change differs. As the digital revolution unfolds, it is driving profound changes in the global economy, labor markets, old institutions, and society as a whole. It is enabling spectacular innovation and unprecedented wealth creation. At the same time, growing social inequality, the decline of the middle class, and pernicious unemployment and underemployment are fueling unrest. Networks enable outsourcing, offshoring, and the globalization of labor markets.

Government architectures and policies have not evolved, and fiscal crises and threats to the industrial age social safety net loom everywhere. Data, a new asset class, has been captured by powerful forces. A handful of companies capture the largesse of the digital age asymmetrically, eroding personal privacy and prosperity in the process. Climate change is threatening our biosphere, with huge displacement and other disruptions just beginning to be felt.

Blockchain can enable people to improve this situation as we discussed in chapters 7, 8, and 9. However, within the Fourth Industrial Revolution—one centered on blockchain, machine learning, robotics, the Internet of Things, and even biotechnology—many core functions of knowledge work, many companies and many industries, are in jeopardy and many of our institutions will be shaken.[1]

Industrial age organizations for solving global problems, those based on the Bretton Woods model of global institutions, have stalled. The upshot is that the *social contract*—the agreements, laws, and appropriate behaviors that people, companies, civil society, and governments adhere to by consensus—no longer serves us well.

In the book, we called for nothing less than a new social contract, which we called a "Declaration of *Inter*dependence"—a phrase aptly coined during the Great Depression—because of the need for new multistakeholder approaches, whereby governments, the private sector, civil society, and individuals could forge and agree on new understandings and new action plans.[2]

After the book was published, we set out to research and write this declaration, which we see as a sort of Manifesto for the Digital Age. The spectacular innovations provide civilization with a new set of opportunities to leap forward rather than decline or even collapse. We believe that the next era of the digital economy will bring epoch-making wealth, with new networked models of global problem solving to realize such a dream. We can even conceive of a new, achievable set of rights for individuals everywhere and future generations to claim:

- *Security of personhood*: the right to our own identities, privacy of our personal information, and our reputations. Blockchain enables this but it requires leadership and profound social and political change.
- *Education*: the right to access affordable digital infrastructures, student-centered education, media literacy at any level, and lifelong learning. Blockchain is a new platform, but old institutions are tough to change.
- *Vocation*: the right to work, change jobs, create a business, monetize our own assets and the data we generate, contribute productively to society, and have that contribution valued economically.
- *Health*: the right to health care, including access to safe, affordable, and nutritious food, medicine, medical treatment, and ongoing care in the event of prolonged illness, disease, disability, or old age.
- *Economic safety net*: the right to a basic income that sustains ourselves and our families—but let's focus on the right to a job, whether created by the private sector, the state, civil society, or philanthropy. Yes, the Fourth Industrial Revolution will wipe out entire industries and job types but will bring new ones.
- *Climate stability*: a right to clean air, safe water, and ecologically sound homelands in perpetuity. Blockchain can help us bring back science, validate the truth, and both codify and incentivize green behavior, in turn mobilizing societies to solve environmental problems. Consider the tokenizing of carbon credits to motivate every human being to make green choices.
- *Peace with dignity*: a right to live a life free of national, religious, or tribal conflict, terrorism, and other types of violence or oppression. Blockchain can build a more transparent, distributed, and high-metabolism military supply chain to fight, say, terrorists, but more important, to help build a more prosperous and just world with fewer terrorists. "All we are saying is give peace a chance."
- *Political and institutional accountability*: the right to transparency online and in increasingly automated processes, where elected representatives are answerable to their voters through smart

contracts, corporate leaders are accountable to the communities in which they operate, and strong codes of ethics guide technological development in the interests of all humanity.

To recognize, realize, and enforce these rights, we need profound changes to our industrial institutions and infrastructures including education, health care, labor unions, food supply chains, transportation and energy systems, and, above all, governments. Networks enable citizens to participate fully in their own governance, and we can now move to a second era of democracy based on a culture of public deliberation and active citizenship. Mandatory voting encourages active, engaged, and responsible citizens.

In the name of global competitiveness and short-term shareholder value, we have let business off the hook for far too long. It's time for business leaders to come to the table as responsible and active participants in the new social contract—for both their own long-term interests as well as in the interest of a healthy society and healthy economy overall. Even—or especially—in a time of exploding information online, we need scientists, researchers, and a professional Fourth Estate of journalists to search for truth, examine options, and inform the ongoing public discourse.

A far cry from blockchain? We think not, as this second era of the Internet will cause disruption on the one hand and new possibilities on the other.

Are these expectations overly ambitious or even utopian? We don't think so, as we consider the alternative trajectory that humanity now faces. We have the unprecedented opportunity either to achieve universal prosperity—or to stoke our current economic inequality into nothing less than class warfare.

We have no delusions of grandeur that our project for a new social contract can provide a strategy for instituting and enforcing each of these rights. Rather our hope is to help catalyze investigation, debate, and action.

Do join the discussion at www.blockchainresearchinstitute.org/social contract.

NOTES

Preface to the Paperback Edition

1. The S&P 500 is an American stock market index based on the market capitalizations of five hundred large companies having common stock listed on the NYSE or NASDAQ. The S&P 500 index components and their weightings are determined by S&P Dow Jones Indices. Typically, the smallest member of the index has a market capitalization of around $7 to $9 billion.
2. The value of cryptoassets changes constantly. We've plugged in values from April 24, 2018, the date we finished the new material for the paperback. For a sense of how volatile this asset class can be, compare this figure with the market value today.
3. Jim Epstein, "Bitcoin Sends Elite Economists into Glorious Fits of Confusion," *Reason.com*, Reason Foundation, November 30, 2017; reason.com/blog/2017/11/30/bitcoin-joseph-stiglitz-crypto-outlaw, accessed March 8, 2018. Both Joseph Stiglitz and Paul Krugman have expressed skepticism about bitcoin.
4. Noah Buhayar, "Munger Calls Bitcoin a 'Noxious Poison' Government Should Tackle," *Bloomberg Markets*, February 14, 2018; www.bloomberg.com/news/articles/2018-02-14/munger-calls-bitcoin-a-noxious-poison-government-should-tackle, accessed March 8, 2018.
5. Vitalik "Not giving away ETH" Buterin @VitalikButerin, "So Total Cryptocoin Market Cap Just Hit $0.5T Today. But Have We *Earned* It?" Twitter, 4:46 PM, December 12, 2017; twitter.com/VitalikButerin/status/940744724431982594, accessed March 8, 2018.
6. Vitalik "Not giving away ETH" Buterin @VitalikButerin, "How Many Unbanked People Have We Banked?" Twitter, 4:46 PM, December 12, 2017; twitter.com/VitalikButerin/status/940744820406013954, accessed March 8, 2018.
7. Vitalik "Not giving away ETH" Buterin @VitalikButerin, "How Much Value Is Stored in Smart Contracts That Actually Do Anything Interesting?" Twitter, 4:50 PM, December 12, 2017; twitter.com/VitalikButerin/status/940745607047548928, accessed March 8, 2018.
8. Vitalik "Not giving away ETH" Buterin @VitalikButerin, "The Answer to All of These Questions Is Definitely Not Zero, and in Some Cases It's Quite Significant. But Not Enough to Say It's $0.5T Levels of Significant. Not Enough," Twitter, 4:53 PM, December 12, 2017; twitter.com/VitalikButerin/status/940746391256678400, accessed March 8, 2018.
9. Tommy Wilkes and Fanny Potkin, "Twitter to Ban Cryptocurrency Ads from Tuesday as Online Crackdown," Reuters, Thomson Reuters, March 26, 2018,

www.reuters.com/article/us-crypto-currencies-twitter/twitter-to-ban-crypto currency-ads-from-tuesday-as-online-crackdown-widens-idUSKBN1H222H.

10. Joseph Lubin, interviewed via telephone by Don Tapscott, March 3, 2018.
11. JPMorgan Chase & Co., "FORM 10-K Annual Report Pursuant to Section 13 or 15(d) of the Securities Exchange Act of 1934, for the Fiscal Year Ended 31 Dec. 2017, Commission File Number 1-5805," Investor Relations, JPMorgan Chase & Co., February 28, 2018, 25; investor.shareholder.com/jpmorganchase/secfiling .cfm?filingID=19617-18-57&CIK=19617#CORP10K2017_HTM_S440D 20F00AA0567AADC9B36846A275C5, accessed March 8, 2018.
12. For more detail, check out *Cryptoassets* by Chris Burniske and Jack Tatar.
13. Michael del Castillo, "JPMorgan Integrates Zcash Privacy Tech into Quorum Blockchain," *CoinDesk*, October 18, 2017; www.coindesk.com/jpmorgan-inte grates-zcash-privacy-tech-enterprise-blockchain, accessed March 8, 2018.
14. https://metronome.io, accessed March 14, 2018.
15. Nick Szabo, "Winning Strategies for Smart Contracts," Blockchain Research Institute, December 4, 2017.
16. Charles Bovaird, "The #Flippening: Will Ether 'Pass' Bitcoin and What Would It Mean?" *CoinDesk*, June 14, 2017; www.coindesk.com/flippening-will-ether-pass -bitcoin-will-mean, accessed March 13, 2018.
17. Joseph Lubin, interviewed via telephone by Don Tapscott, March 3, 2018.
18. Scott Nelson, e-mail to Don Tapscott, March 12, 2018.
19. Primavera De Filippi, et al., "Regulatory Framework for Token Sales: An Overview of Relevant Laws and Regulations in Different Jurisdictions," Blockchain Research Institute and COALA, April 2018.
20. Michael Casey, "The Token Economy: When Money Becomes Programmable," Blockchain Research Institute, September 28, 2017.
21. "The New Digital Gold Standard," *RMG*, Royal Mint Ltd., n.d.; rmg.royalmint .com, accessed March 8, 2018.
22. Casey, "The Token Economy."
23. Marcia Dunn, "Elon Musk Hails 'Silly but Fun' SpaceX Rocket Launch," *Independent.ie*, February 7, 2018; www.independent.ie/world-news/elon-musk-hails -silly-but-fun-spacex-rocket-launch-36576997.html, accessed March 8, 2018.
24. Scott Reyburn, "Will Cryptocurrencies Be the Art Market's Next Big Thing?" *The New York Times*, January 13, 2018; www.nytimes.com/2018/01/13/arts/crypto currency-art-market.html.
25. DADA overview, dada.nyc/presale, and DADA white paper, dada.nyc/presale, accessed March 8, 2018.
26. Rachel A. J. Pownall, "TEFAF 2017 Global Art Market Report," *TEFAF.com*, March 6, 2017, p. 8; tinyurl.com/ycexbc6x, accessed March 14, 2018; http://1uyx qn3lzdsa2ytyzj1asxmmmpt.wpengine.netdna-cdn.com/wp-content/uploads /2017/03/TEFAF-Art-Market-Report-20173.pdf.
27. Charles "Chuck" Bair, "Some Good Things About the Petro," *Bair Dot Com*, March 8, 2018; bair.com/2018/03/08/good-things-about-the-petro, accessed March 14, 2018.
28. Jack Karsten and Darrell M. West, "Venezuela's 'Petro' Undermines Other Cryptocurrencies—and International Sanctions," Brookings Institution, March 9, 2018; www.brookings.edu/blog/techtank/2018/03/09/venezuelas-petro-undermines-oth er-cryptocurrencies-and-international-sanctions, accessed March 14, 2018.
29. Gabriel Abed, interviewed by Alex Tapscott, March 13, 2018.
30. Timothy B. Lee, "Why Experts Are Worried About Tether, a Dollar-Pegged Cryptocurrency," *Ars Technica*, February 5, 2018; arstechnica.com/tech-policy

/2018/02/tether-says-its-cryptocurrency-is-worth-2-billion-but-its-audit-failed, accessed March 13, 2018.

31. Gabriel Abed, interviewed by Alex Tapscott, March 13, 2018.
32. www.linuxfoundation.org/press-release/linux-foundation-unites-industry-leaders-to-advance-blockchain-technology/#.WZ8FmCiG, accessed March 8, 2018.
33. www.hyperledger.org/about, accessed March 8, 2018.
34. Olga Kharif, "Next-Generation Crypto-Ledgers Take the Block Out of Blockchain," *Bloomberg Technology*, February 14, 2018; www.bloomberg.com/news/articles/2018-02-14/next-generation-crypto-ledgers-take-the-block-out-of-blockchain, accessed March 13, 2018.
35. www.hyperledger.org/projects, accessed March 8, 2018.
36. www.hyperledger.org/community/projects, accessed March 8, 2018.
37. Gary Wolf and Kevin Kelly, "About the Quantified Self," *Quantified Self*, November 1, 2016; quantifiedself.com/about, accessed February 22, 2018.
38. Doc Searls, "Sovereign-Source vs. Administrative Identity," *ProjectVRM*, March 25, 2012; blogs.harvard.edu/vrm/2012/03/25/sovereign-source-vs-administrative-identity, accessed February 21, 2018.
39. Moxie Marlinspike, "What Is 'Sovereign Source Authority'?" *The Moxy Tongue*, February 15, 2012; www.moxytongue.com/2012/02/what-is-sovereign-source-authority.html, accessed February 21, 2018.
40. UNICEF, "The Births of Around One Fourth of the Global Population of Children Under Five Have Never Been Registered," *UNICEF Data: Monitoring the Situation of Children and Women*, January 2018; data.unicef.org/topic/child-protection/birth-registration, accessed February 21, 2018.
41. Ibid.
42. Jean-Luc Lemahieu and Angela Me, directors, UN Office on Drugs and Crime, "Global Report on Trafficking in Persons 2016," 25; www.unodc.org/documents/data-and-analysis/glotip/2016_Global_Report_on_Trafficking_in_Persons.pdf, accessed February 21, 2018.
43. UNICEF, "The Births of Around One Fourth of the Global Population of Children Under Five Have Never Been Registered."
44. "Principles on Identification for Sustainable Development: Toward the Digital Age," The World Bank Group, February 27, 2017; documents.worldbank.org/curated/en/213581486378184357/Principles-on-identification-for-sustainable-development-toward-the-digital-age, accessed February 22, 2018.
45. Samer Aburass, "Syrian Refugees' Documentation Crisis," *NRC* (Norwegian Refugee Council), January 26, 2017; www.nrc.no/news/2017/january/syrian-refugees-documentation-crisis, accessed February 22, 2018.
46. Bronwen Manby, Jonathan Marskell, and Julia Clark, "'Papers Please?' The Importance of Refugees and Other Forcibly-Displaced Persons Being Able to Prove Identity," *dev4peace Blog*, June 20, 2017; blogs.worldbank.org/dev4peace/papers-please-importance-refugees-and-other-forcibly-displaced-persons-being-able-prove-identity, accessed February 22, 2018.
47. "Identification for Development (ID4D)," The World Bank, 2018; www.worldbank.org/en/programs/id4d, accessed February 22, 2018.
48. Special Correspondent, "Aadhaar Covers 99% of Adults in India: Prasad," *The Hindu*, January 17, 2017; www.thehindu.com/business/Aadhaar-covers-99-of-adults-in-India-Prasad/article17104609.ece, accessed January 17, 2018.
49. Unique Identification Authority of India, "[Aadhaar Authentication] Operation

Model," Government of India, n.d.; https://uidai.gov.in/authentication/authentication-overview/operation-model.html, accessed 13 Jan. 2018.

50. "Aadhaar Card Data Security Issues: UIDAI Says System Fully Safe, Never Breached or Leaked," *The Financial Express*, November 20, 2017; www.financialexpress.com/money/aadhaar-card-data-security-issues-uidai-says-system-fully-safe-never-breached-or-leaked/940459, accessed February 22, 2018.
51. Joseph Lubin, "This Is How the Blockchain Will Disrupt the Election System," ConsenSys Media, November 7, 2016; media.consensys.net/this-is-how-the-blockchain-will-disrupt-the-election-system-4e988d1d7d8f, accessed March 13, 2018.
52. Brendan Pierson, "Anthem to Pay Record $115 Million to Settle U.S. Lawsuits over Data Breach," *Reuters*, June 23, 2017; www.reuters.com/article/us-anthem-cyber-settlement/anthem-to-pay-record-115-million-to-settle-u-s-lawsuits-over-data-breach-idUSKBN19E2ML, accessed February 22, 2018.
53. Jessica Davis, "Anthem: Insider Theft Exposes Data of 18,000 Medicare Members," *Healthcare IT News*, October 26, 2017; www.healthcareitnews.com/news/anthem-insider-theft-exposes-data-18000-medicare-members, accessed February 22, 2018. The thief was an employee of Anthem's Medicare insurance coordinator, LaunchPoint Ventures.
54. Casey, "The Token Economy." See also "Bootstrapping Identity," *OpenStack Docs*, February 20, 2018; docs.openstack.org/keystone/pike/admin/identity-bootstrap.html.
55. Robert Hackett, "Can This Man Build a Better Bitcoin?" *Fortune*, December 18, 2017; fortune.com/2017/12/18/jp-morgan-bitcoin-zcash-wilcox, accessed February 23, 2018.
56. Christopher Allen, "The Path to Self-Sovereign Identity," *CoinDesk*, May 1, 2016; www.coindesk.com/path-self-sovereign-identity, accessed February 21, 2018.
57. "OpenPDS/SafeAnswers: The Privacy-Preserving Personal Data Store," MIT Human Dynamics Group, Massachusetts Institute of Technology School of Architecture + Planning, n.d.; openpds.media.mit.edu. See also Yves-Alexandre de Montjoye, Erez Shmueli, Samuel S. Wang, and Alex "Sandy" Pentland, "OpenPDS: Protecting the Privacy of Metadata Through SafeAnswers," *PLOS ONE*, July 9, 2014; journals.plos.org/plosone/article?id=10.1371%2Fjournal.pone.0098790, accessed February 22, 2018.
58. https://gomedici.com/22-companies-leveraging-blockchain-for-identity-management-and-authentication.
59. Christian Lundkvist, Rouven Heck, Joel Torstensson, Zac Mitton, and Michael Sena, "uPort: A Platform for Self-Sovereign Identity," white paper, uPort.me, February 21, 2017; https://whitepaper.uport.me/uPort_whitepaper_DRAFT20170221.pdf, accessed March 2, 2018.
60. "Decentralized Identity Foundation," n.d.; identity.foundation, accessed February 21, 2018.
61. "Working Groups," n.d.; identity.foundation/working-groups, accessed February 21, 2018.
62. Alyssa Hertig, "The Father of the ICO Is All About Identity Now," *CoinDesk*, December 5, 2017; www.coindesk.com/forget-token-sales-the-father-of-the-ico-is-all-about-identity-now, accessed February 28, 2018.
63. Elizabeth M. Pierce, "Designing a Data Governance Framework to Enable and Influence IQ Strategy," presentation, Proceedings of the MIT 2007 Information Quality Industry Symposium, Cambridge, MA, July 19, 2007; http://mitiq.mit.edu/IQIS/Documents/CDOIQS_200777/Papers/01_08_1C.pdf, accessed December 7, 2017.

64. Ibid.
65. David A. Jaffray, interviewed via telephone by Kirsten Sandberg, March 13, 2018.
66. Tom McCarthy, "How Russia Used Social Media to Divide Americans," *The Guardian*, October 2017, www.theguardian.com/us-news/2017/oct/14/russia-us-politics-social-media-facebook.
67. Yuval Noah Harari, "On Big Data, Google and the End of Free Will," *Financial Times*, August 26, 2016, www.ft.com/content/50bb4830-6a4c-11e6-ae5b-a7cc5dd5a28c.
68. "Cambridge Analytica Receives Top Honor in the 2017 ARF David Ogilvy Awards," *PR Newswire*, March 21, 2017, www.prnewswire.com/news-releases/cambridge-analytica-receives-top-honor-in-the-2017-arf-david-ogilvy-awards-300426997.html.
69. Harari, "On Big Data, Google and the End of Free Will."
70. Marcelo Gleiser, "Biometric Data and the Rise of Digital Dictatorship," National Public Radio, February 28, 2018, www.npr.org/sections/13.7/2018/02/28/589477976/biometric-data-and-the-rise-of-digital-dictatorship.
71. Szabo, "Winning Strategies for Smart Contracts."
72. Eric Jaffe, "Old World, High Tech," *Smithsonian Magazine*, December 1, 2006; www.smithsonianmag.com/science-nature/old-world-high-tech-141284744/?no-ist=&page=2, accessed February 21, 2018.
73. Maki Shiozawa, "16 Things You Didn't Know About Vending Machines," *Coca-Cola Journey*, February 20, 2015; www.coca-colacompany.com/stories/16-things-you-didnt-know-about-vending-machines-in-japan-and-around-the-world, accessed February 28, 2018.
74. Primavera De Filippi and Aaron Wright, *Blockchain and the Law: The Rule of Code* (Boston: Harvard University Press, 2018): 79.
75. For a good overview of jurisdictions, see "Can smart contracts be legally binding contracts? An R3 and Norton Rose Fulbright White Paper," http://www.nortonrosefulbright.com/files/r3-and-norton-rose-fulbright-white-paper-full-report-144581.pdf.
76. Nick Szabo, "Money, Blockchains, and Social Scalability," *Unenumerated*, February 9, 2017; https://unenumerated.blogspot.com/2017/02/money-blockchains-and-social-scalability.html, accessed September 27, 2017.
77. Szabo, "Winning Strategies for Smart Contracts."
78. Lucinda Shen, "The 10 Biggest Business Scandals of 2017," *Fortune*, December 31, 2017; fortune.com/2017/12/31/biggest-corporate-scandals-misconduct-2017-pr, accessed February 28, 2018.
79. Szabo, "Winning Strategies for Smart Contracts."
80. De Filippi and Wright, *Blockchain and the Law: The Rule of Code*, 29.
81. Alan Majer, "Slock.it: Enabling IoT and the Universal Sharing Network," Blockchain Research Institute, December 21, 2017.
82. Stephan Tual, "Share&Charge Launches Its Mobile App, On-boards over 1,000 Charging Stations on the Blockchain," *Slock.it Blog*, May 1, 2017; https://blog.slock.it/share-charge-launches-its-app-on-boards-over-1-000-charging-stations-on-the-blockchain-ba8275390309, accessed November 22, 2017.
83. Ibid.
84. Jon Buck, "UN Branch Using Ethereum-Based Smart Contracts for Public Confidence," *Cointelegraph*, August 7, 2017; cointelegraph.com/news/un-branch-using-ethereum-based-smart-contracts-for-public-confidence, accessed February 28, 2018.
85. Majer, "Slock.it: Enabling IoT and the Universal Sharing Network." Indeed, some have blamed the Solidity language itself for the difficulty in identifying anti-patterns like the ones that caused the attack on the DAO.

86. Ibid.
87. Tom Serres and Bettina Warburg, "Introducing Asset Chains: The Cognitive, Friction-free, and Blockchain-enabled Future of Supply Chains," Blockchain Research Institute, November 28, 2017.
88. Ibid.
89. Scott Nelson, e-mail to Don Tapscott, March 12, 2018.
90. Serres and Warburg, "Introducing Asset Chains."
91. Irving Wladawsky-Berger, "Building a Framework for Blockchain Adoption: What CEOs Should Know," Blockchain Research Institute, October 26, 2017.
92. Ibid.
93. Ibid.
94. Ibid.
95. Klaus Schwab, *The Fourth Industrial Revolution* (Geneva: World Economic Forum, 2016).
96. Oliver T. Bussmann, "Blockchain and the Global CIO: How Distributed Ledger Technology Will Transform Enterprise Architecture and the CIO Role," Blockchain Research Institute, January 3, 2018.
97. Vlad Gheorghiu, Sergey Gorbunov, Michele Mosca, and Bill Munson, "Quantum-Proofing the Blockchain," Blockchain Research Institute, November 23, 2017.
98. Open MIC Project, *Breaking the Mold: Investing in Racial Diversity in Tech; A Report for Investors*, February 2017; http://breakingthemold.openmic.org, accessed February 28, 2018.
99. Alberto Felice De Toni and Giovanni De Zan, "The Complexity Dilemma: Three Tips for Dealing with Complexity in Organizations," *Emergence: Complexity and Organization* 3–4 (December 2016): H1. Pindar Wong, interviewed via telephone by Don Tapscott, April 7, 2017.
100. Wong interview, April 7, 2017.
101. Andy Spence, "Blockchain and the Chief Human Resources Officer: Transforming the HR Function and the Market for Skills, Talent, and Training," Blockchain Research Institute, January 29, 2018.
102. Ibid.
103. Jeremy Epstein, "Blockchain and the CMO: The Next Era of Marketing," Blockchain Research Institute, September 28, 2017.
104. Ibid.
105. Paul H. Madore, "IOTA Update: The Tangled Web of Home-Rolled Cryptography," *Hacked.com*, September 10, 2017; hacked.com/iota-update-tangled-web-home-rolled-cryptography, accessed February 28, 2018.
106. Ethan Heilman, Neha Narula, Thaddeus Dryja, and Madars Virza, "IOTA Vulnerability Report: Cryptanalysis of the Curl Hash Function Enabling Practical Signature Forgery Attacks on the IOTA Cryptocurrency," *GitHub*, September 7, 2017; github.com/mit-dci/tangled-curl/blob/master/vuln-iota.md#iota-vulnerability-report-cryptanalysis-of-the-curl-hash-function-enabling-practical-signature-forgery-attacks-on-the-iota-cryptocurrency, accessed March 13, 2018.
107. Joichi Ito, "Our Response to 'A Cryptocurrency Without a Blockchain Has Been Built to Outperform Bitcoin,'" MIT Media Lab, December 20, 2017; www.media.mit.edu/posts/iota-response, accessed February 28, 2018.
108. Thomas M. Isaacson, "Patents and Blockchain Innovation: Strategic Approaches to Intellectual Property," Blockchain Research Institute, January 29, 2018.
109. Choe Sang-hun, "Samsung's Leader Is Indicted on Bribery Charges," *The New York*

Times, February 28, 2017; www.nytimes.com/2017/02/28/world/asia/lee-jae-yong-samsung.html. Donna Borak, "Former Equifax Executive Charged with Insider Trading," *CNN Money*, March 14, 2018; money.cnn.com/2018/03/14/news/companies/equifax-insider-trading/index.html. Osamu Tsukimori and Aaron Sheldrick, "Mitsubishi Materials Finds More Products with Falsified Data," *Reuters*, December 19, 2017; www.reuters.com/article/us-mitsubishi-ma-scandal/mitsubishi-materials-finds-more-products-with-falsified-data-idUSKBN1ED14A. Ken Sweet, "Wells Fargo Says 3.5 Million Accounts Involved in Scandal," *AP News*, August 31, 2017; apnews.com/c3de75ac78004f04be8291b1b76c2cd0. All articles accessed March 14, 2018.

110. Global Solution Networks Program, a program conducted by The Tapscott Group for the Rotman School of Management, 2014–16; http://gsnetworks.org, accessed June 20, 2017.
111. The Muskoka Group, "Manifesto on Building a Healthy Blockchain Ecosystem," www.muskokagroup.org.
112. "Realizing the Potential of Blockchain: A Multi-stakeholder Approach to the Stewardship of Blockchain and Cryptocurrencies," www3.weforum.org/docs/WEF_Realizing_Potential_Blockchain.pdf.
113. Global Solution Networks, http://gsnetworks.org.
114. Don Tapscott, Hilary Carter, and Jill Rundle, "The Networked Hotbeds of Blockchain: Creating Global Hubs for the Internet's Second Era," Blockchain Research Institute, January 15, 2018.
115. Ibid.
116. Shiwen Yap, "Singapore Emerges as Third Largest Global ICO Hub," *Deal Street Asia*, November 27, 2017; www.dealstreetasia.com/stories/singapore-emerges-asia-ico-hub-86574, accessed December 22, 2017.
117. Sonal Chokshi, Chris Dixon, Denis Nazarov, Jesse Walden, and Ali Yahya, "Crypto Canon: Blockchain and Cryptocurrencies," Andreessen Horowitz, March 7, 2018; a16z.com/2018/02/10/crypto-readings-resources, accessed March 13, 2018.
118. See 99bitcoins.com/who-accepts-bitcoins-payment-companies-stores-take-bitcoins for a list.

Chapter 1: The Trust Protocol

1. https://www.technologyreview.com/s/419452/moores-outlaws/.
2. https://cryptome.org/jya/digicrash.htm.
3. "How DigiCash Blew Everything," translated from Dutch into English by Ian Grigg and colleagues and e-mailed to Robert Hettinga mailing list, February 10, 1999. Cryptome.org. John Young Architects. Web. July 19, 2015. https://cryptome.org/jya/digicash.htm. "How DigiCash Alles Verknalde" www.nextmagazine.nl/ecash.htm. *Next! Magazine*, January 1999. Web. July 19, 2015. https://web.archive.org/web/19990427/http://nextmagazine.nl/ecash.htm.
4. http://nakamotoinstitute.org/the-god-protocols/.
5. Brian Fung, "Marc Andreessen: In 20 Years, We'll Talk About Bitcoin Like We Talk About the Internet Today," *The Washington Post*, May 21, 2014; www.washingtonpost.com/blogs/the-switch/wp/2014/05/21/marc-andreessen-in-20-years-well-talk-about-bitcoin-like-we-talk-about-the-internet-today/, accessed January 21, 2015.
6. Interview with Ben Lawsky, July 2, 2015.
7. www.economist.com/news/leaders/21677198-technology-behind-bitcoin-could-transform-how-economy-works-trust-machine.

8. www.coindesk.com/bitcoin-venture-capital/.
9. Fung, "Marc Andreessen."
10. www.coindesk.com/bank-of-england-economist-digital-currency/.
11. Leigh Buchanan reports on the Kauffman Foundation research in "American Entrepreneurship Is Actually Vanishing," www.businessinsider.com/927-people-own-half-of-the-bitcoins-2013-12.
12. This definition was developed in Don Tapscott and David Ticoll, *The Naked Corporation* (New York: Free Press, 2003).
13. www.edelman.com/news/trust-institutions-drops-level-great-recession/.
14. www.gallup.com/poll/1597/confidence-institutions.aspx.
15. Interview with Carlos Moreira, September 3, 2015.
16. Don Tapscott is a member of the WISeKey Advisory Board.
17. Don Tapscott has been one of many authors to write about the dark side potential of the digital age, for example in *The Digital Economy: Promise and Peril in the Age of Networked Intelligence* (New York: McGraw Hill, 1995).
18. Interview with Carlos Moreira, September 3, 2015.
19. Tom Peters, "The Wow Project," *Fast Company*, Mansueto Ventures LLC, April 30, 1999; http://www.fastcompany.com/36831/wow-project.
20. Interview with Carlos Moreira, September 3, 2015.
21. "The Virtual You" is a term popularized by Ann Cavoukian and Don Tapscott in *Who Knows: Safeguarding Your Privacy in a Networked World* (New York: McGraw-Hill, 1997).
22. Scott McNealy, then CEO of Sun Microsystems, was the first in 1999.
23. Interview with Andreas Antonopoulos, July 20, 2015.
24. Interview with Joe Lubin, July 30, 2015.
25. Eventually sophisticated personal data query services will not even be able to read that data because it will be given to them encrypted. Still they will be able to answer questions regarding that data by asking those questions of the encrypted data itself using homomorphic encryption techniques.
26. Leading thinkers have a broad view of prosperity that goes beyond GDP growth. Harvard's Michael Porter has created a social progress imperative http://www.socialprogressimperative.org. Economist Joseph Stiglitz and others have researched measures beyond GDP—http://www.insee.fr/fr/publications-et-services/dossiers_web/stiglitz/doc-commission/RAPPORT_anglais.pdf. There are other efforts that try to improve GDP but stay closer to home—http://www.forbes.com/sites/realspin/2013/11/29/beyond-gdp-get-ready-for-a-new-way-to-measure-the-economy/.
27. Interview with Vitalik Buterin, September 30, 2015.
28. Luigi Marco Bassani, "Life, Liberty and . . . : Jefferson on Property Rights," *Journal of Libertarian Studies* 18(1) (Winter 2004): 58.
29. Interview with Hernando de Soto, November 27, 2015.
30. Ibid.
31. www.theguardian.com/music/2013/feb/24/napster-music-free-file-sharing, accessed August 12, 2015.
32. www.inc.com/magazine/201505/leigh-buchanan/the-vanishing-startups-in-decline.html.
33. *Naked City* was a police drama series that aired from 1958 to 1963 on the ABC television network.
34. An October 2015 World Economic Forum report suggests it will not become mainstream until 2027.
35. Interview with David Ticoll, December 12, 2015.

Chapter 2: Bootstrapping the Future: Seven Design Principles of the Blockchain Economy

1. Interview with Ann Cavoukian, September 2, 2015.
2. Guy Zyskind, Oz Nathan, and Alex "Sandy" Pentland, "Enigma: Decentralized Computation Platform with Guaranteed Privacy," white paper, Massachusetts Institute of Technology, 2015. June 10, 2015. Web. October 3, 2015, arxiv.org/pdf/1506.03471.pdf.
3. Interview with Ann Cavoukian, September 2, 2015.
4. Ibid.
5. Interview with Austin Hill, July 22, 2015.
6. Interview with Ann Cavoukian, September 2, 2015.
7. Vitalik Buterin, "Proof of Stake: How I Learned to Love Weak Subjectivity," *Ethereum blog*, Ethereum Foundation, November 25, 2014. Web. October 3, 2015, blog.ethereum.org/2014/11/25/proof-stake-learned-love-weak-subjectivity.
8. Dino Mark Angaritis, e-mail attachment, November 27, 2015. He arrived at his calculation by "assuming hashrate of 583,000,000 Gh/s. (Gh/s = billion hashes /s). There are 600 seconds in 10 minutes. 600*583,000,000 = 349,800,000,000 billion hashes in 10 minutes. That's 350 Quintillion / 350,000,000,000,000,000,000 / 350 million million billion."
9. *Proof of burn* calls for miners to send their coins to a dead-end address where they become unredeemable. In exchange for burning these coins, miners gain entrance to a lottery where, presumably, they win back more than they burned. It's not a consensus mechanism but a trust mechanism.
10. Interview with Paul Brody, July 7, 2015.
11. Franklin Delano Roosevelt, "Executive Order 6102—Requiring Gold Coin, Gold Bullion and Gold Certificates to Be Delivered to the Government," *The American Presidency Project*, ed. Gerhard Peters and John T. Woolley, April 5, 1933, www.presidency.ucsb.edu/ws/?pid=14611, accessed December 2, 2015.
12. Interview with Josh Fairfield, June 1, 2015.
13. Allusion to Bandai's digital toy designed so that users would take care of it and protect it. It died if no one tended to it.
14. Joseph E. Stiglitz, "Lessons from the Global Financial Crisis of 2008," *Seoul Journal of Economics* 23(3) (2010).
15. Ernst & Young LLP, "The Big Data Backlash," December 2013, www.ey.com/UK/en/Services/Specialty-Services/Big-Data-Backlash; http://tinyurl.com/ptfm4ax.
16. The type of attack was named after "Sybil," a pseudonym of a woman diagnosed with dissociative identity disorder described in a 1973 book of that name. Cat-loving computer scientist John "JD" Douceur popularized the name in a 2002 paper.
17. Satoshi Nakamoto, "Bitcoin: A Peer-to-Peer Electronic Cash System," www.bitcoin.org, November 1, 2008; www.bitcoin.org/bitcoin.pdf, section 6, "Incentive."
18. Nick Szabo. "Bit gold." Unenumerated. Nick Szabo. December 27, 2008. Web. October 3, 2015. http://unenumerated.blogspot.com/2005/12/bit-gold.html.
19. Interview with Austin Hill, July 22, 2015.
20. Neal Stephenson, *Snow Crash* (1992). An allusion to *Snow Crash*'s virtual world, of which Hiro Protagonist is the protagonist and hero. Hiro was one of the top hackers of the Metaverse. Kongbucks are like bitcoin: the franchulates (corporate states, from the combination of *franchise* and *consulate*) issue their own money.
21. Ernest Cline, *Ready Player One* (New York: Crown, 2011).
22. Interview with Austin Hill, July 22, 2015.

23. John Lennon. "Imagine." *Imagine*. Producers John Lennon, Yoko Ono, and Phil Spector. October 11, 1971. www.lyrics007.com/John%20Lennon%20Lyrics/Imagine%20Lyrics.html.
24. Andy Greenberg. "Banking's Data Security Crisis." *Forbes*. November 2008. Web. October 3, 2015. www.forbes.com/2008/11/21/data-breaches-cybertheft-identity08-tech-cx_ag_1121breaches.html.
25. Ponemon Institute LLC, "2015 Cost of Data Breach Study: Global Analysis," sponsored by IBM, May 2015, www-03.ibm.com/security/data-breach.
26. Ponemon Institute LLC, "2014 Fifth Annual Study on Medical Identity Theft," sponsored by Medical Identity Fraud Alliance, February 23, 2015, Medidfraud.org/2014-fifth-annual-study-on-medical-identity-theft.
27. Interview with Andreas Antonopoulos, July 20, 2015.
28. Michael Melone, "Basics and History of PKI," *Mike Melone's blog*, Microsoft Corporation, March 10, 2012. Web. October 3, 2015. http://tinyurl.com/ngxuupl.
29. "Why Aren't More People Using Encrypted Email?," *Virtru blog*, Virtru Corporation, January 24, 2015. Web. August 8, 2015. www.virtru.com/blog/aren't-people-using-email-encryption, August 8, 2015.
30. Interview with Andreas Antonopoulos, July 20, 2015.
31. Interview with Austin Hill, July 22, 2015.
32. Ibid.
33. Interview with Ann Cavoukian, September 2, 2015.
34. Ibid.
35. David McCandless, "Worlds Biggest Data Breaches," *Information Is Beautiful*, David McCandless, October 2, 2015. Web. October 3, 2015. www.informationisbeautiful.net/visualizations/worlds-biggest-data-breaches-hacks/.
36. Interview with Haluk Kulin, June 9, 2015.
37. Interview with Austin Hill, July 22, 2015.
38. Coinbase privacy policy, www.coinbase.com/legal/privacy, November 17, 2014, accessed July 15, 2015.
39. See Don Tapscott and David Ticoll, *The Naked Corporation: How the Age of Transparency Will Revolutionize Business* (New York: Simon & Schuster, 2003).
40. Interview with Haluk Kulin, June 9, 2015.
41. ProofofExistence.com, September 2, 2015; www.proofofexistence.com/about/.
42. Interview with Steve Omohundro, May 28, 2015.
43. Interview with Andreas Antonopoulos, July 20, 2015.
44. Ibid.
45. Interview with Stephen Pair, June 11, 2015.
46. Edella Schlarger and Elinor Ostrom, "Property-Rights Regimes and Natural Resources: A Conceptual Analysis," *Land Economics* 68(3) (August 1992): 249–62; www.jstor.org/stable/3146375.
47. Interview with Haluk Kulin, June 9, 2015.
48. John Paul Titlow, "Fire Your Boss: Holacracy's Founder on the Flatter Future of Work," *Fast Company*, Mansueto Ventures LLC, July 9, 2015; www.fastcompany.com/3048338/the-future-of-work/fire-your-boss-holacracys-founder-on-the-flatter-future-of-work.
49. World Bank, September 2, 2015; www.worldbank.org/en/news/press-release/2015/04/15/massive-drop-in-number-of-unbanked-says-new-report.
50. "Bitcoin Powers New Worldwide Cellphone Top-Up Service," *CoinDesk*, February 15, 2015; www.coindesk.com/bitcoin-powers-new-worldwide-cellphone-top-service/, accessed August 26, 2015. FAQs, BitMoby.com, mHITs Ltd., n.d.; www.bitmoby.com/faq.html, accessed November 14, 2015.

51. Interview with Gavin Andresen, June 8, 2015.
52. Interview with Austin Hill, July 22, 2015.
53. Jakob Nielsen, "Nielsen's Law of Internet Bandwidth," Nielsen Norman Group, April 5, 1998; www.nngroup.com/articles/law-of-bandwidth/, accessed August 26, 2015.
54. Matthew Weaver, "World Leaders Pay Tribute at Auschwitz Anniversary Ceremony," *The Guardian*, Guardian News and Media Limited, January 27, 2015. Web. September 5, 2015, http://www.theguardian.com/world/2015/jan/27/-sp-watch-the-auschwitz-70th-anniversary-ceremony-unfold.

Chapter 3: Reinventing Financial Services

1. Estimates range from $87.5 million to $112 million (IMF).
2. https://ripple.com/blog/the-true-cost-of-moving-money/.
3. Interview with Vikram Pandit, August 24, 2015.
4. www.nytimes.com/2015/07/12/business/mutfund/putting-the-public-back-in-public-finance.html.
5. www.worldbank.org/en/topic/poverty/overview.
6. http://hbswk.hbs.edu/item/6729.html.
7. Interview with Hernando de Soto, November 27, 2015.
8. http://corporate.westernunion.com/About_Us.html.
9. Interview with Erik Voorhees, June 16, 2015.
10. Paul A. David, "The Dynamo and the Computer: An Historical Perspective on the Modern Productivity Paradox," *Economic History of Technology* 80(2) (May 1990): 355–61.
11. Joseph Stiglitz, "Lessons from the Global Financial Crisis," revised version of a lecture presented at Seoul National University, October 27, 2009.
12. www.finextra.com/finextra-downloads/newsdocs/The%20Fintech%202%200%20Paper.pdf.
13. www.bloomberg.com/news/articles/2015-07-22/the-blockchain-revolution-gets-endorsement-in-wall-street-survey.
14. www.swift.com/assets/swift_com/documents/about_swift/SIF_201501.pdf.
15. https://lightning.network/.
16. Interview with Chris Larsen, July 27, 2015.
17. Interview with Austin Hill, July 22, 2015.
18. Interview with Blythe Masters, July 27, 2015.
19. Ibid.
20. Ibid.
21. Ibid.
22. https://bitcoinmagazine.com/21007/nasdaq-selects-bitcoin-startup-chain-run-pilot-private market-arm/.
23. Interview with Austin Hill, July 22, 2015.
24. July 2015 by Greenwich Associates; www.bloomberg.com/news/articles/2015-07-22/the-blockchain-revolution-gets-endorsement-in-wall-street-survey.
25. Blythe Masters, from Exponential Finance keynote presentation: www.youtube.com/watch?v=PZ6WR2R1MnM.
26. https://bitcoinmagazine.com/21007/nasdaq-selects-bitcoin-startup-chain-run-pilot-private-market-arm/.
27. Interview with Jesse McWaters, August 13, 2015.
28. Interview with Austin Hill, July 22, 2015.
29. https://blog.ethereum.org/2015/08/07/on-public-and-private-blockchains/.
30. Interview with Chris Larsen, July 27, 2015.

31. Interview with Adam Ludwin, August 26, 2015.
32. Interview with Blythe Masters, July 27, 2015.
33. Interview with Eric Piscini, July 13, 2015.
34. Interview with Derek White, July 13, 2015.
35. Ibid.
36. Later, Bank of America, BNY Mellon, Citi, Commerzbank, Deutsche Bank, HSBC, Mitsubishi UFJ Financial Group, Morgan Stanley, National Australia Bank, Royal Bank of Canada, SEB, Société Générale, and Toronto Dominion Bank; www.ft.com/intl/cms/s/0/f358ed6c-5ae0-11e5-9846-de406ccb37f2.html#axzz3mf3orbRX; www.coindesk.com/citi-hsbc-partner-with-r3cev-as-blockchain-project-adds-13-banks/.
37. http://bitcoinnewsy.com/bitcoin-news-mike-hearn-bitcoin-core-developer-joins-r3cev-with-5-global-banks-including-wells-fargo/.
38. http://www.linuxfoundation.org/news-media/announcements/2015/12/linux-foundation-unites-industry-leaders-advance-blockchain.
39. www.ifrasia.com/blockchain-will-make-dodd-frank-obsolete-bankers-say/21216014.article.
40. http://appft.uspto.gov/netacgi/nph-Parser?Sect1=PTO2&Sect2=HITOFF&p=1&u=%2Fnetahtml%2FPTO%2Fsearch-bool.html&r=1&f=G&l=50&co1=AND&d=PG01&s1=20150332395&OS=20150332395&RS=20150332395?p=cite_Brian_Cohen_or_Bitcoin_Magazine.
41. www.youtube.com/watch?v=A6kJfvuNqtg.
42. Interview with Jeremy Allaire, June 30, 2015.
43. Ibid.
44. Ibid.
45. Ibid.
46. Heralded as another sign the industry was "growing up"; www.wsj.com/articles/goldman-a-lead-investor-in-funding-round-for-bitcoin-startup-circle-1430363042.
47. Interview with Jeremy Allaire, June 30, 2015.
48. Interview with Stephen Pair, June 11, 2015.
49. Alex Tapscott has consulted to Vogogo Inc.
50. Interview with Suresh Ramamurthi, September 28, 2015.
51. E-mail correspondence with Blythe Masters, December 14, 2015.
52. Interview with Tom Mornini, July 20, 2015.
53. These ideas were originally explored in *The Naked Corporation* by Don Tapscott and David Ticoll.
54. Ibid.
55. www.accountingweb.com/aa/auditing/human-errors-the-top-corporate-tax-and-accounting-mistakes.
56. Ibid.
57. Interview with Simon Taylor, July 13, 2015.
58. Ibid.
59. Interview with Jeremy Allaire, June 30, 2015.
60. Interview with Christian Lundkvist, July 6, 2015.
61. Interview with Austin Hill, July 22, 2015.
62. Interview with Eric Piscini, July 13, 2015.
63. www2.deloitte.com/us/en/pages/about-deloitte/articles/facts-and-figures.html.
64. Interview with Eric Piscini, July 13, 2015.
65. Ibid.
66. Interview with Tom Mornini, July 20, 2015.

67. Ibid.
68. www.calpers.ca.gov/docs/forms-publications/global-principles-corporate-governance.pdf.
69. Interview with Izabella Kaminska, August 5, 2015.
70. http://listedmag.com/2013/06/robert-monks-its-broke-lets-fix-it/.
71. The Right to Be Forgotten Movement is gaining steam, particularly in Europe: http://ec.europa.eu/justice/data-protection/files/factsheets/factsheet_data_protection_en.pdf.
72. www.bloomberg.com/news/articles/2014-10-07/andreessen-on-finance-we-can-reinvent-the-entire-thing.
73. http://www.nytimes.com/2015/12/24/business/dealbook/banks-reject-new-york-city-ids-leaving-unbanked-on-sidelines.html.
74. Interview with Patrick Deegan, June 6, 2015.
75. Ibid.
76. https://btcjam.com/.
77. Interview with Erik Voorhees, June 16, 2015.
78. www.sec.gov/about/laws/sa33.pdf.
79. http://www.wired.com/2015/12/sec-approves-plan-to-issue-company-stock-via-the-bitcoin-blockchain/.
80. http://investors.overstock.com/mobile.view?c=131091&v=203&d=1&id=2073583.
81. https://bitcoinmagazine.com/21007/nasdaq-selects-bitcoin-startup-chain-run-pilot-private-market-arm/.
82. James Surowiecki, *The Wisdom of Crowds: Why the Many Are Smarter Than the Few and How Collective Wisdom Shapes Business, Economies, Societies and Nations* (New York: Doubleday, 2014).
83. www.augur.net.
84. From e-mail exchange with the Augur team: Jack Peterson, Core Developer; Joey Krug, Core Developer; Peronet Despeignes, SpecialOps.
85. Interview with Andreas Antonopoulos, December 8, 2014.
86. Interview with Barry Silbert, September 22, 2015.
87. Interview with Benjamin Lawsky, July 2, 2015.

Chapter 4: Re-architecting the Firm: The Core and the Edges

1. Interview with Joe Lubin, July 13, 2015.
2. Companies like Apple and Spotify will be able to use the new platform as well. The goal is that it will be owned by many entities in the music industry, especially artists. It will likely be easier to earn tokens if you create content than if you just resell someone else's content.
3. https://slack.com/is.
4. https://github.com.
5. Coase wrote: "A firm has a role to play in the economic system if . . . [t]ransactions can be organized within the firm at less cost than if the same transactions were carried out through the market. The limit to the size of the firm [is reached] when the costs of organizing additional transactions within the firm [exceed] the costs of carrying the same transactions in the market." As cited in Oliver Williamson and Sydney G. Winter, eds., *The Nature of the Firm* (New York and Oxford: Oxford University Press, 1993), 90.

6. Oliver Williamson, "The Theory of the Firm as Governance Structure: From Choice to Contract," *The Journal of Economic Perspectives* 16(3) (Summer 2002) 171–95.
7. Ibid.
8. Peter Thiel with Blake Masters, *Zero to One: Notes on Startups, or How to Build the Future* (New York: Crown Business, 2014).
9. Lord Wilberforce, *The Law of Restrictive Trade Practices and Monopolies* (Sweet & Maxwell, 1966), 22.
10. Interview with Yochai Benkler, August 26, 2015.
11. John Hagel and John Seely Brown, "Embrace the Edge or Perish," *Bloomberg*, November 28, 2007; www.bloomberg.com/bw/stories/2007-11-28/embrace-the-edge-or-perishbusinessweek-business-news-stock-market-and-financial-advice.
12. Interview with Vitalik Buterin, September 30, 2015.
13. Interview with Andreas Antonopoulos, July 20, 2015.
14. The one exception is the Way Back Machine, which allows you to get a deeper history.
15. Oliver E. Williamson, "The Theory of the Firm as Governance Structure: From Choice to Contract," *Journal of Economic Perspectives* 16 (3), Summer 2002.
16. Ibid.
17. Michael C. Jensen and William H. Meckling, "Theory of the Firm: Managerial Behavior, Agency Costs and Ownership Structure," *Journal of Financial Economics* 305 (1976): 310–11 (arguing that the corporation—or, more generally, a firm—is a collection of consensual relationships among shareholders, creditors, managers, and perhaps others); see also generally, Frank H. Easterbrook and Daniel R. Fischel, *The Economic Structure of Corporate Law* (Cambridge, Mass.: Harvard University Press, 1991).
18. Vitalik Buterin, "Bootstrapping a Decentralized Autonomous Corporation: Part I," *Bitcoin Magazine*, September 19, 2013; https://bitcoinmagazine.com/7050/bootstrapping-a-decentralized-autonomous-corporation-part-i/.
19. Nick Szabo, "Formalizing and Securing Relationships on Public Networks," http://szabo.best.vwh.net/formalize.html.
20. http://szabo.best.vwh.net/smart.contracts.html.
21. Interview with Aaron Wright, August 10, 2015.
22. Cryptographers started using "Alice" and "Bob" instead of "Party A" and "Party B" as a convenient way to describe exchanges between them, lending some clarity and familiarity to discussions about computational encryption. The practice is said to date from Ron Rivest's 1978 work, "Security's Inseparable Couple," *Communications of the ACM. Network World*, February 7, 2005; www.networkworld.com/news/2005/020705widernetaliceandbob.html.
23. GitHub.com, January 3, 2012; https://github.com/bitcoin/bips/blob/master/bip-0016.mediawiki, accessed September 30, 2015.
24. www.coindesk.com/hedgy-hopes-tackle-bitcoin-volatility-using-multi-signature-technolog/.
25. https://books.google.ca/books?id=VXIDgGjLHVgC&pg=PA19&lpg=PA19&dq=a+workman+moves+from+department+Y+to+department+X&source=bl&ots=RHb0qrpLz_&sig=LaZFqatLYllrBW8ikPn4PEZ9_7U&hl=en&sa=X&ved=0ahUKEwjgyuO2gKfKAhUDpB4KHb0JDcAQ6AEIITAB#v=onepage&q=a%20workman%20moves%20from%20department%20Y%20to%20department%20X&f=false.
26. Elliot Jaques, "In Praise of Hierarchy," *Harvard Business Review*, January–February 1990.

27. Interview with Yochai Benkler, August 26, 2015.
28. Tapscott and Ticoll, *The Naked Corporation*.
29. Werner Erhard and Michael C. Jensen, "Putting Integrity into Finance: A Purely Positive Approach," November 27, 2015, Harvard Business School NOM Unit Working Paper No. 12-074; Barbados Group Working Paper No. 12-01; European Corporate Governance Institute (ECGI)—Finance Working Paper No. 417/2014.
30. Bank of America's Average Return on Equity since December 31, 2009, is less than 2 percent; https://ycharts.com/companies/BAC/return_on_equity.
31. Interview with Steve Omohundro, May 28, 2015.
32. E-mail interview with David Ticoll, December 9, 2015.
33. Interview with Melanie Swan, September 14, 2015.
34. https://hbr.org/1990/05/the-core-competence-of-the-corporation.
35. Michael Porter, "What Is Strategy?," *Harvard Business Review*, November–December 1996.
36. Interview with Susan Athey, November 20, 2015.

Chapter 5: New Business Models: Making It Rain on the Blockchain

1. To prevent spam, it's possible that new public keys (personas) with low reputation will have to pay a certain fee to list. The fee can be moved to an escrow contract and be returned once the persona has successfully rented their property out, or perhaps after some period of time elapses and they decide to delete their listing. Large data items like pictures will be kept on IPFS or Swarm, but the hash of the data and the information identifying the persona that owns the data will be kept on the blockchain inside the bAirbnb contract.
2. Perhaps using the Whisper protocol.
3. Formatted and annotated with the Hypertext Markup Language (HTML).
4. David McCandless, "World's Biggest Data Breaches," *Information Is Beautiful*, October 2, 2015; www.informationisbeautiful.net/visualizations/worlds-biggest-data-breaches-hacks/, accessed November 27, 2015.
5. As defined by Vitalik Buterin: "Cryptoeconomics is a technical term roughly meaning 'it's decentralized, it uses public key cryptography for authentication, and it uses economic incentives to ensure that it keeps going and doesn't go back in time or incur any other glitch.'" From "The Value of Blockchain Technology, Part I," https://blog.ethereum.org/2015/04/13/visions-part-1-the-value-of-blockchain-technology/.
6. www.youtube.com/watch?v=K2fhwMKk2Eg.
7. http://variety.com/2015/digital/news/netflix-bandwidth-usage-internet-traffic-1201507187/.
8. Interview with Bram Cohen, August 17, 2015.
9. Stan Franklin and Art Graesser, "Is It an Agent, or Just a Program? A Taxonomy for Autonomous Agents," www.inf.ufrgs.br/~alvares/CMP124SMA/IsItAnAgentOrJustAProgram.pdf.
10. Ibid., 5.
11. Vitalik Buterin, https://blog.ethereum.org/2014/05/06/daos-dacs-das-and-more-an-incomplete-terminology-guide/. "Autonomous agents are on the other side of the automation spectrum; in an autonomous agent, there is no necessary specific human involvement at all; that is to say, while some degree of human effort might

be necessary to build the hardware that the agent runs on, there is no need for any humans to exist that are aware of the agent's existence."

12. Ibid.
13. Technical detail: Because storing data directly on blockchains is very expensive, it's more likely that there is a hash of the data, and the data itself will be on some other decentralized data storage network like Swarm or IPFS.
14. Interview with Vitalik Buterin, September 30, 2015.
15. Interview with Andreas Antonopoulos, July 20, 2015.
16. Ibid.
17. Don Tapscott and Anthony D. Williams, *Wikinomics: How Mass Collaboration Changes Everything* (New York: Portfolio/Penguin, 2007). *Wikinomics* defined seven such business models. The list has been extended here.
18. Commons-based Peer Production is a term developed by Harvard Law professor Yochai Benkler in the seminal article "Coase's Penguin," *The Yale Law Journal*, 2002; www.yale.edu/yalelj/112/BenklerWEB.pdf.
19. http://fortune.com/2009/07/20/information-wants-to-be-free-and-expensive/.
20. Interview with Yochai Benkler, August 26, 2015.
21. Interview with Dino Mark Angaritis, August 7, 2015.
22. Andrew Lih, "Can Wikipedia Survive?," *The New York Times*, June 20, 2015; www.nytimes.com/2015/06/21/opinion/can-wikipedia-survive.html.
23. http://techcrunch.com/2014/05/09/monegraph/.
24. http://techcrunch.com/2015/06/24/ascribe-raises-2-million-to-ensure-you-get-credit-for-your-art/.
25. www.nytimes.com/201%4/15/technology/15twitter.html?_r=0.
26. http://techcrunch.com/2014/05/09/monegraph/.
27. www.verisart.com/.
28. http://techcrunch.com/2015/07/07/verisart-plans-to-use-the-blockchain-to-verify-the-authencity-of-artworks/.
29. Interview with Yochai Benkler, August 26, 2015.
30. Interview with David Ticoll, August 7, 2015.
31. Interview with Yochai Benkler, August 26, 2015.
32. www.nytimes.com/2013/07/21/opinion/sunday/friedman-welcome-to-the-sharing-economy.html?pagewanted=1&_r=2&partner=rss&emc=rss&.
33. Sarah Kessler, "The Sharing Economy Is Dead and We Killed It," *Fast Company*, September 14, 2015; www.fastcompany.com/3050775/the-sharing-economy-is-dead-and-we-killed-it#1.
34. "Prosumers" is a term invented by Alvin Toffler in *Future Shock* (1980). In *The Digital Economy* (1994) Don Tapscott developed the concept and notion of "prosumption."
35. Interview with Robin Chase, September 2, 2015.
36. https://news.ycombinator.com/item?id=9437095.
37. This scenario was originally explained by Don Tapscott in "The Transparent Burger," *Wired*, March 2004; http://archive.wired.com/wired/archive/12.03/start.html?pg=2%3ftw=wn_tophead_7.
38. Interview with Yochai Benkler, August 26, 2015.
39. Called "the wiki workplace" in *Wikinomics*.
40. CAPTCHA stands for "Completely Automated Public Turing Test to Tell Computers and Humans Apart."
41. Interview with Joe Lubin, July 13, 2015.
42. Ibid.

Chapter 6: The Ledger of Things: Animating the Physical World

1. Not their real names. This story is based on discussions with individuals familiar with the situation.
2. Primavera De Filippi, "It's Time to Take Mesh Networks Seriously (and Not Just for the Reasons You Think)," *Wired*, January 2, 2014.
3. Interview with Eric Jennings, July 10, 2015.
4. Ibid.
5. Interview with Lawrence Orsini, July 30, 2015.
6. Don predicted the development of such networks in Don Tapscott and Anthony Williams, *Macrowikinomics: New Solutions for a Connected Planet* (New York: Portfolio/Penguin, 2010, updated 2012).
7. Interview with Lawrence Orsini, July 30, 2015.
8. Puja Mondal, "What Is Desertification? Desertification: Causes, Effects and Control of Desertification," *UNEP: Desertifcation*, United Nations Environment Programme, n.d.; https://desertification.wordpress.com/category/ecology-environment/unep/, accessed September 29, 2015.
9. www.internetlivestats.com/internet-users/, as of December 1, 2015.
10. Cadie Thompson, "Electronic Pills May Be the Future of Medicine," CNBC, April 21, 2013; www.cnbc.com/id/100653909; and Natt Garun, "FDA Approves Edible Electronic Pills That Sense When You Take Your Medication," *Digital Trends*, August 1, 2012; www.digitaltrends.com/home/fda-approves-edible-electronic-pills/.
11. Mark Jaffe, "IOT Won't Work Without Artificial Intelligence," *Wired*, November 2014; www.wired.com/insights/2014/11/iot-wont-work-without-artificial-intelligence/.
12. IBM, "Device Democracy," 2015, 4.
13. Allison Arieff, "The Internet of Way Too Many Things," *The New York Times*, September 5, 2015.
14. IBM, "Device Democracy," 10.
15. Interview with Dino Mark Angaritis, August 11, 2015.
16. Interview with Carlos Moreira, September 3, 2015.
17. Ibid.
18. Interview with Michelle Tinsley, June 25, 2015.
19. Ibid.
20. McKinsey Global Institute, "The Internet of Things: Mapping the Value Beyond the Hype," June 2015.
21. Interview with Eric Jennings, July 10, 2015.
22. IBM Institute for Business Value, "The Economy of Things: Extracting New Value from the Internet of Things," 2015.
23. Cadie Thompson, "Apple Has a Smart Home Problem: People Don't Know They Want It Yet," *Business Insider*, June 4, 2015; www.businessinsider.com/apple-homekit-adoption-2015-6.
24. McKinsey Global Institute, "The Internet of Things."
25. Interview with Eric Jennings, July 10, 2015.
26. IBM, "Device Democracy," 9.
27. Ibid., 13.
28. McKinsey Global Institute, "The Internet of Things." MGI defined nine settings with value potential.

29. www.wikihow.com/Use-Uber.
30. http://consumerist.com/tag/uber/page/2/.
31. Mike Hearn, "Future of Money," Turing Festival, Edinburgh, Scotland, August 23, 2013, posted September 28, 2013; www.youtube.com/watch?v=Pu4PAMFPo5Y&feature=youtu.be.
32. McKinsey, "An Executive's Guide to the Internet of Things," August 2015; www.mckinsey.com/Insights/Business_Technology/An_executives_guide_to_the_Internet_of_Things?cid=digital-eml-alt-mip-mck-oth-1508.

Chapter 7: Solving the Prosperity Paradox: Economic Inclusion and Entrepreneurship

1. http://datatopics.worldbank.org/financialinclusion/country/nicaragua.
2. www.budde.com.au/Research/Nicaragua-Telecoms-Mobile-and-Broadband-Market-Insights-and-Statistics.html.
3. "Property Disputes in Nicaragua," U.S. Embassy, http://nicaragua.usembassy.gov/property_disputes_in_nicaragua.html. There are an estimated thirty thousand properties in dispute.
4. Interview with Joyce Kim, June 12, 2015.
5. Ibid.
6. Ibid.
7. www.worldbank.org/en/news/press-release/2015/04/15/massive-drop-in-number-of-unbanked-says-new-report; and C. K. Prahalad, *The Fortune at the Bottom of the Pyramid: Eradicating Poverty Through Profits* (Philadelphia: Wharton School Publishing, 2009). This figure is an estimate.
8. Interview with Joyce Kim, June 12, 2015.
9. www.ilo.org/global/topics/youth-employment/lang—en/index.htm.
10. Thomas Piketty, *Capital in the Twenty-First Century* (Cambridge, Mass.: Belknap Press, 2014).
11. www.brookings.edu/~/media/research/files/papers/2014/05/declining%20business%20dynamism%20litan/declining_business_dynamism_hathaway_litan.pdf.
12. Ruth Simon and Caelainn Barr, "Endangered Species: Young U.S. Entrepreneurs," *The Wall Street Journal*, January 2, 2015; www.wsj.com/articles/endangered-species-young-u-s-entrepreneurs-1420246116.
13. World Bank Group, Doing Business, www.doingbusiness.org/data/exploretopics/starting-a-business.
14. Interview with Hernando de Soto, November 27, 2015.
15. www.tamimi.com/en/magazine/law-update/section-6/june-4/dishonoured-cheques-in-the-uae-a-criminal-law-perspective.html.
16. www.worldbank.org/en/topic/poverty/overview. To be precise, it was 1.91 billion in 1990.
17. http://digitalcommons.georgefox.edu/cgi/viewcontent.cgi?article=1003&context=gfsb.
18. http://reports.weforum.org/outlook-global-agenda-2015/top-10-trends-of-2015/1-deepening-income-inequality/.
19. Ibid.
20. Interview with Tyler Winklevoss, June 9, 2015.
21. Congo, Chad, Central African Republic, South Sudan, Niger, Madagascar, Guinea,

Cameroon, Burkina Faso, Tanzania; http://data.worldbank.org/indicator/FB.CBK.BRCH.P5?order=wbapi_data_value_2013+wbapi_data_value+wbapi_data_value-last&sort=asc.
22. www.aba.com/Products/bankcompliance/Documents/SeptOct11CoverStory.pdf.
23. http://www.nytimes.com/2015/12/24/business/dealbook/banks-reject-new-york-city-ids-leaving-unbanked-on-sidelines.html.
24. E-mail correspondence with Joe Lubin, August 6, 2015.
25. David Birch, *Identity Is the New Money* (London: London Publishing Partnership, 2014), 1.
26. E-mail correspondence with Joe Lubin, August 6, 2015.
27. Interview with Joyce Kim, June 12, 2015.
28. Interview with Hernando de Soto, November 27, 2015.
29. Interview with Haluk Kulin, June 9, 2015.
30. E-mail correspondence with Joe Lubin, August 6, 2015.
31. Interview with Balaji Srinivasan, May 29, 2014.
32. www.doingbusiness.org/data/exploretopics/starting-a-business.
33. Interview with Haluk Kulin, June 9, 2015.
34. Analie Domingo agreed to let us shadow her as she went through her normal routine of sending money home to her mother in the Philippines. Analie has been an employee of Don Tapscott and Ana Lopes for twenty years and is also a close friend.
35. www12.statcan.gc.ca/nhs-enm/2011/dp-pd/prof/details/page.cfm?Lang=E&Geo1=PR&Code1=01&Data=Count&SearchText=canada&SearchType=Begins&SearchPR=01&A1=All&B1=All&Custom=&TABID=1.
36. https://remittanceprices.worldbank.org/sites/default/files/rpw_report_june_2015.pdf.
37. The remittance market is $500 billion; average fees of 7.7 percent translate to $38.5 billion in fees.
38. Dilip Ratha, "The Impact of Remittances on Economic Growth and Poverty Reduction," *Migration Policy Institute* 8 (September 2013).
39. Adolf Barajas, et al., "Do Workers' Remittances Promote Economic Growth?," IMF Working Paper, www10.iadb.org/intal/intalcdi/pe/2009/03935.pdf.
40. "Aid and Remittances from Canada to Select Countries," Canadian International Development Platform, http://cidpnsi.ca/blog/portfolio/aid-and-remittances-from-canada/.
41. World Bank Remittance Price Index, https://remittanceprices.worldbank.org/en.
42. 2011 National Household Survey Highlights, Canadian Census Bureau, www.fin.gov.on.ca/en/economy/demographics/census/nhshi11-1.html.
43. https://support.skype.com/en/faq/FA1417/how-much-bandwidth-does-skype-need.
44. Interview with Eric Piscini, July 13, 2015.
45. http://corporate.westernunion.com/Corporate_Fact_Sheet.html.
46. At the time of writing, Abra had not opened its doors in Canada. However, we were able to test Abra's technology with Analie and her mother successfully with Abra's help.
47. Interview with Bill Barhydt, August 25, 2015.
48. Ibid.
49. Ibid.
50. "Foreign Aid and Rent-Seeking, *The Journal of International Economics*, 2000, 438; http://conferences.wcfia.harvard.edu/sites/projects.iq.harvard.edu/files/gov2126/files/1632.pdf.

51. Ibid.
52. www.propublica.org/article/how-the-red-cross-raised-half-a-billion-dollars-for-haiti-and-built-6-homes.
53. "Mortality, Crime and Access to Basic Needs Before and After the Haiti Earthquake," *Medicine, Conflict and Survival* 26(4) (2010).
54. http://unicoins.org/.
55. Jeffrey Ashe with Kyla Jagger Neilan, *In Their Own Hands: How Savings Groups Are Revolutionizing Development* (San Francisco: Berrett-Koehler Publishers, 2014).
56. E. Kumar Sharma, "Founder Falls," *Business Today* (India), December 25, 2011; www.businesstoday.in/magazine/features/vikram-akula-quits-sks-microfiance-loses-or-gains/story/20680.html.
57. Ning Wang, "Measuring Transaction Costs: An Incomplete Survey," *Ronald Coase Institute Working Papers* 2 (February 2003); www.coase.org/workingpapers/wp-2.pdf.
58. www.telesurtv.net/english/news/Honduran-Movements-Slam-Repression-of-Campesinos-in-Land-Fight-20150625-0011.html.
59. USAID, the Millennium Challenge Corporation, and the UN Food and Agriculture Organization.
60. Paul B. Siegel, Malcolm D. Childress, and Bradford L. Barham, "Reflections on Twenty Years of Land-Related Development Projects in Central America: Ten Things You Might Not Expect, and Future Directions," Knowledge for Change Series, International Land Coalition (ILC), Rome, 2013; http://tinyurl.com/oekhzos, accessed August 26, 2015.
61. Ibid.
62. Ambassador Michael B. G. Froman, US Office of the Trade Representative, "2015 National Trade Estimate Report on Foreign Trade Barriers," USTR.gov, April 1, 2015; https://ustr.gov/sites/default/files/files/reports/2015/NTE/2015%20NTE%20Honduras.pdf.
63. Interview with Hernando de Soto, November 27, 2015.
64. http://in.reuters.com/article/2015/05/15/usa-honduras-technology-idINKBN0O01V720150515.
65. Interview with Kausik Rajgopal, August 10, 2015.
66. World Bank, "Doing Business 2015: Going Beyond Efficiencies," Washington, D.C.: World Bank, 2014; DOI: 10.1596/978-1-4648-0351-2, License Creative Commons Attribution CC BY 3.0 IGO.
67. "ITU Releases 2014 ICT Figures," www.itu.int/net/pressoffice/press_releases/2014/23.aspx#.VEfalovF_Kg.
68. www.cdc.gov/healthliteracy/learn/understandingliteracy.html.
69. www.proliteracy.org/the-crisis/adult-literacy-facts.
70. CIA World Factbook, literacy statistics, www.cia.gov/library/publications/the-world-factbook/fields/2103.html#136.

Chapter 8: Rebuilding Government and Democracy

1. http://europa.eu/about-eu/countries/member-countries/estonia/index_en.htm; http://www.citypopulation.de/Canada-MetroEst.html.
2. In-person conversation between Estonian president Toomas Hendrik Ilves and Don Tapscott at the World Economic Forum's Global Agenda Council meeting in Abu Dhabi, United Arab Emirates, October 2015.

3. www.socialprogressimperative.org/data/spi#data_table/countries/com6/dim1,dim2,dim3,com9,idr35,com6,idr16,idr34.
4. https://e-estonia.com/the-story/the-story-about-estonia/. Estonia is very proud of its e-Estonia initiatives and has published a lot of information on the Web. All of the information and statistics used in this section came from the Government of Estonia Web site.
5. "Electronic Health Record," e-Estonia.com, n.d.; https://e-estonia.com/component/electronic-health-record/, accessed November 29, 2015.
6. "e-Cabinet," e-Estonia.com, n.d.; https://e-estonia.com/component/e-cabinet/, accessed November 29, 2015.
7. "Electronic Land Register," e-Estonia.com, n.d.; https://e-estonia.com/component/electronic-land-register/, accessed November 29, 2015.
8. Charles Brett, "My Life Under Estonia's Digital Government," *The Register*, www.theregister.co.uk/2015/06/02/estonia/.
9. Interview with Mike Gault, August 28, 2015.
10. "Keyless Signature Infrastructure," e-Estonia.com, n.d.; https://e-estonia.com/component/keyless-signature-infrastructure/, accessed November 29, 2015.
11. Olga Kharif, "Bitcoin Not Just for Libertarians and Anarchists Anymore," *Bloomberg Business*, October 9, 2014; www.bloomberg.com/bw/articles/2014-10-09/bitcoin-not-just-for-libertarians-and-anarchists-anymore. To be sure, there is a strong libertarian trend in the American population as a whole. According to the Pew Research Center, 11 percent of Americans describe themselves as libertarian and know the definition of the term. "In Search of Libertarians," www.pewresearch.org/fact-tank/2014/08/25/in-search-of-libertarians/.
12. "Bitcoin Proves the Libertarian Idea of Paradise Would Be Hell on Earth," *Business Insider*, www.businessinsider.com/bitcoin-libertarian-paradise-would-be-hell-on-earth-2013-12#ixzz3kQqSap00.
13. Human Rights Watch, "World Report 2015: Events of 2014," www.hrw.org/sites/default/files/wr2015_web.pdf.
14. Interview with Hernando de Soto, November 27, 2015.
15. Seymour Martin Lipset, *Political Man: The Social Bases of Politics*, 2nd ed. (London: Heinemann, 1983), 64.
16. Interview with Hernando de Soto, November 27, 2015.
17. Hernando de Soto, "The Capitalist Cure for Terrorism," *The Wall Street Journal*, October 10, 2014; www.wsj.com/articles/the-capitalist-cure-for-terrorism-1412973796, accessed November 27, 2015.
18. Interview with Hernando de Soto, November 27, 2015.
19. Interview with Carlos Moreira, September 3, 2015.
20. Melanie Swan, *Blockchain: Blueprint for a New Economy* (Sebastopol, Calif.: O'Reilly Media, January 2015), 45.
21. Emily Spaven, "UK Government Exploring Use of Blockchain Recordkeeping," *CoinDesk*, September 1, 2015; www.coindesk.com/uk-government-exploring-use-of-blockchain-recordkeeping/.
22. J. P. Buntinx, "'Blockchain Technology' Is Bringing Bitcoin to the Mainstream," Bitcoinist.net, August 29, 2015; http://bitcoinist.net/blockchain-technology-bringing-bitcoin-mainstream/.
23. Melanie Swan, quoted in Adam Stone, "Unchaining Innovation: Could Bitcoin's Underlying Tech Be a Powerful Tool for Government?," *Government Technology*, July 10, 2015; www.govtech.com/state/Unchaining-Innovation-Could-Bitcoins-Underlying-Tech-be-a-Powerful-Tool-for-Government.html.

24. See for example www.partnerships.org.au/ and www.in-control.org.uk/what-we-do.aspx.
25. Interview with Perianne Boring, August 7, 2015. See also Joseph Young, "8 Ways Governments Could Use the Blockchain to Achieve 'Radical Transparency,'" *CoinTelegraph*, July 13, 2015; http://cointelegraph.com/news/114833/8-ways-governments-could-use-the-blockchain-to-achieve-radical-transparency.
26. www.data.gov.
27. www.data.gov.uk.
28. Ben Schiller, "A Revolution of Outcomes: How Pay-for-Success Contracts Are Changing Public Services," *Co.Exist*, www.fastcoexist.com/3047219/a-revolution-of-outcomes-how-pay-for-success-contracts-are-changing-public-services. Also see: www.whitehouse.gov/blog/2013/11/20/building-smarter-more-efficient-government-through-pay-success.
29. R. C. Porter, "Can You 'Snowden-Proof' the NSA?: How the Technology Behind the Digital Currency—Bitcoin—Could Stop the Next Edward Snowden," *Fortuna's Corner*, June 3, 2015; http://fortunascorner.com/2015/06/03/can-you-snowden-proof-the-nsa-how-the-technology-behind-the-digital-currency-bitcoin-could-stop-the-next-edward-snowden/.
30. Elliot Maras, "London Mayoral Candidate George Galloway Calls for City Government to Use Block Chain for Public Accountability," *Bitcoin News*, July 2, 2015; www.cryptocoinsnews.com/london-mayoral-candidate-george-galloway-calls-city-government-use-block-chain-public-accountability/.
31. Tapscott, *The Digital Economy*, 304.
32. Al Gore, speech to the We Media conference, October 6, 2005; www.fpp.co.uk/online/05/10/Gore_speech.html.
33. Ibid.
34. "The Persistence of Conspiracy Theories," *The New York Times*, April 30, 2011; www.nytimes.com/2011/05/01/weekinreview/01conspiracy.html?pagewanted=all&_r=0.
35. www.nytimes.com/2014/07/06/upshot/when-beliefs-and-facts-collide.html?module=Search&mabReward=relbias:w;%201RI:6%20%3C{:}%3E.
36. "Plain Language: It's the Law," Plain Language Action and Information Network, n.d.: www.plainlanguage.gov/plLaw/, accessed November 30, 2015.
37. https://globalclimateconvergence.org/news/nyt-north-carolinas-election-machine-blunder.
38. http://users.encs.concordia.ca/~clark/papers/2012_fc.pdf.
39. http://link.springer.com/chapter/10.1007%2F978-3-662-46803-6_16.
40. http://blogs.wsj.com/digits/2015/07/29/scientists-in-greece-design-cryptographic-e-voting-platform/.
41. http://nvbloc.org/.
42. http://cointelegraph.com/news/114404/true-democracy-worlds-first-political-app-blockchain-party-launches-in-australia.
43. www.techinasia.com/southeast-asia-blockchain-technology-bitcoin-insights/.
44. Ibid.
45. www.washingtonpost.com/news/wonkblog/wp/2014/08/06/a-comprehensive-investigation-of-voter-impersonation-finds-31-credible-incidents-out-of-one-billion-ballots-cast/.
46. www.eac.gov/research/election_administration_and_voting_survey.aspx.
47. http://america.aljazeera.com/opinions/2015/7/most-americans-dont-vote-in-elections-heres-why.html.

48. Interview with Eduardo Robles Elvira, September 10, 2015.
49. www.chozabu.net/blog/?p=78.
50. https://agoravoting.com/.
51. Interview with Eduardo Robles Elvira, September 10, 2015.
52. http://cointelegraph.com/news/111599/blockchain_technology_smart_contracts_and_p2p_law.
53. Patent Application of David Chaum, "Random Sample Elections," June 19, 2014; http://patents.justia.com/patent/20140172517.
54. https://blog.ethereum.org/2014/08/21/introduction-futarchy/.
55. Federico Ast (@federicoast) and Alejandro Sewrjugin (@asewrjugin), "The CrowdJury, a Crowdsourced Justice System for the Collaboration Era," https://medium.com/@federicoast/the-crowdjury-a-crowdsourced-court-system-for-the-collaboration-era-66da002750d8#.e8yynqipo.
56. http://crowdjury.org/en/.
57. The entire process is described in Ast and Sewrjugin, "The CrowdJury."
58. A brief description of the jury selection process in early Athens is described at www.agathe.gr/democracy/the_jury.html.
59. See full report and recommendations here, including a description of models worldwide: www.judiciary.gov.uk/reviews/online-dispute-resolution/.
60. http://blog.counter-strike.net/index.php/overwatch/.
61. Environmental Defense Fund, www.edf.org/climate/how-cap-and-trade-works.
62. Swan, *Blockchain: Blueprint for a New Economy.*
63. Interview with Andreas Antonopoulos, July 20, 2015.

Chapter 9: Freeing Culture on the Blockchain: Music to Our Ears

1. "2015 Women in Music Honours Announced," *M Online*, PRS for Music, October 22, 2015; www.m-magazine.co.uk/news/2015-women-in-music-honours-announced/, accessed November 21, 2015.
2. Interview with Imogen Heap, September 16, 2015.
3. David Byrne, "The Internet Will Suck All Creative Content Out of the World," *The Guardian*, June 20, 2014; www.theguardian.com/music/2013/oct/11/david-byrne-internet-content-world, accessed September 20, 2015.
4. Interview with Imogen Heap, September 16, 2015.
5. In-person conversation between Paul Pacifico and Don Tapscott at the home of Imogen Heap, November 8, 2015.
6. "Hide and Seek," performed by Ariana Grande, YouTube, Love Ariana Grande Channel, October 17, 2015; www.youtube.com/watch?v=2SDVDd2VpP0, accessed November 21, 2015.
7. Interview with Imogen Heap, September 16, 2015.
8. David Byrne, et al., "Once in a Lifetime," *Remain in Light*, Talking Heads, February 2, 1981.
9. Interview with Imogen Heap, September 16, 2015.
10. Johan Nylander, "Record Labels Part Owner of Spotify," *The Swedish Wire*, n.d.; www.swedishwire.com/jobs/680-record-labels-part-owner-of-spotify, accessed September 23, 2015. According to Nylander, Sony had 5.8 percent, Universal 4.8 percent, and Warner 3.8 percent. Before its sell-off, EMI had a 1.9 percent stake.
11. Interview with Imogen Heap, September 16, 2015.
12. David Johnson, "See How Much Every Top Artist Makes on Spotify," *Time*,

November 18, 2014; http://time.com/3590670/spotify-calculator/, accessed September 25, 2015.
13. Micah Singleton, "This Was Sony Music's Contract with Spotify," *The Verge*, May 19, 2015; www.theverge.com/2015/5/19/8621581/sony-music-spotify-contract, accessed September 25, 2015.
14. Stuart Dredge, "Streaming Music: What Next for Apple, YouTube, Spotify . . . and Musicians?," *The Guardian*, August 29, 2014; www.theguardian.com/technology/2014/aug/29/streaming-music-apple-youtube-spotify-musicians, accessed August 14, 2015.
15. Ed Christman, "Universal Music Publishing's Royalty Portal Now Allows Writers to Request Advance," *Billboard*, July 20, 2015; www.billboard.com/articles/business/6634741/universal-music-publishing-royalty-window-updates, accessed November 24, 2015.
16. Robert Levine, "Data Mining the Digital Gold Rush: Four Companies That Get It," *Billboard* 127(10) (2015): 14–15.
17. Interview with Imogen Heap, September 16, 2015.
18. Imogen Heap, "Panel Session," *Guardian Live*, "Live Stream: Imogen Heap Releases Tiny Human Using Blockchain Technology, Sonos Studio London," October 2, 2015; www.theguardian.com/membership/2015/oct/02/live-stream-imogen-heap-releases-tiny-human-using-blockchain-technology. Passage edited by Imogen Heap, e-mail, November 27, 2015.
19. Ibid.
20. Interview with Andreas Antonopoulos, July 20, 2015.
21. Interview with Imogen Heap, September 16, 2015.
22. Ibid.
23. Stuart Dredge, "How Spotify and Its Digital Music Rivals Can Win Over Artists: 'Just Include Us,'" *The Guardian*, October 29, 2013; www.theguardian.com/technology/2013/oct/29/spotify-amanda-palmer-songkick-vevo, accessed August 14, 2015.
24. George Howard, "Bitcoin and the Arts: An Interview with Artist and Composer, Zoe Keating," *Forbes*, June 5, 2015; www.forbes.com/sites/georgehoward/2015/06/05/bitcoin-and-the-arts-and-interview-with-artist-and-composer-zoe-keating/, accessed August 14, 2015.
25. Ibid.
26. Joseph Young, "Music Copyrights Stored on the Bitcoin BlockChain: Rock Band 22HERTZ Leads the Way," *CoinTelegraph*, May 6, 2015; http://cointelegraph.com/news/114172/music-copyrights-stored-on-the-bitcoin-blockchain-rock-band-22hertz-leads-the-way, accessed August 14, 2015.
27. Press release, "Colu Announces Beta Launch and Collaboration with Revelator to Bring Blockchain Technology to the Music Industry," *Business Wire*, August 12, 2015.
28. Gideon Gottfried, "How 'the Blockchain' Could Actually Change the Music Industry, *Billboard*, August 5, 2015; www.billboard.com/articles/business/6655915/how-the-blockchain-could-actually-change-the-music-industry.
29. PeerTracks Inc., September 24, 2015; http://peertracks.com/.
30. "About Us," Artlery: Modern Art Appreciation, September 3, 2015; https://artlery.com.
31. Ellen Nakashima, "Tech Giants Don't Want Obama to Give Police Access to Encrypted Phone Data," *Washington Post*, WP Company LLC, May 19, 2015; www.washingtonpost.com/world/national-security/tech-giants-urge-obama-to-resist-backdoors-into-encrypted-communications/2015/05/18/11781b4a-fd69-11e4-833c-a2de05b6b2a4_story.html.

32. David Kaye, "Report of the Special Rapporteur on the Promotion and Protection of the Right to Freedom of Opinion and Expression," Human Rights Council, United Nations, Twenty-ninth session, Agenda item 3, advance edited version, May 22, 2015; www.ohchr.org/EN/Issues/FreedomOpinion/Pages/CallForSub mission.aspx, accessed September 25, 2015.
33. The UN report refers readers to the Centre for International Governance Innovation and Chatham House, *Toward a Social Compact for Digital Privacy and Security: Statement by the Global Commission on Internet Governance* (2015).
34. The Social Progress Imperative, *Social Progress Index 2015*, April 14, 2015; www .socialprogressimperative.org/data/spi#data_table/countries/com9/dim1,dim2,dim3 ,com9, accessed September 24, 2015. Our ranking is derived from the component scores, not from the overall opportunity score.
35. "Regimes Seeking Ever More Information Control," *2015 World Press Freedom Index*, Reporters Without Borders, 2015; http://index.rsf.org/#!/themes/regimes -seeking-more-control.
36. Reporters Without Borders, "Has Russia Gone So Far as to Block Wikipedia?," August 24, 2015; https://en.rsf.org/russia-has-russia-gone-so-far-as-to-block-24 -08-2015,48253.html, accessed September 25, 2015.
37. Scott Neuman, "China Arrests Nearly 200 over 'Online Rumors,'" August 30, 2015; www.npr.org/sections/thetwo-way/2015/08/30/436097645/china-arrests -nearly-200-over-online-rumors.
38. GetGems.org, September 2, 2015; http://getgems.org/.
39. "Factom: Business Processes Secured by Immutable Audit Trails on the Blockchain," www.factom.org/faq.
40. Interview with Stephen Pair, June 11, 2015.
41. Miguel Freitas, About Twister. http://twister.net.co/?page_id=25.
42. Mark Henricks, "The Billionaire Dropout Club," *CBS MarketWatch*, CBS Interactive Inc., January 24, 2011, updated January 26, 2011; www.cbsnews.com/news/ the-billionaire-dropout-club/, accessed September 20, 2015.
43. Interview with Joichi Ito, August 24, 2015.
44. Ibid.
45. Interview with Melanie Swan, September 14, 2015.
46. Ibid.
47. "Introducing UNESCO: What We Are." Web. Accessed November 28, 2015; http://www.unesco.org/new/en/unesco/about-us/who-we-are/introducing-unesco.

Chapter 10: Overcoming Showstoppers: Ten Implementation Challenges

1. Lev Sergeyevich Termen, "Erhöhung der Sinneswahrnehmung durch Hypnose [Increase of Sense Perception Through Hypnosis]," *Erinnerungen an A. F. Joffe*, 1970. "Theremin, Léon," *Encyclopedia of World Biography*, 2005, *Encyclopedia.com*, www.encyclopedia.com, accessed August 26, 2015.
2. Maciej Ceglowski, "Our Comrade the Electron," speech given at Webstock 2014, St. James Theatre, Wellington, New Zealand, February 14, 2014; www.webstock .org.nz/talks/our-comrade-the-electron/, accessed August 26, 2015. Ceglowski's talk inspired the opening of this chapter.
3. Interview with Andreas Antonopoulos, July 20, 2015.
4. Interview with Tyler Winklevoss, June 9, 2015.
5. Satoshi Nakamoto, P2pfoundation.ning.com, February 18, 2009.

6. Ken Griffith and Ian Grigg, "Bitcoin Verification Latency: The Achilles Heel for Time Sensitive Transactions," white paper, February 3, 2014; http://iang.org/papers/BitcoinLatency.pdf, accessed July 20, 2015.
7. Interview with Izabella Kaminska, August 5, 2015.
8. Ibid.
9. Primavera De Filippi and Aaron Wright, "Decentralized Blockchain Technology and the Rise of Lex Cryptographia," Social Sciences Research Network, March 10, 2015, 43.
10. Interview with Josh Fairfield, June 1, 2015.
11. Izabella Kaminska, "Bitcoin's Wasted Power—and How It Could Be Used to Heat Homes," FT Alphaville, *Financial Times*, September 5, 2014.
12. CIA, "The World Factbook," www.cia.gov, 2012; http://tinyurl.com/noxwvle, accessed August 28, 2015. Note that Cyprus's carbon gas emissions in the same period were 8.801 million metric megatons (2012).
13. "After the Bitcoin Gold Rush," *The New Republic*, February 24, 2015; www.newrepublic.com/article/121089/how-small-bitcoin-miners-lose-crypto-currency-boom-bust-cycle, accessed May 15, 2015.
14. Interview with Bob Tapscott, July 28, 2015.
15. Interview with Gavin Andresen, June 8, 2015.
16. Interview with Eric Jennings, July 10, 2015.
17. Interview with Stephen Pair, June 11, 2015.
18. Interview with Erik Voorhees, June 16, 2015.
19. Sangjin Han, "On Fair Comparison Between CPU and GPU," blog, February 12, 2013; www.eecs.berkeley.edu/~sangjin/2013/02/12/CPU-GPU-comparison.html, accessed August 28, 2015.
20. Interview with Bob Tapscott, July 28, 2015.
21. Interview with Valery Vavilov, July 24, 2015.
22. Hass McCook, "Under the Microscope: Economic and Environmental Costs of Bitcoin Mining," CoinDesk Ltd., June 21, 2014; www.coindesk.com/microscope-economic-environmental-costs-bitcoin-mining/, accessed August 28, 2015.
23. Interview with Bob Tapscott, July 28, 2015.
24. my-mr-wanky, eBay.com, May 8, 2014; www.ebay.com/itm/3-Cointerra-Terra Miner-IV-Bitcoin-Miner-1-6-TH-s-ASIC-Working-Units-in-Hand-/331192098368, accessed July 25, 2015.
25. "PC Recycling," *MRI of Australia*, MRI (Aust) Pty Ltd. Web. August 28, 2015; http://www.mri.com.au/pc-recycling.shtml.
26. Interview with Gavin Andresen, June 8, 2015.
27. Vitalik Buterin, "Proof of Stake: How I Learned to Love Weak Subjectivity," *Ethereum blog*, November 25, 2014; https://blog.ethereum.org/2014/11/25/proof-stake-learned-love-weak-subjectivity/.
28. Stefan Thomas and Evan Schwartz, "Ripple Labs' W3C Web Payments," position paper, March 18, 2014; www.w3.org/2013/10/payments/papers/webpayments2014-submission_25.pdf.
29. Interview with Austin Hill, July 22, 2015.
30. Interview with Roger Ver, April 30, 2015.
31. Satoshi Nakamoto, "Re: Bitcoin P2P E-cash Paper," *The Mail Archive*, November 7, 2008; www.mail-archive.com/, http://tinyurl.com/oofvok7, accessed July 13, 2015.
32. Interview with Josh Fairfield, June 1, 2015.
33. Interview with Stephen Pair, June 11, 2015.

34. Interview with Jerry Brito, June 29, 2015.
35. Ibid.
36. Interview with Josh Fairfield, June 1, 2015.
37. Interview with Andreas Antonopoulos, July 20, 2015.
38. Interview with Izabella Kaminska, August 5, 2015.
39. Interview with Stephen Pair, June 11, 2015.
40. Andrew Vegetabile, "An Objective Look into the Impacts of Forking Blockchains Due to Malicious Actors," The Digital Currency Council, July 9, 2015; www.digitalcurrencycouncil.com/professional/an-objective-look-into-the-impacts-of-forking-blockchains-due-to-malicious-actors/.
41. Interview with Keonne Rodriguez, May 11, 2015.
42. Vegetabile, "An Objective Look."
43. Peter Todd, "Re: [Bitcoin-development] Fwd: Block Size Increase Requirements," *The Mail Archive*, June 1, 2015; www.mail-archive.com/, http://tinyurl.com/pk4ordw, accessed August 26, 2015.
44. Satoshi Nakamoto, "Re: Bitcoin P2P E-cash Paper," Mailing List, *Cryptography*, Metzger, Dowdeswell & Co. LLC, November 11, 2008. Web. July 13, 2015, www.metzdowd.com/mailman/listinfo/cryptography.
45. Pascal Bouvier, "Distributed Ledgers Part I: Bitcoin Is Dead," *FiniCulture blog*, August 4, 2015; http://finiculture.com/distributed-ledgers-part-i-bitcoin-is-dead/, accessed August 28, 2015.
46. Western Union, "Company Facts," Western Union, Western Union Holdings, Inc., December 31, 2014. Web. January 13, 2016; http://corporate.westernunion.com/Corporate_Fact_Sheet.html.
47. Interview with Gavin Andresen, June 8, 2015.
48. Ibid.
49. Interview with Austin Hill, July 22, 2015.
50. Interview with Gavin Andresen, June 8, 2015.
51. Andreas Antonopoulos, "Bitcoin as a Distributed Consensus Platform and the Blockchain as a Ledger of Consensus States," interview with Andreas Antonopoulos, December 9, 2014.
52. Andy Greenberg, "Hackers Remotely Kill a Jeep on the Highway—with Me in It," *Wired*, July 21, 2015.
53. International Joint Conference on Artificial Intelligence, July 28, 2015, Buenos Aires, Argentina; http://futureoflife.org/AI/open_letter_autonomous_weapons#signatories.
54. Lisa Singh, "Father of the Internet Vint Cerf's Forecast for 'Internet of Things,'" *Washington Exec*, August 17, 2015.
55. Interview with Keonne Rodriguez, May 11, 2015.
56. Ceglowski, "Our Comrade the Electron."
57. Interview with Ann Cavoukian, September 2, 2015.
58. Ceglowski, "Our Comrade the Electron."
59. http://www.lightspeedmagazine.com/nonfiction/interview-marc-goodman/.
60. Marc Goodman, *Future Crimes: Everything Is Connected, Everyone Is Vulnerable, and What We Can Do About It* (New York, Doubleday, 2015).
61. Interview with Steve Omohundro, May 28, 2015.
62. *The Silver Stallion*, chapter 26; www.cadaeic.net/cabell.htm, accessed October 2, 2015.
63. Interview with Yochai Benkler, August 26, 2015.

Chapter 11: Leadership for the Next Era

1. Stephan Tual, "Announcing the New Foundation Board and Executive Director," *Ethereum blog*, Ethereum Foundation, July 30, 2015; https://blog.ethereum.org/2015/07/30/announcing-new-foundation-board-executive-director/, accessed December 1, 2015.
2. *Ethereum: The World Computer*, produced by Ethereum, YouTube, July 30, 2015; www.youtube.com/watch?v=j23HnORQXvs, accessed December 1, 2015.
3. Interview with Vitalik Buterin, September 30, 2015.
4. Ibid.
5. Ibid.
6. Ibid.
7. *Henry VI*, part 2, act 4, scene 2.
8. E-mail correspondence with Vitalik Buterin, October 1, 2015.
9. David D. Clark, "A Cloudy Crystal Ball," presentation, IETF, July 16, 1992; http://groups.csail.mit.edu/ana/People/DDC/future_ietf_92.pdf.
10. Interview with Brian Forde, June 26, 2015.
11. Interview with Erik Voorhees, June 16, 2015; interview with Andreas Antonopolous, July 20, 2015.
12. Interview with Erik Voorhees, June 16, 2015.
13. Interview with Jim Orlando, September 28, 2015.
14. http://www.coindesk.com/bitcoin-venture-capital/.
15. E-mail correspondence with Tim Draper, August 3, 2015.
16. Interview with Gavin Andresen, June 8, 2015.
17. Ibid.
18. Interview with Brian Forde, June 26, 2015.
19. Interview with Joichi Ito, August 24, 2015.
20. Interview with Jerry Brito, June 29, 2015.
21. Ibid.
22. www.cryptocoinsnews.com/us-colleges-universities-offering-bitcoin-courses-fall/.
23. Interview with Adam Draper, May 31, 2015.
24. Interview with Benjamin Lawsky, July 2, 2015.
25. Interview with Perianne Boring at Money 2020, October 26, 2015.
26. Interview with Joichi Ito, August 24, 2015.
27. Interview with Blythe Masters, July 29, 2015.
28. For a full list of all the major victories Lawsky achieved while superintendent of NYDFS, please visit www.dfs.ny.gov/reportpub/2014_annualrep_summ_mea.htm.
29. Interview with Benjamin Lawsky, July 2, 2015.
30. Ibid.
31. Ibid.
32. Interview with Jerry Brito, June 29, 2015.
33. Interview with Benjamin Lawsky, July 2, 2015.
34. Ibid.
35. Required reading for anyone looking for a fresh take by a typically conservative government body: www.parl.gc.ca/Content/SEN/Committee/412/banc/rep/repl2jun15-e.pdf.
36. Ibid.
37. Interview with Senator Doug Black of Canada, July 8, 2015.
38. Ibid.

39. Ibid.
40. Ibid.
41. Ibid.
42. Interview with Aaron Wright, August 10, 2015.
43. Interview with Josh Fairfield, June 1, 2015.
44. The Federal Reserve was not the first national bank in the United States. The First National Bank, brought into existence by Congress in 1791, and architected by the United States' first secretary of the treasury, Alexander Hamilton, was far more limited in scope and President Andrew Jackson ultimately dismantled its successor, the Second National Bank, in 1836.
45. Interview with Carolyn Wilkins, August 27, 2015.
46. http://qz.com/148399/ben-bernanke-bitcoin-may-hold-long-term-promise/.
47. In Canada: www.bankofcanada.ca/wpcontent/uploads/2010/11/regulation_canadian_financial.pdf; in the United States: www.federalreserve.gov/pf/pdf/pf_5.pdf.
48. Interview with Carolyn Wilkins, August 27, 2015.
49. "Money in a Digital World," remarks by Carolyn Wilkins, Senior Deputy Governor of the Bank of Canada, Wilfred Laurier University, Waterloo, Ontario, November 13, 2014.
50. Interview with Carolyn Wilkins, August 27, 2015.
51. Ibid.
52. Interview with Jerry Brito, June 29, 2015.
53. Interview with Steve Beauregard, April 30, 2015.
54. Interview with Jerry Brito, June 29, 2015.
55. Don Tapscott and Lynne St. Amour, "The Remarkable Internet Governance Network—Part I," Global Solution Networks Program, Martin Prosperity Institute, University of Toronto, 2014.
56. E-mail correspondence with Vint Cerf, June 12, 2015.
57. www.w3.org/Payments/.
58. www.intgovforum.org/cms/wks2015/index.php/proposal/view_public/239.
59. www.internetsociety.org/inet-bangkok/speakers/mr-pindar-wong.
60. Adam Killick, "Knowledge Networks," Global Solution Networks Program, Martin Prosperity Institute, University of Toronto, 2014.
61. Interview with Jerry Brito, June 29, 2015.
62. Interview with Tyler Winklevoss, June 9, 2015.
63. Interview with Joichi Ito, August 24, 2015.
64. http://coala.global/?page_id=13396.
65. www.digitalchamber.org/.
66. https://blog.coinbase.com/2014/10/13/welcome-john-collins-to-coinbase/.
67. http://www.digitalchamber.org/assets/press-release---g7---for-website.pdf.
68. Anthony Williams, "Platforms for Global Problem Solving," Global Solution Networks Program, Martin Prosperity Institute, University of Toronto 2013.
69. Interview with Brian Forde, June 26, 2015.
70. Interview with Gavin Andresen, June 8, 2015.
71. www3.weforum.org/docs/WEF_GAC15_Technological_Tipping_Points_report_2015.pdf, 7.
72. Interview with Constance Choi, April 10, 2015.
73. The digital revolution has moved on to "the second half of the chessboard"—a clever phrase coined by the American inventor and author Ray Kurzweil. He tells a story of the emperor of China being so delighted with the game of chess that he offered the game's inventor any reward he desired. The inventor asked for rice. "I

would like one grain of rice on the first square of the chessboard, two grains of rice on the second square, four grains of rice on the third square, and so on, all the way to the last square," he said. Thinking this would add up to a couple bags of rice, the emperor happily agreed. He was misguided. While small at the outset, the amount of rice escalates to more than two billion grains halfway through the chessboard. The final square would require nine billion billion grains of rice—enough to cover all of Earth.

74. E-mail interview with Timothy Draper, August 3, 2015.
75. Interview with Hernando de Soto, November 27, 2015.

Afterword to the Paperback Edition

1. Klaus Schwab, *The Fourth Industrial Revolution* (Geneva: World Economic Forum, 2016).
2. Henry Wallace, the thirty-third vice president of the United States, used the phrase to describe the New Deal farm program when he was secretary of agriculture in 1933. See John C. Culver and John Hyde, *American Dreamer: The Life and Times of Henry A. Wallace* (New York: W. W. Norton & Company, 2001), 193.

INDEX